AF293580

Kevin D. Altria

Analysis of Pharmaceuticals by Capillary Electrophoresis

CHROMATOGRAPHIA CE Series

Edited by Kevin D. Altria, Glaxo Wellcome R&D, UK

There are currently a number of general textbooks covering Capillary Electrophoresis where information on a range of applications and techniques can be found. Readers who are interested in a specific area of CE struggle to find truly comprehensive treatments of their areas of interest. The CHROMATOGRAPHIA CE series has been established to allow comprehensive books to be produced covering individual topics. The books are written by well known authors in their specialist application areas and cover CE topics such as DNA analysis, analysis of pharmaceuticals, chiral separations, MECC, carbohydrate analysis, biomedical applications and troubleshooting in CE.

- **Volume 1**: C. Heller (Ed.), Analysis of Nucleic Acids by Capillary Electrophoresis
- **Volume 2**: K. D. Altria, Analysis of Pharmaceuticals by Capillary Electrophoresis

Kevin D. Altria

Analysis of Pharmaceuticals by Capillary Electrophoresis

ISBN-13: 978-3-322-85013-3   e-ISBN-13: 978-3-322-85011-9
DOI: 10.1007/978-3-322-85011-9

Preface

During the 1980's the analysis of pharmaceuticals was dominated by the use of High Performance Liquid Chromatography (HPLC). Other separative techniques such as Gas Chromatography and Thin Layer Chromatography offered alternatives but their quantitative capabilities and/or solute range could not approach that of HPLC.

The majority of pharmaceuticals are ionic and it would be reasonable to assume that electrophoresis may be useful in the analysis of pharmaceuticals. However, the electrophoretic instruments available in the 1980's were labour intensive and employed post-separation detection procedures. During the late 1980's and early 1990's extensive research was conducted into the possibilities of conducting electrophoretic separations in capillaries. This approach allowed on-line detection and could be performed on fully automated equipment. This research led to the advent of modern day capillary electrophoresis (CE) instruments which offer similar performance and automation levels to that of HPLC. Research was also focused on developing applications for CE and particular attention was paid to applications within the pharmaceutical analysis area. These applications proved that CE could be applied to a wide range of drug types including water insoluble and neutral compounds. The ability to achieve efficient chiral separations of drugs also increased the popularity of the technique. CE with indirect UV detection has become established as a simple and effective alternative to ion-exchange chromatography for the determination of small inorganic or organic ions.

Routine CE methods are now used in many industrial pharmaceutical companies and applications include determination of related impurities, main component assay, chiral separations and drug residue determinations. Research into CE continues to broaden the application range in drug analysis. In particular current research is focused in the area of non-aqueous solvent systems and the developing technique of capillary electrochromatography (CEC). In CEC the capillaries used in CE are packed with HPLC stationary phase material and a high voltage is used to achieve separations by combining electrokinetic and chromatographic processes. All of these application and development areas are covered in individual chapters within this book.

CE is still a rapidly developing technique and new applications and developments appear on a weekly basis. This book broadly reflects the current state of the art for CE analysis of pharmaceuticals and contains several hundred references to specific applications and methods. The main purpose of the book is to present the application possibilities of CE but I hope the extensive use of tabulated application data should also make the book useful as a reference point for specific applications.

London 1998   Kevin D Altria

Acknowledgement

I would like to extend my appreciation to Dr. Angelika Schulz from Vieweg Publishing who has kindly supported the creation of the Chromatographia CE Series throughout its continued development. My thanks are also extended once again to my family and friends who have helped me to retain? a little sanity during the preparation of this volume. In particular my gratitude is extended to my wife Fatima and my colleagues Dave Rudd and Simon Bryant.

Contact details

Adress: Kevin D Altria, Pharmaceutical Development , GlaxoWellcome R&D,
             Park Road, Ware, Herts. SG12 ODP, UK
Phone :   ++44-1920-883616
Fax :   ++44-1920-883873
E-mail: KDA8029@ggr.co.uk
Website: http://dspace.dial.pipex.com/town/terrace/ho81/kda.htm

Contents

1 Introduction to CE and the Use of CE in Pharmaceutical Analysis

1.1 Capillary Electrophoresis (CE) Theory and Background

Electrophoresis refers to the migration of charged electrical species when dissolved, or suspended, in an electrolyte through which an electric current is passed. Cations migrate toward the negatively charged electrode (cathode) and anions are attracted toward the positively charged electrode (anode). Neutral solutes are not attracted to either electrode. Conventionally electrophoresis has been performed on layers of gel or paper. The traditional electrophoresis equipment offered a low level of automation and long analysis times. Detection of the separated bands was performed by post-separation visualisation. The analysis times were long as only relatively low voltages could be applied before excessive heat formation caused loss of separation.

The advantages of conducting electrophoresis in capillaries was highlighted in the early 1980's by the work of Jorgenson and Lukacs (1) who popularised the use of CE. Performing electrophoretic separations in capillaries was shown to offer the possibility of automated analytical equipment, fast analysis times and on-line detection of the separated peaks. Heat generated inside the capillary was effectively dissipated through the walls of the capillary which allowed high voltages to be used to achieve rapid separations. The capillary was inserted through the optical centre of a detector which allowed on-capillary detection.

Capillary electrophoresis has grown to become a collection of a range of separation techniques which involve the application of high voltages across buffer filled capillaries to achieve separations. The variations include separation based on size and charge differences between analytes (termed Capillary Zone Electrophoresis, CZE, or Free Solution CE, FSCE), separation of neutral compounds using surfactant micelles (Micellar electrokinetic capillary chromatography, MECC or sometimes referred to as MEKC) sieving of solutes through a gel network (Capillary Gel Electrophoresis, CGE), and separation of zwitterionic solutes within a pH gradient (Capillary Isoelectric Focusing, CIEF). Capillary electrochromatography (CEC) is an associated electrokinetic separation technique which involves applying voltages across capillaries filled with silica gel stationary phases. Separation selectivity in CEC is a combination of both electrophoretic and chromatographic processes. Many of the CE separation techniques rely on the presence of an electrically induced flow of solution (electroosmotic flow, EOF) within the capillary to pump solutes towards the detector. The basis to EOF is discussed later in this chapter.

FSCE and MECC are the most frequently used separation techniques in pharmaceutical analysis. GCE and CIEF are of importance for the separation of biomolecules such as DNA and proteins respectively and are becoming of increasing importance as development of biotechnology derived drugs is becoming more frequent. Generally CE is performed using aqueous based electrolytes, however there is a growing use of non-aqueous solvents in CE.

Operation of a CE system involves application of a high voltage (typically 10–30 kV) across a narrow bore (25–100 μm) capillary. The capillary is filled with electrolyte solution which conducts current through the inside of the capillary. The ends of the capillary are dipped into reservoirs filled with the electrolyte. Electrodes made of an inert material such as platinum are also inserted into the electrolyte reservoirs to complete the electrical circuit. A small volume of sample is injected into one end of the capillary. The capillary passes through a detector, usually a UV absorbance detector, at the opposite end of the capillary from the injection. Application of a voltage causes movement of sample ions along the capillary and passing through the detector. A plot of detector response with time is generated which is termed an electropherogram. A flow of electrolyte, known as electroendosmotic flow, EOF, (discussed later in this chapter) results in a flow of the solution along the capillary usually towards the detector. This flow can significantly reduce analysis times or force an ion to overcome its migration tendency towards the electrode it is being attracted to by the sign of its charge (discussed more fully in the FSCE section below).

Detailed treatments of the background theory and non-pharmaceutical based applications can be obtained from a number of reference books (2–5)

Commercially available CE instruments (Figure 1.1) are PC controlled and consist of a buffer filled capillary passing through the optical centre of a detector, a means of introducing the sample into the capillary, a high voltage power supply and an autosampler.

The typical voltages used are in the range of 5–30 kV which results in currents in the range of 10–100 μA. Higher currents than this can cause problems of heating inside the capillary which can broaden peaks resulting in loss of resolution.

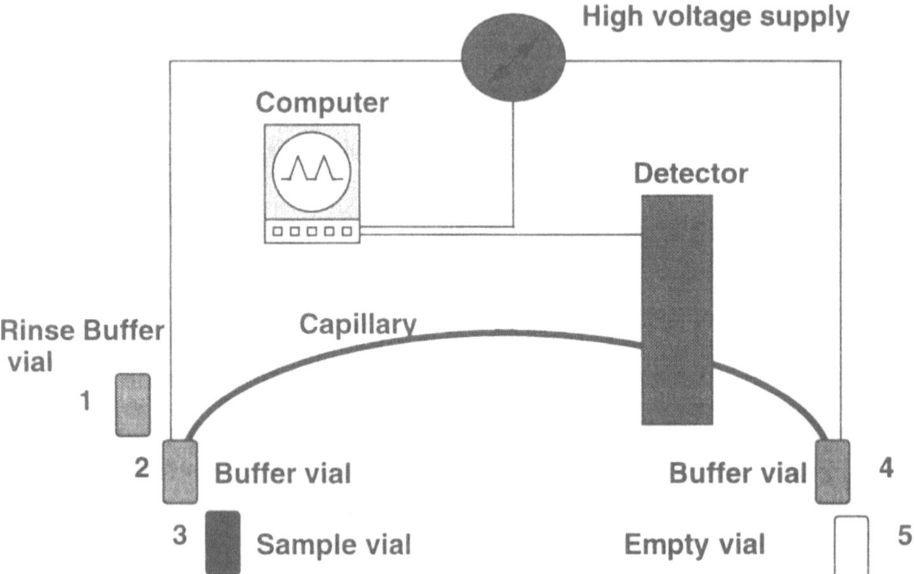

Figure 1.1   Typical CE separation system

1.2 CE Instrumentation

1.2.1 Capillaries

The capillaries used are normally fused silica capillaries covered with an external polyimide protective coating to give them increased mechanical strength as bare fused silica is extremely fragile. A small portion of this coating is removed to form a window for detection purposes. The window is aligned in the optical centre of the detector. Capillaries are typically 25–100 cm long with 50 and 75 micron being the most commonly employed inner diameters. On standard commercial CE instruments the capillary is often held in a housing device, such as a cartridge, to facilitate ease of capillary insertion into the instrument and to protect the delicate detection window area. The inner surface of the capillary can be chemically modified by covalently binding (coating) different substances onto the capillary wall. These coatings are used for a variety of purposes such as to reduce sample adsorption or to change the ionic charge on the capillary wall.

1.2.2 Temperature control

It is important to regulate the temperature of the environment around the capillary to ensure consistent separations. To achieve this capillaries are often inserted into cartridges which are placed in the CE instrument. Temperature controlled air or liquid coolant is then forced through the cartridge to regulate the temperature.

1.2.3 Sample introduction

Sample solution is forced into the end of the capillary furthest from the detector. Typical injection volumes are 10–100 nl. The most frequently used injection mode is to dip the capillary into the sample solution vial. The vial is then pressurised causing a volume of solution to be forced into the capillary. An alternative, less popular sample injection procedure, is to dip the capillary and electrode into the sample solution vial and to apply a voltage. If the sample is ionised and the appropriate voltage polarity is used then sample ions will migrate into the capillary. This type of injection is known as electrokinetic sampling.

1.2.4 Detectors

The most frequently used detector is a UV absorbance detector which is standard on commercial CE instruments. The majority of instruments also have UV diode array detectors available. Alternative detector modes commercially available include fluorescence, laser induced fluorescence, conductivity and indirect detection. The hyphenation of CE and mass spectrometers is frequently used to give structural information on the resolved peaks.

The detectors can be interfaced with data acquisition devices to calculate results. Integrated peak areas are routinely used for quantitation as these give increased dynamic ranges compared to use of peak heights (6).

1.3 Band Broadening Effects in CE

The capillary format employed in CE is advantageous as it minimises or eliminates most sources of band broadening that occur in conventional electrophoresis or in HPLC.

1.3.1 Heat dissipation

In conventional slab gel electrophoresis the Joule heat associated with the generation of current during separation can cause problems of peak dispersion. This Joule heat causes the formation of convention currents within the gel which mixes the zones during separation and results in band broadening and peak dispersion. Heat generation therefore restricts the operating voltages that can be used in slab gel electrophoresis which produces longer analysis times. Performing electrophoresis in a capillary allows the heat to be effectively dissipated through the capillary walls which reduces any convection related band broadening. This improved heat dissipation means that higher operating voltages can be used in CE which can produce significantly faster analysis times.

1.3.2 Electroendosmotic flow

The walls of a fused silica capillary contain silanols which ionise in contact with a high pH electrolyte solution (Figure 1.2). This dissociation produces a negatively charged wall. A layer of metal ions is then established at the wall to preserve electroneutrality. When a voltage is applied these metal ions and their associated solvating water molecules migrate towards the cathode. This movement of ions and their associated water molecules results in a flow of solution towards the detector. This flow effectively pumps solute ions along the capillary generally towards the detector and could be considered as an "electrically-driven pump". At low pH the silanols are unionised and therefore the flow rate is much reduced or can become zero at very low pH values.

The extent of the flow is related to the charge on the capillary, the buffer viscosity and dielectric constant of the buffer:

The magnitude of the EOF is given by (1).

$$\mu_{eo} = \left(\varepsilon \zeta / 4 \pi \eta\, r \right) \tag{1}$$

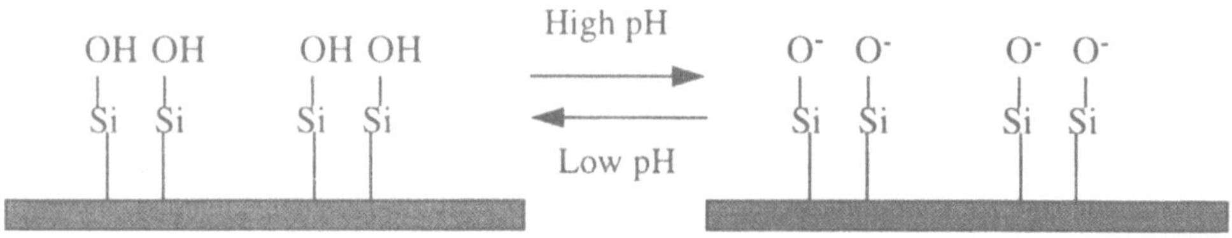

Figure 1.2   Silanol dissociation process

Where: μ_{eo} = "EOF mobility" (rate of EOF), η = viscosity, ζ = zeta potential (charge on capillary surface), r = capillary radius

The velocity of the electroosmotic flow, V_{eo}, is described by equation 2:

$$V_{eo} = \mu_{eo}(V/L) \tag{2}$$

where V is the applied voltage, and L is the capillary length.

The level of EOF is highly dependent upon electrolyte pH as the zeta potential is largely governed by the ionisation of the acidic silanols on the capillary wall. Below pH 4 the ionisation is small and the EOF flow rate is therefore not significant, above ~pH 9 the silanols are fully ionised and EOF is strong. The level of EOF decreases with increased electrolyte concentration as the zeta potential is reduced.

The EOF is generated by the entire length of the capillary and is therefore produces constant flow rate at all distance along the capillary. This means that the flow profile of EOF is plug-like in nature (Figure 1.3) and that the solutes are being swept along at the same rate throughout their transport along the capillary which minimises sample dispersion. This is an advantage compared to the laminar flow encountered in pumped systems such as HPLC. In laminar flow the solution is pushed from one end of the column and the solution at the edges of the column is moving slower than the solution in the middle of the column which results in different solute speeds across the column. Therefore laminar flow broadens the peaks as they travel along the column.

1.3.3 On-capillary detection

One of band broadening processes that occurring in HPLC is post-separation mixing of peaks during transfer from the end of the column to the detector. This is avoided in CE as a portion of the capillary is used as the detector which eliminates the possibility for such problems.

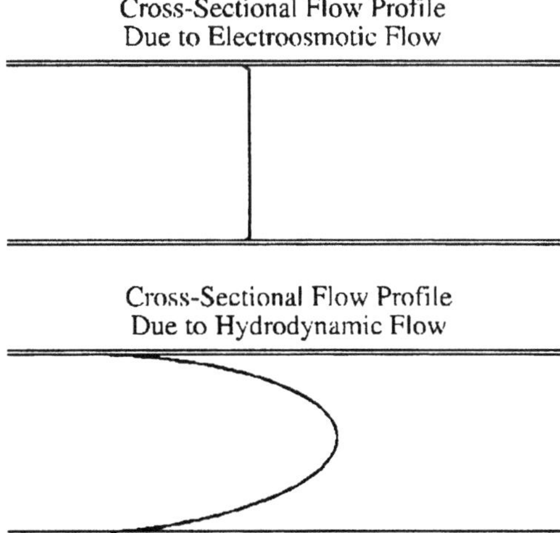

Figure 1.3 Profile of EOF and laminar flow

1.3.4 Molecular diffusion

A major dispersive effect remaining in CE is that of molecular diffusion of the solute as it passes along the capillary. This diffusion is lowest for large molecules such as proteins which have small diffusion coefficients. Therefore it is possible to obtain theoretical plates counts (N) of several million for biomolecules such as proteins and nucleotides in CE. The theoretical plates count can be calculated:

$$N = \frac{(\mu_{ep} + \mu_{eo})V}{2D} \tag{3}$$

in which D is the diffusion coefficient and μ_{ep} is the electrophoretic mobility of the sample ion (discussed later in Section 1.4.1)

1.3.5 Injection related broadening

In addition to molecular diffusion the other major band broadening factor remaining in CE is the length of the injection volume which may a few mm in length. This is a significant length given that the total capillary length may be 25 cm and the detection window may be 0.1mm. The starting zone length of the sample injection can be reduced by utilisation of a process termed "stacking". This stacking reduces the width of the sample zone before separation and this results in an improved sensitivity (as the sample becomes more concentrated on-capillary) and increased peak efficiency. Stacking occurs when the sample is dissolved in a lower ionic strength solution than that of the separation electrolyte. Under these circumstances the field strength is higher in the sample zone than in the rest of the capillary which is filled with electrolyte (Figure 1.4). The sample ions move forwards rapidly in the sample zone until they encounter the electrolyte boundary where they experience a lower applied field and their migration rate slows down. In this way the sample zone is focused and this can lead to up a 10 fold reduction in the starting peak width. Stacking is optimised if the sample is dissolved in pure water or a 1 to 10 dilution of the run electrolyte.

1.4 Separation Modes Available

1.4.1 Free Solution Capillary Electrophoresis (FSCE)

The separation of ions in the simplest form of CE is often termed Free Solution Capillary Electrophoresis (FSCE). The separations rely principally on the pH controlled dissociation of acidic groups on the solute or the protonation of basic functions on the solute. These ionic species are separated based on differences in their charge-to-mass ratios. For example basic drugs are separated at low pH as cations whilst acidic drugs are separated as anions at high pH. In FSCE all neutral compounds are swept, unresolved, through the detector together. Separation of neutrals is generally achieved by Micellar Electrokinetic Capillary Chromatography (MECC) which is discussed later in this chapter.

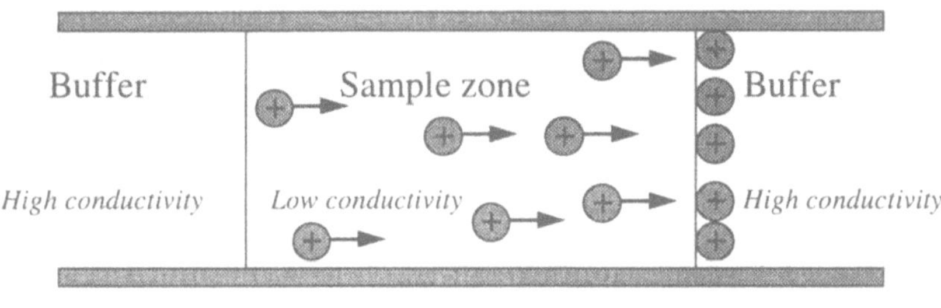

Figure 1.4 Principles of stacking

Under the influence of an applied electric field sample ions will move towards their appropriate electrode. Cations towards the cathode and anions towards the anode. The speed of their movement towards the electrode is governed by their size and number of appropriate charges. Smaller molecules with a large number of charges will move more quickly than larger or less charged compounds. The speed of movement, known as the electrophoretic mobility, is characteristic of the solute and can be derived from the following expression:

$$\mu_{ep} = q/(6\pi\eta r) \tag{4}$$

where q = number of ionic charges, η = solution viscosity and r = ionic radius.

Therefore, when a mixture of cations is separated (Figure 1.5) the smaller imidazole (relative molecular mass: RMM 68) is detected first followed by the larger aminobenzoate (RMM 137) and the considerably larger aspartame (RMM 294) is detected last.

Equation 5 shows that the rate of movement, V_{ep}, is directly related to the magnitude of the strength of the applied field, E, per cm of the capillary length (V/cm).

$$V_{ep} = \mu_{ep}(V/L) \tag{5}$$

where V is the applied voltage, and L is the capillary length.

The time taken by a solute to migrate the entire length of the capillary of a solute, t, is related to capillary length and both the velocity of the EOF and the electrophoretic velocity:

$$t = \frac{L}{(V_{ep} + V_{eo})} = \frac{L2}{(\mu_{ep} + \mu_{eo})} \tag{6}$$

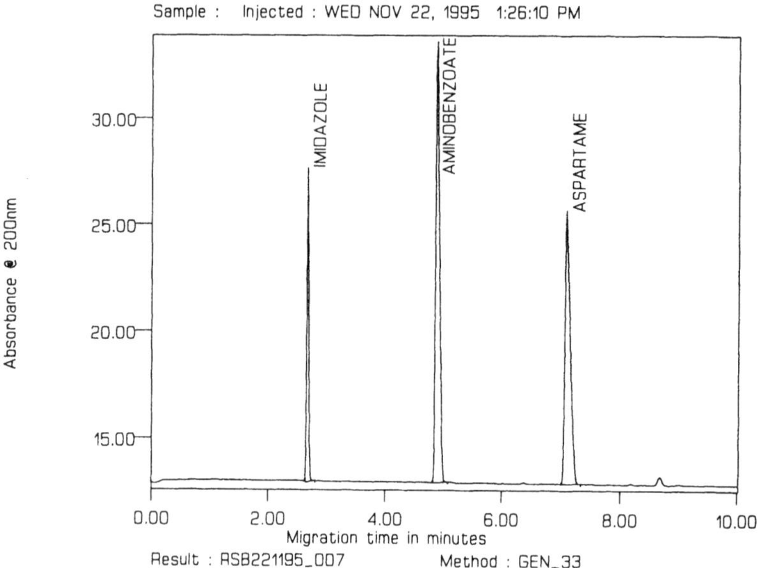

Figure 1.5   Separation of a range of cations at low pH

Therefore to obtain rapid separations high voltages should be applied across short capillaries. The migration speed is related to the temperature through a viscosity term. At higher temperatures the buffer solution is less viscous and the sample ions meet less resistance to their passage through the solution towards the electrode and therefore have a faster migration rate.

The mobility of a species can also be changed by complexing the ion as it moves along the capillary. For example additives such as cyclodextrins can be used to complex with drug enantiomers to achieve chiral separations.

The presence of EOF allows the separation and detection of both cations and anions within a single analysis. For example the EOF is sufficiently strong at pH 7, and above, to sweep anions to the cathode regardless of their charge. Figure 1.6 shows an electropherogram of a neutral compound (Nimbex) which migrates on top of the EOF front and three singly charged anions which migrate against the EOF and are therefore detected after the EOF front. The separation was performed at pH 9.5 where the EOF is strong. The besylate is the smallest of the anions (RMM 122) and so is able to migrate most effectively against the EOF. The largest anion beta-naphthoxy acetic acid (RMM 202) is less able to migrate against the EOF and is therefore detected first.

Therefore pH is the major operating parameter in FSCE as it affects the separation of ionic species by controlling both the solute charged state and the level of EOF.

The overall migration time of a solute is therefore related to both the mobility of the solute and EOF. The term apparent mobility (μA) is measured from the migration time, and is a sum of both μE and μEOF.

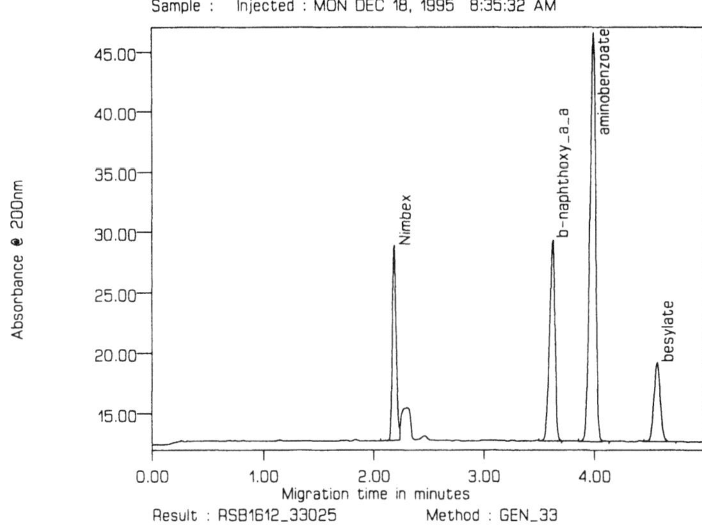

Figure 1.6   Separation of a neutral compound and 3 acids at pH 9.5

The time, t, necessary for a solute to migrate the entire length of the capillary is represented by the equation:

$$t = \frac{L}{(V_{ep} + V_{eo})} \tag{7}$$

There are a number of variables in FSCE that can be used in the optimisation of FSCE methods. These include the operating pH, electrolyte type and concentration, capillary dimensions, temperature and injection volume. Electrolyte such as ion-pair reagents and chiral substances can also be employed in order to manipulate selectivity. Highly efficient chiral CE separations can obtained (Chapter 4) by the addition of chirally selective substances, such as cyclodextrins or crown ethers, into the electrolyte.

1.4.2 Micellar electrokinetic capillary chromatography (MECC)

MECC was initially developed (7) for the resolution of uncharged compounds which cannot be separated using simple free solution CE. The separation conditions generally involve use of a high pH electrolyte containing relatively high levels of surfactant such as sodium dodecyl sulphate (SDS). Above a specific surfactant concentration, the critical micelle concentration (CMC), the surfactant molecules begins to self-aggregate, forming micelles in which the hydrophilic head groups form an outer shell and the hydrophobic tail groups form a nonpolar core into which solutes can partition. SDS micelles have a negative charge and migrate against the EOF. However, the EOF is sufficiently strong to force the micelles to eventually

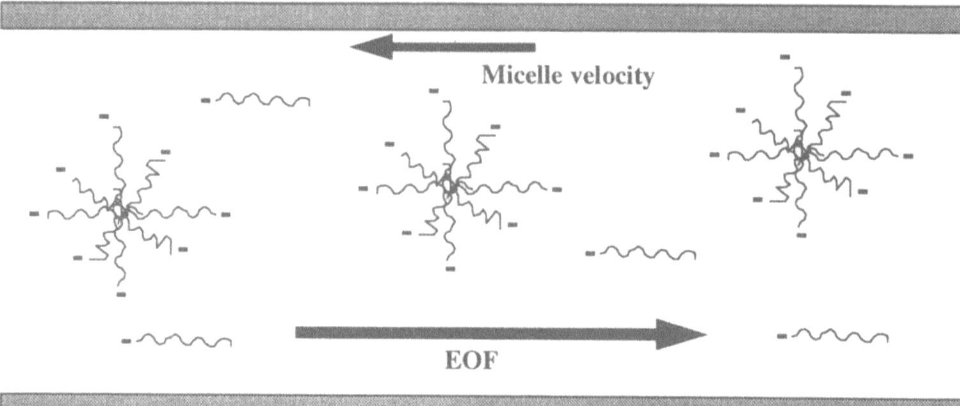

Figure 1.7 Schematic of the principles of MECC

pass through the detector. Figure 1.7 shows the separation principles. Sample species can partition into the interior of the micelle in a fashion similar to retention on a stationary phase in HPLC. The differential partitioning between the buffered aqueous mobile phase and the micellar (pseudo) stationary phase is the sole basis for separation for neutral molecules .
The retention time, t_r for a neutral species is always between t_0 and t_{mc} :

$$t_r = \left(\frac{1+k'}{1+\dfrac{t_0}{t_{mc}} k'} \right) t_0 \qquad (8)$$

Where t_0 is the time required for an unretained substance to travel through the capillary (from injection point to detection window), t_{mc} is the time required for a micelle to traverse the capillary.

Solutes being highly retained by the micelle will be eluted later (Figure 1.8) whilst solutes have only a limited interaction with the micelle will be eluted near to the EOF front (t_0). Extremely hydrophobic compounds may be totally included into the micelle and would be detected at t_{mc}. Figure 1.8 shows separation of a test mixture, peak 1 represents methanol which is not retained by the micelle and elutes at t_0 with the EOF flow. Sudan III is a hydrophobic dye which is totally included into the micelle (t_{mc}) and is widely used to mark the migration time of the micelle.

The capacity factor k' for a neutral species can be calculated in MECC using the equation:

$$k' = \frac{\dfrac{t_r}{t_0} - 1}{1 - \dfrac{t_r}{t_{mc}}} \qquad (9)$$

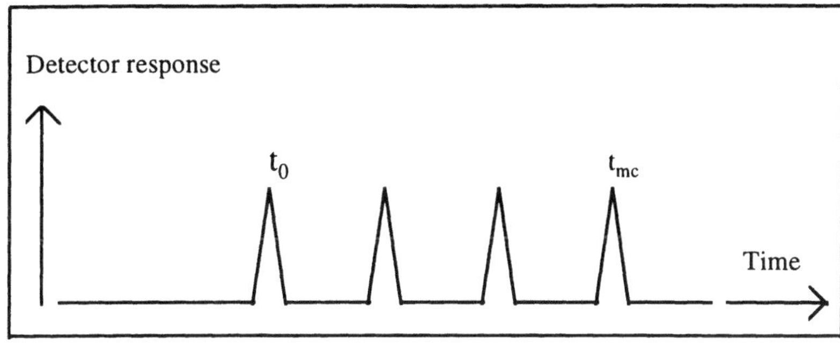

Figure 1.8 Elution window for solutes in MECC

A solute which elutes with the solvent front has a capacity factor of zero and a solute eluting with at the t_{mc} would be considered to have an infinite capacity factor. A solute spending equal time in the aqueous solute and in the micelle would have a capacity factor of 1.

Method development in MECC concentrates on the type and concentration of surfactant in the first instance as both of these factors affect solute capacity factors and selectivity. The most common surfactant used for MECC is sodium dodecylsulfate (an anionic surfactant). Others include cetyltrimethylammonium bromide (cationic surfactant) and bile salts (anionic surfactants). Mixtures of surfactants can be used including neutral surfactants such as Tween and Brij. Separations are invariably conducted at high pH where there is an appreciable EOF. Organic solvents and ion-pair reagents can also be added to the MECC buffer to adjust the capacity factors, just as in reverse-phase HPLC separations.

MECC is especially useful for the resolution of water insoluble, neutral compounds such as steroids. Figure 1.9 shows a separation of a range of steroids using MECC. In this example the micelles used are sodium cholate. There are a number of bile salts such as sodium cholate which can be used in MECC. The bile salts are chiral and can be used in chiral separations (Chapter 4) – alternatively chiral separations can be achieved through use of combinations of SDS micelles and cyclodextrins.

Since both HPLC and MECC are chromatographic based techniques the separation profiles are similar to reverse-phase HPLC. However, solute partitioning is different between MECC and HPLC, which results in different profiles. Additionally, if species are charged then they will be separated in MECC based on the sum of both their electrophoretic mobility and partitioning. Therefore, MECC is useful for determination of drug related impurities where mixtures of charged and uncharged components may be resolved. The other feature of MECC is that all components injected into the capillary, provided that they are sufficiently soluble in the electrolyte, will migrate between t_0 and t_{mc}. This is unlike HPLC where some components may be irreversibly adsorbed onto the stationary phase. MECC separations are performed on the same equipment as FSCE and employ capillaries of similar dimensions.

1.4.3 Capillary Gel Electrophoresis (CGE)

When separating DNA species problems occur as their electrophoretic mobilities may be very similar as they have very similar charge to mass ratios as any increases in the number of

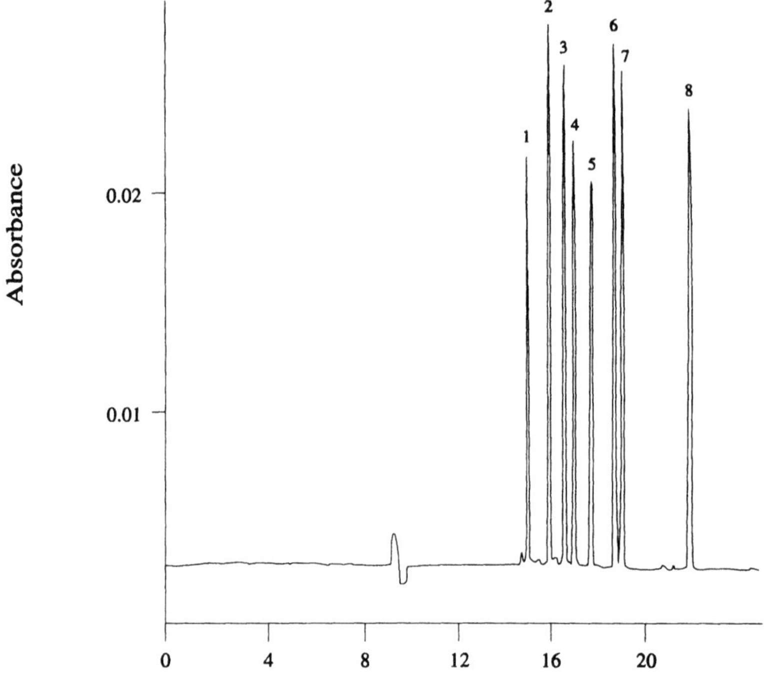

Figure 1.9   Separation of Corticosteroids by MECC. Buffer, 100 mM sodium cholate, 100 mM borate, pH 8.45. 1, triamcinolone; 2, hydrocortisone; 3, betamethasone; 4, hydrocortisone acetate; 5, dexamethasone acetate; 6, triamcinolone acetonide; 7, fluocinolone acetanide; 8, fluocinolone

charges is directly related to the increased size of the solute. Therefore FSCE is often insufficient for adequate resolution. In this case separations are performed in capillaries filled with a gel solution. In Capillary Gel Electrophoresis (CGE) a sieving effect occurs as solutes of various sizes migrate through the gel filled capillary towards the detector. The smaller ions are able to migrate quickly through the gel whilst the larger ions become entangled in the gel matrix and their migration rate is reduced. This is a similar process to Gel Permeation Chromatography.

Initially the gels used in CGE involved polyacrylamide covalently bonded to the capillary wall. However these fixed gels suffered from problems of shrinkage and blockage and could have relatively short lifetimes. In addition, if sample matrix components contaminated the gel then it would often need to be discarded. Therefore there has been a recent tendency towards the use of pumpable gel solutions which can be used to fill the capillary with a liquid gel matrix. Pumpable gels include the use of derivatised celluloses dissolved in the run buffer. Use of the liquid gels also allows replacement of the gel in-between injections to reduce the contamination problems encountered with fixed gels.

1.4.4 Capillary Isoelectric Focusing (CIEF)

This CE technique is used for the separation of zwitterionic species which contain both positive and negative charged groups. The analytes separated by CIEF tend to be proteins and peptides. Zwitterionic species have a pI value which is pH value at which the number of positive charges exactly equal the number of negative charges. The solute is therefore neutral at its pI value and does not migrate. In CIEF the capillary is filled with a pH gradient which remains fixed during the application of a separation voltage (Figure 1.10). The sample is injected at the low pH end of the capillary and as it is positively charged it will therefore migrate along the capillary. When the solute reaches it pI position along the pH gradient it will become uncharged and will stop migrating and remain in a fixed position. The various solutes in a mixture will have different pI values and will therefore be separated at various positions along the gradient. Once the focusing of all the zones is completed a pressure is applied across the capillary whilst the voltage is being applied. The pressure forces the pH gradient along the capillary and through the detector which allows detection of the separated zones.

1.4.5 Capillary electrochromatography (CEC)

Capillary electrochromatography is a rapidly evolving hybrid technique between HPLC and CE. In essence, CE capillaries are packed with HPLC packing and a voltage is applied across the packed capillary which generates an electro-osmotic flow (EOF). The EOF transports solutes along the capillary towards the detector. Both differential partitioning and electrophoretic migration of the solutes occurs during their transportation towards the detector which leads to CEC separations. It is therefore possible to obtain unique separation selectivities using CEC compared to both HPLC and CE. The beneficial flow profile of EOF reduces flow related band broadening and separation efficiencies of several hundred thousand plates per metre are often obtained in CEC. Generally, carrier electrolytes containing high levels (40–80%) of organic solvents such as methanol or acetonitrile are employed in CEC. Therefore resolution of both water insoluble and neutral solutes is readily achieved in CEC whilst these separations are more difficult to achieve in CE. Chapter 12 extensively covers the background to CEC and pharmaceutical analysis applications.

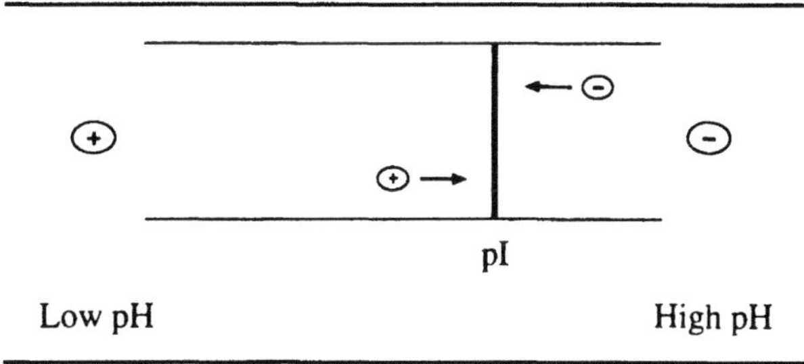

Figure 1.10 Principles of Capillary Isoelectric Focusing

1.5 Application of CE to Specific Drug Classes

Several reports (Table 1.1) have shown that CE methods have been developed (8–46) to cover testing of a range of similar drugs or drugs from the same class. These general analysis methods can be used for assay and identity confirmation of drugs in the formulated products or in the drug substance form. Use of these general separation conditions can often significantly reduce method development efforts. However when attempting to resolve complex mixtures it is important to optimise all factors in order to achieve the required selectivity. This would include the choice of buffering species used as electrolyte. For example a range of operating conditions were investigated for the separation of cardiovasculars (21) before the final separation conditions of 50 mM SDS, 100 mM borate, 15% acetone were obtained.

The concentration of the buffer can also be a critical parameter when separating complex mixtures of closely related compounds. For example (19) a 320 mM citrate concentration was needed to resolve 10 beta-blocker drugs at pH 2.0 This separation had been impossible to achieve using a phosphate buffer of the same pH.

Table 1.1 Separation of drug types by CE

Sample	Electrolyte	Ref	Comment
10 Acidic drugs and excipients	Borate pH 9.5	8	Validated method for identification and quantitation
5 Analgesics	MECC, SDS	9	4 minute separation of test mix
13 Aminoglycoside antibiotics	Imidazole-acetate pH 5, FC135	10	Indirect detection at 214 nm
6 Analgesics	MECC, SDS	11	100 % recovery obtained in validation
8 Analgesics	MECC, SDS, MeOH	12	Separation within 5 minutes
9 Antihistamines	MECC, SDS, CD, ion-pair	13	Commercial samples analysed
9 Barbiturates	MECC, SDS, pH 8.5	14	Correlation between hydrophobacity and migration order
12 Basic drugs and excipients	Phosphate pH 2.5	15	Validated method for identification and quantitation
22 Basic drugs in urine	Phosphate pH 2	16	Injection precision less than 2 % RSD
17 Basic drugs	Phosphate pH 2.4	17	Migration time precision ~1 %
9 Beta-blockers	CTAB, MECC	18	Fundamental study on parameter effects
10 Beta blockers	Citrate buffer, low pH	19	Concentration affected selectivity
10 Benzodiazepines	MECC, SDS, MeOH	20	Glycine-triethanolamine pH 9 buffer
6 Cardiovascular drugs	MECC, SDS, solvents	21	Range of solvents used including acetone
Cardiovascular drugs	Various separation conditions	22	Review paper on the use of CE in the analysis of cardiovascular drugs
5 Cephalosporins	MECC, SDS, ion-pair	23	Identity confirmation by migration time
9 Cephalosporins	pH 7, phosphate-borate	14	pH 7 optimal

Table 1.1 continued

Sample	Electrolyte	Ref	Comment
9 Cephalosporins	MECC, SDS, ion-pair	24	TMAB best ion-pair reagent evaluated
12 Cold medicine ingredients	LMT, MECC (or bile salts)	25	Plate counts up to 350,000
14 Cold medicine ingredients	MECC, bile salt	26	Capsules and granules analysed
8 Cocaine related substances	CTAB, phosphate-borate pH 8.6 + 7.5% acetonitrile	27	Drug seizure profiling
7 Corticosteroids	Borate, SDS, MECC	28	Experimental design optimisation. Cortisol measured in plasma
8 Corticosteroids	MECC, SDS, CD	29	γ-CD shown to be optimal MECC additive
8 Corticosteroids	MECC bile salt	30	Steroid cream analysed
17 Corticosteroids	Bile salt MECC	31	Various mixtures of bile salts and SDS
15 Diuretics in blood and urine	60mM CAPS pH 10.6	32	SPE, relative migration times used for identification purposes
Diuretic drugs	Various operating conditions	33	Review of the use of CE to determine diuretic drugs
18 Common drugs of abuse	85 mM SDS 8.5 mM phosphate / 8.5 mM borate / 15 % acetonitrile pH 8.5	34	Used to analyse heroin seizures
20 Drugs of abuse	MECC and FSCE	35	Used to analyse range of illicit drug samples
12 Ecdysteroids	MECC, SDS, MeOH	36	Plant extract analysis
10 Mycotoxins	MECC, SDS or bile salt	37	Mycotoxins identified in samples by RMT
15 NSAID's	MECC, SDS, ACN	38	Acetonitrile more effective than methanol addition
7 Penicillins	SDS, MECC	39	Analysed in plasma
8 Penicillins	MECC, SDS	40	Clear separation within 14 minutes
14 Phenothiazines	MECC, CTAB	41	Related compounds resolved by MECC
Polymyxin antibiotics	Phosphate pH 2.5 with PAPS	42	Complex mixtures separated – identity confirmed
14 Quinolone anti-bacterials	MECC, SDC, ion-pair	43	Baseline separation of test mixture within 14 minutes
9 Sulphonamides	pH 7, phosphate	17	pH optimised for separation
16 Sulphonamides	pH 7, imidazole-acetate	44	Quantitation in pork meat extracts
9 Theophylline tablet components	MECC, SDS	45	Simultaneous quantitation of all components
Xanthine derivatives	MECC, SDS	46	Tablet formulation assayed

Key to **Table 1.1**

CAPS = cyclohexaneamino sulphonic acid
CD = cyclodextrin
CTAB = cetyltrimethylammonium bromide
NSAID's = non steroidal anti-inflammatories
RMT = relative migration time

RSD = relative standard deviation
PAPS = phosphoadenosine phosphosulfate
SDC = sodium deoxycholate
SPE = solid phase extraction

1.6 The Role of CE in Pharmaceutical Analysis

Traditionally electrophoresis has been used for the separation of biomolecules such as proteins and DNA. Chromatographic techniques such as HPLC predominate pharmaceutical analysis as they offer the possibility of highly selective, sensitive and highly automated methods. The automation capabilities of HPLC are a key feature as the number of samples routinely analysed is very high in pharmaceutical analysis laboratories. Previously the time consuming nature of slab gel electrophoresis and relatively long analysis times had prevented the widespread use of electrophoresis in drug analysis where large number of samples are analysed and the need for automated sample analysis is a fundamental requirement.

The pioneering papers in CE were published in the early 1980's principally by Jorgenson et al (1) and Terabe et al (7). These papers demonstrated that highly efficient and rapid electrokinetic separations could be obtained using capillaries and that the equipment arrangement could be readily automated. Initial work in the area of CE was concerned with the separation of biomolecules such as DNA, proteins, and peptides. The first reports of the analysis of pharmaceuticals by CE were in the late 1980's (11, 47). Commercial CE instrumentation became available in the late 1980's. The nature of these instruments allowed unattended automated operation and has enabled CE to become established as a complimentary and alternative technique to HPLC which previously dominated pharmaceutical analysis laboratories.

A survey of CE users in the US and Europe (48) showed (Figure 1.11) that pharmaceutical analysis was the most common application of CE.

The range of application areas within pharmaceutical analysis are very similar for both CE and HPLC. A survey of a number of major pharmaceutical companies in both the US and UK on their use of CE confirmed (49) that CE, like HPLC, is used for determination of related impurities (Chapter 3), chiral separations (Chapter 4), identity confirmation (Chapter 2) and main component assay (Chapter 2). The use of indirect UV detection in CE allows quantitation of small ions such as inorganic anions and metal ions which is routinely used in many laboratories as an alternative to ion-exchange chromatography. These indirect UV methods are used for the determination of drug counter-ions (Chapter 5) and raw materials/excipients (Chapter 7). Figure 1.12 shows the percentage of use for each application type. The range of application areas is more diverse than those mentioned in the survey and separate chapters covering the extended range are included such as vitamin analysis (Chapter 9), dissolution testing (Chapter 8), clinical determinations (Chapter 10), forensic applications (Chapter 15) and trace level determinations (Chapter 6) are also included. There may be many factors to optimise in CE method development and the use of experimental designs and chemometrics is covered in Chapter 14. Conventionally CE has been performed using

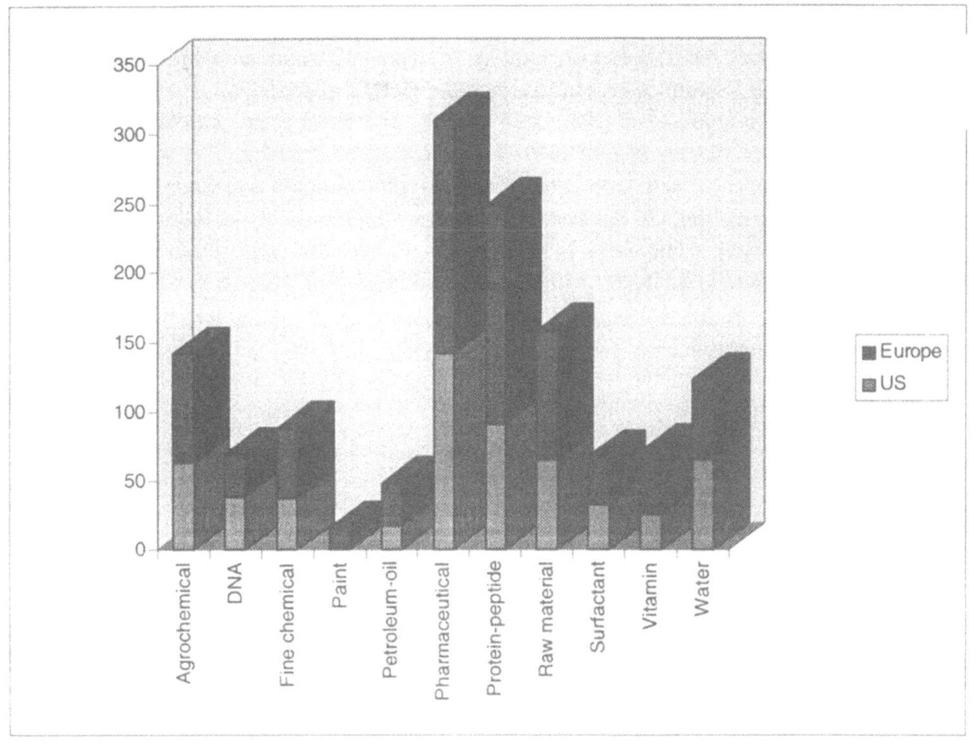

Figure 1.11 Application areas of CE across all industries. (Reproduced with permission from ref. 48)

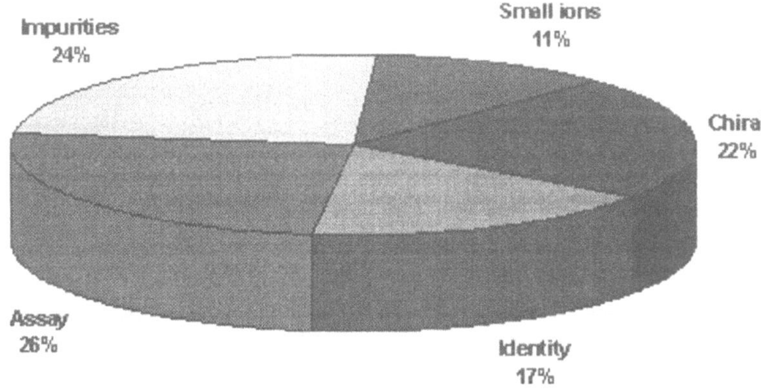

Figure 1.12 Application areas of CE in the pharmaceutical industry. (Reproduced with permission from ref. 49)

aqueous buffers but there is a increase in the use of non-aqueous electrolytes (Chapter 13). The advent of capillary electrochromatography (Chapter 12) increases the importance of capillary electrodriven separation techniques in pharmaceutical analysis.

Methods have been successfully validated in many industrial pharmaceutical analysis laboratories and Chapter 11 provides an overview on method validation. The attitude of regulatory authorities towards a new technique of considerable concern and Chapter 16 contains a section which confirms that CE has been well received by regulatory authorities.

The general intention of this book is to highlight the possible application areas of CE in the area of pharmaceutical analysis. Although the compound of interest to each reader may not be included there are many examples and references throughout the text which may steer readers in the right direction.

References

1. Jorgenson J W and Lukacs K D, Zone electrophoresis in open-tubular glass capillaries, *Anal. Chem.*, 53 (**1981**) 1298-1302.

2. Li SFY, in Capillary Electrophoresis, principles, practice and applications, Elsevier press, **1992**.

3. Kuhn R and Hoffstetter-Kuhn, in Capillary Electrophoresis: Principles and Practice, Springer-Verlag Press, Berlin, **1993**.

4. Camilleri P, in Capillary Electrophoresis: Theory and Practice, CRC Press, Boca Raton, **1993**.

5. Altria KD, Ed. Capillary Electrophoresis Guidebook – Principles, Operation and applications, Humana Press, Totowa, **1995**.

6. H Watzig, Appropriate calibration functions for capillary electrophoresis I. Precision and sensitivity using peak areas and heights, *J. Chromatogr. A*, 700 (**1995**) 1-7.

7. Terabe S, Otsuka K and Ando T, Electrokinetic chromatography with micellar solution and open-tubular capillary, *Anal. Chem.*, 57 (**1985**) 834-841.

8. Altria KD, Bryant SM and Hadgett T, Validated capillary electrophoresis method for the assay of a range of acidic drugs and excipients, *J. Pharm. Biomed. Analysis.*, 15 (**1997**) 1091-1101.

9. Swartz ME, Method development and selectivity control for small molecule pharmaceutical separations by capillary electrophoresis, *J. Liq. Chromatogr.*, 14 (**1991**) 923-938.

10. Ackermans MT, Everaerts FM and Beckers JL, Determination of aminogycloside antibiotics in pharmaceuticals by capillary zone electrophoresis with indirect UV detection coupled with micellar electrokinetic capillary chromatography, *J.Chromatogr.*, 606 (**1992**) 229-235.

11. Fujiwara S and Honda S, Determination of ingredients of antipyretic analgesic preparations by micellar electrokinetic capillary chromatography, *Anal. Chem.*, 59 (**1987**) 2773-2776.

12. McLaughlin GM, Nolan JA, Lindahl JL, Morrison JA and Bronzert TJ, Practical drug separations by HPCE : practical considerations, *J. Liquid Chromatogr.*, 15 (**1992**) 961-1021.

13. Ong CP, Ng CL, Lee HK, Li SFY, Determination of antihistamines in pharmaceuticals by capillary electrophoresis, *J. Chromatogr.*, 588 (**1991**) 335-339.

14. Wainwright A, Capillary electrophoresis applied to the analysis of pharmaceutical compounds, *J. Microcol Sep.*, 2 (**1990**) 166-175.

15. Altria KD, Frake P, Gill I, Hadgett T, Kelly MA, and Rudd D R, Validated capillary electrophoresis method for the assay of a range of basic drugs and excipients, *J. Pharm. Biomed. Analysis*, 13 (**1995**) 951-957.

16. Gonzalez E and Laserna JJ, Capillary zone electrophoresis for the rapid screening of banned drugs in sport, *Electrophoresis*, 15 (**1994**) 240-243.

17. Chee GL and Wan TSM, Reproducible and high-speed separation of basic drugs by capillary zone electrophoresis, *J. Chromatogr.*, 612 (**1993**) 172-177.

18. Lukkari P, Ennelin A, Siren H and Riekkola M L, Effect of temperature, effective capillary length, and applied voltage on the migration of 9 ß-blockers in micellar electrokinetic capillary chromatography, *J. Liq. Chromatogr.*, 16 (**1993**) 2069-2079.

19. Lin C-E, Chang C-C, Lin W-C Lin and Lin EC, Capillary zone electrophoretic separation of β-blockers using citrate buffer at low pH, *J. Chromatogr. A*, 753 (**1996**) 133-138.

20. Bechet I, Fillet M, Hubert P and Crommen J, Determination of benzodiazeprines by micellar electrokinetic chromatography, *Electrophoresis*, 15 (**1994**) 1316-1321.

21. Bretnall AE and Clarke GS, Investigation and optimisation of the use of micellar electrokinetic chromatography for the analysis of six cardiovascular drugs, *J.Chromatogr.*, 700 (**1995**) 173-178.

22. Nguyen NT and Siegler RW, Capillary electrophoresis of cardiovascular drugs, *J. Chromatogr. A*, 735 (**1996**) 123-150.

23. Sciacchitano CJ, Mopper B, and Specchio JJ, Identification and separation of five cephalosporins by micellar electrokinetic capillary chromatography, *J. Chromatogr.*, 657 (**1994**) 395-399.

24. Nishi H, Tsumagari N and Terabe S, Effect of tetraalkylammonium salts on micellar electrokinetic chromatography of ionic substances, *Anal. Chem.*, 61 (**1989**) 2434-2439.

25. Nishi H, Fukayama T, Matsuo M and Terabe S, Effect of surfactant structures on the separation of cold medicine ingredients by micellar electrokinetic chromatography, *J. Pharm. Sci.*, 79 (**1990**) 519-523.

26. Nishi H, Fukayama T, Matsuo M and Terabe S, Separation and determination of the ingredients of a cold medicine by micellar electrokinetic chromatography with bile salts, *J. Chromatogr.*, 498 (**1990**) 313-323.

27. Trenerry V C, Robertson J and Wells R J, The determination of cocaine and related substances by micellar electrokinetic capillary chromatography, *Electrophoresis*, 15 (**1994**) 103-108.

28. Jumppanen J H, Wiedmer S K, Siren H, Riekkola M L and Haario H, Optimized separation of seven corticosteroids by micellar electrokinetic chromatography, *Electrophoresis*, 15 (**1994**) 1267-1272.

29. Nishi H and Matsuo M, Separation of corticosteroids and aromatic hydrocarbons by cyclodextrin-modified micellar electrokinetic chromatography, *J. Liq. Chromatogr.*, 14 (**1991**) 973-986.

30. Nishi H, Fukayama T, Matsuo M and Terabe S, Separation and determination of lipophilic corticosteroids and benzothiazepin analogues by micellar electrokinetic chromatography with bile salts, *J. Chromatogr.*, 513 (**1990**) 279-295.

31. Bumgarner J G and Khaledi M G, Mixed micellar electrokinetic chromatography of corticosteroids, *Electrophoresis*, 15 (**1994**) 1260-1266.

32. Jumppanen J, Siren H and Riekkola M L, Screening for diuretics in urine and blood serum by capillary zone electrophoresis, *J. Chromatogr.*, 652 (**1993**) 441-450.

33. Riekkola ML and Jumppanen JH, Capillary electrophoresis of diuretics, *J. Chromatogr. A*, 735 (**1996**) 151-164.

34. Weinberger R and Lurie I S Micellar electrokinetic capillary chromatography of illicit drug substances, *Anal. Chem.*, 63 (**1991**) 823-827.

35. Tagliaro F, Smith FP, Turrina S, Equisetto V and Marigo M, Complementary use of capillary zone electrophoresis and micellar electrokinetic capillary chromatography for mutual confirmation of results in forensic drug analysis, *J. Chromatogr. A*, 735 (**1996**) 227-235.

36. Davis P, Lafon R, Large T, Morgan E D and Wilson I D, Micellar capillary electrophoresis of the ecdysteroids, *Chromatographia*, 37 (**1993**) 37-42.

37. Holland R D and Sepaniak M J, Qualitative analysis of mycotoxins using micellar electrokinetic capillary chromatography, *Anal. Chem.*, 65 (**1993**) 1140-1146.

38. Donato M G, Baeyens W, Van-den-Bossche W and Sandra P, The determination of non-steroidal antiinflammatory drugs in pharmaceuticals by capillary zone electrophoresis and micellar electrokinetic capillary chromatography, *J. Pharm. Biomed. Anal.*, 12 (**1994**) 21-26.

39. Nishi H and Terabe S, Application of micellar electrokinetic chromatography to pharmaceutical analysis, *Electrophoresis*, 11 (**1990**) 691-701.

40. Swartz ME, Method development and selectivity control for small molecule pharmaceutical separations by capillary electrophoresis, *J. Liq. Chromatogr.*, 14 (**1991**) 923-938.

41. Muijselaar PGHM, Claessens HA and Cramers CA, Determination of structurally related phenothiazines by capillary zone electrophoresis and micellar electrokinetic chromatography, *J. Chromatogr. A*, 735 (**1996**) 395-402.

42. Kristensen H K and Hansen S H, Separation of polymyxins by micellar electrokinetic capillary chromatography, *J. Chromatogr.*, 628 (**1993**) 309-315.

43. Sun S-W and Chen L-Y, Optimisation of capillary electrophoretic separation of quinolone antibacterials using the overlapping resolution mapping scheme, *J. Chromatogr. A*, 766 (**1997**) 215-224.

44. Ackermans MT, Beckers JL, Everaerts FM, Hoogland H and Tomassen MJH Determination of sulphonamides in pork meat extracts by capillary zone electrophoresis, *J. Chromatogr.*, 596 (**1992**) 101-109.

45. Dang Q-X, Yan L-X, Sun Z-P and Ling D-K, Separation and simultaneous determination of the active ingredients in theophylline tablets by micellar electrokinetic capillary chromatography, *J. Chromatogr.*, 630 (**1993**) 363-369.

46. Korman M, Vindevogel J and Sandra P, Application of micellar electrokinetic chromatography to the quality control of pharmaceutical formulations: The analysis of xanthine derivatives, *Electrophoresis*, 15 (**1994**) 1304-1309.

47. Altria KD and Simpson CF, Analysis of some pharmaceuticals by capillary zone electrophoresis, *J. Pharm. Biomed. Anal.*, 6 (**1988**) 801-805.

48. Altria KD and Bryant S, An investigation into the current applications and status of capillary electrophoresis, *LC-GC Int.*, (**1997**) 26- 30.

49. Altria KD and Kersey M, Capillary electrophoresis and pharmaceutical analysis: a survey of the industrial application and their status of in the united states and united kingdom, *LC-GC*, Jan. **1995** 40-46.

2 Main Component Assay by CE

2.1 Introduction

FSCE and MECC methods have been employed to quantify the level of drugs in a large variety of pharmaceutical preparations. Combinations of CE and HPLC are frequently employed in cross-validation studies during method validation. The choice between adopting CE or HPLC for a particular application is dependent upon the relative merits of each technique to the individual assay. Often CE can have advantages in terms of reduced sample pre-treatment, reduced consumable costs and reduced analysis time.

Table 2.1 shows that CE has been used to analyse both liquid formulations such as syrups and injection solutions and also to assay solid preparations such as tablets and creams. Various selectivity options have been used to separate water-soluble and water-insoluble drugs and compounds which are ionisable or unionisable. The various options are briefly discussed in this chapter with details of standard electrolyte compositions. The most significant performance measurements in quantitative assay are those of precision and accuracy. The factors affecting CE precision (and to a lesser extent accuracy) are widely different to HPLC which is currently the predominantly employed analytical technique for drug assay. The main difference is that HPLC employs a well controlled fixed volume injector system whilst a time and pressure dependent hydrodynamic injection is generally employed in CE. In addition, sample solution viscosity is an important factor in CE as it greatly effects the volume of sample injected into the capillary whilst viscosity has no influence on HPLC injection volumes. Internal standards are widely used in CE to improve precision (Table 2.1). The accuracy of CE has been widely demonstrated by comparison of CE data with label claim and/or HPLC data. A section is included which describes results from a number of these studies. Validation of CE methods is similar to HPLC and several reports have shown successful validation of CE methods and details from a selection of these reports are covered.

Given that CE methods are a viable alternative to HPLC based methods then often a choice between the two is possible for particular analyses. The relative merits of CE and HPLC are discussed with regards to operational parameters such as analysis time, sample pretreatment requirements, and ease of operation.

2.2 Reported Applications

The reported applications largely fall into two main categories – those compounds soluble or insoluble in water. Water-insoluble drugs are generally analysed by Micellar Electrokinetic Capillary Chromatography (MECC) whilst water soluble ionisable drugs can be analysed by Free Solution CE (FSCE) or MECC as appropriate. Generally basic compounds can be analysed by FSCE at pH values 2–4 whilst acidic compounds have been analysed by FSCE using electrolytes at pH 6–9.

Table 2.1 Quantitative pharmaceutical applications of CE

	Electrolyte	Sample	IS	Ref.
Low pH FSCE				
Basic drugs – various	FSCE, pH 2.5	Various formulations	Yes	1
Bronchodilators	FSCE, pH 5	Various formulations	No	2
Domperidone	FSCE, pH 4	Tablets	Yes	3
Dothiepin	pH 5, CD	Tablets	No	4
Eminase	FSCE pH 2.5, cellulose	Injection solutions	No	5
Isoxsuprine	FSCE, pH 6	Tablets	Yes	6
Loxiglumide	FSCE, pH 6	Injections and tablets	Yes	7
Minoxidol	FSCE, pH 3	Tablets	Yes	8
Non-benzodiapenic anxiolytic agents	FSCE, pH 3	Drug substance samples	Yes	9
Opiods	FSCE, pH 4 with CD	Crude opium	No	10
Quinolone & impurities	FSCE, pH 2	Drug substance	No	11
Ranitidine & impurities	FSCE, low pH	Syrup formulation	No	12
Sumatriptan	FSCE, pH 2	Formulations	Yes	13
High pH FSCE				
Acidic drugs	FSCE, pH 9.5	Formulations	Yes	14
Benzylpenicillin	High pH FSCE	Formulations	Yes	15
Cimetidine	FSCE, pH 7	Formulations	Yes	16
Cefotaxime	pH8 FSCE	Tablets	No	17
Enalapril	FSCE pH 9	Tablets	No	18
Gentamycin and related impurities	FSCE, pH 9.4	Injection solutions	No	19
NSAIDs	Glycine – triethanolamine pH 9.1	Various formulations	No	20
Neomycin and aminoglycoside antibiotics	FSCE, various (indirect detection)	Eardrops	Yes	21
MECC				
Analgesics – various	MECC, SDC	Tablets	Yes	22
Analgesics – various	MECC	Tablets	Yes	23
Analgesics – various	MECC	Formulations	Yes	24
Antibiotics	MECC, SDS	Formulations	No	25
Antihistamines	MECC, SDS, TAA	Tablets	No	26
Benzodiazepines	SDS, MeOH	drug substance	Yes	27

Table 2.1 continued

	Electrolyte	Sample	IS	Ref.
Caffeine and derivatives	MECC, SDS	Tablets	No	28
Cold medicine ingredients	MECC, cholate	Tablets	Yes	29
Cefotaxime	MECC, SDS	Drug substance	No	30
Cephradine and its related impurities	MECC, SDS	Drug substance	No	31
Dilitiazem & impurities	MECC, SDC	Tablets and creams	Yes	32
Enalapril maleate	MECC, Brij 35	Tablets	No	33
Hydrochlorothiazide and chlorothiazide	MECC , SDS	Tablets	No	34
Hydroquinone	MECC, SDS	Cream	Yes	35
NSAIDs – various	MECC, SDS	Tablets	No	36
NSAIDs – various	FSCE and MECC	Capsules, suppositories and others	Yes	37
Paracetamol	MECC, SDS	Capsules	Yes	38
Steroids	MECC	Creams	Yes	39
Steroids and benzothiazepins	MECC, bile salts	Tablets, creams, drug substance	Yes	40
Theophylline and derivatives	MECC, SDS	Tablets	Yes	41
Xanthines – various	MECC, SDS	Tablets	No	42

where :

Brij 35	= neutral surfactant	NSAIDs	= non steroidal anti-imflammatories
CD	= cyclodextrin	SDS	= sodium dodecyl sulphate (surfactant)
FSCE	= free solution capillary electrophoresis	SDC	= sodium deoxycholate (surfactant)
MECC	= micellar electrokinetic capillary chromatography	TAA	= tetra-alkylammonium salts (ion-pairing reagent)

2.2.1 Low pH

2.2.1.1 Water soluble basic drugs

A typical standard buffer employed (1, 5, 8, 11, 13) is NaH_2PO_4, pH adjusted to pH 2.5 with concentrated H_3PO_4, for low pH separations. Typically an internal standard such as imidazole or aminobenzoic acid (HCl salt) is employed to give good precision.

Table 2.2   Assay results for CE for tablets and drug substance

3TC content (mg/tablet)		
Label claim		150
HPLC		152.2
CE		155.6
Label claim		100
CE		104.0
Histamine acid content (% w/w)		
CE	Sample 1	100.3, 100.2
	Sample 2	100.8, 100.7
	Average	100.5

Data reproduced with permission from (1)

Use of a simple pH 2.5, 25 mM phosphate buffer has been validated (1) for assay of a wide range of basic drugs. The method was shown to have good sensitivity, linearity (correlations > 0.999) with relative standard deviation (RSD) values of 0.3–2 % for peak area ratios. The robustness of the method to deviations in the method settings was satisfactorily assessed using an experimental design (see Chapter 14 for details on the use of experimental designs). Shelf-lives of electrolyte and sample solutions were assessed (3 months and 14 days, respectively). The analysis was repeated on different days, by different analysts on different equipment using different reagents and samples. This method is routinely used for quantitative analysis of basic drug substances and pharmaceutical formulations. Good precision was obtained using either aminobenzoic acid or imidazole as the internal standard. Figure 2.1 shows a typical separation obtained under these conditions. Good agreement (1) was obtained between CE data and the label claim for tablets (Table 2.2).

Levels of isoxsuprine in tablets were determined by analysing aqueous extracts containing an internal standard (6). Peak area ratio precision was better than 2%, recovery experiments showed a satisfactory 99.3 % result. Similar good performance has been reported in the determination of cimetidine (16).

Levels of domperidone were assayed (3) in various formulations using a citrate-phosphate electrolyte (pH 4) and the assay results were in good agreement with label claim and HPLC data. The use of an internal standard enabled CE peak area ratio RSD values of <2 % RSD. In a similar study, levels of the anti-migraine drug, sumatriptan, in injection solutions, were quantified (13) using a pH 2 phosphate buffer and CE results were in close agreement (Table 2.3) with HPLC data and label claim (12 mg/ml). An internal standard was used for improved CE precision. Figure 2.2 shows the separation of sumatriptan and the internal standard.

Ackermans et al (2) quantified levels of the three bronchodilators, salbutamol, terbutaline and fenoterol, in six different pharmaceutical dosage forms. The CE data was compared successfully (Table 2.4) with label claim and results generated by HPLC. The CE data showed RSD values typically around 1% with linearity data showing correlation greater than 0.9997. In this study internal standards were not employed.

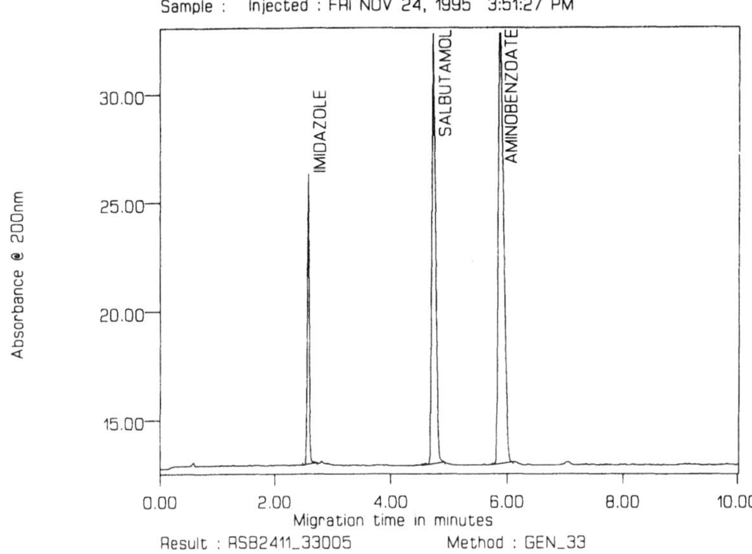

Figure 2.1 Separation of salbutamol and 2 internal standards. Separation conditions: 25 mM phosphate pH 2.5, detection at 200 nm, 27 cm × 50 µm capillary.

Table 2.3 Sumatriptan contents in Injection solutions as determined by CE and HPLC

Sample	Sumatriptan content (mg/ml)	
	CE	HPLC
Batch 2 Sample 1	11.5, 11.6	11.6, 11.6
Batch 2 Sample 2	11.6, 11.6	11.7, 11.7
Batch 3 Sample 1	11.7, 11.8	11.8, 11.8
Batch 3 Sample 2	11.6, 11.6	11.7, 11.7
Batch 4 Sample 1	11.7, 11.8	11.8, 11.8
Batch 4 Sample 2	11.7, 11.6	11.7, 11.7

Data reproduced with permission from (13)

2.2.1.2 Water insoluble basic drugs

Water-insoluble basic drugs can often be analysed by dissolving the sample in either an organic solvent–water mixture or in a dilute acid solution. For example (8) levels of the water insoluble compound minoxidil were determined at pH 3 using a 100 mM NaH_2PO_4 buffer. Samples of tablets and topical ointments were prepared by dissolving in water–ethanol (1:1).

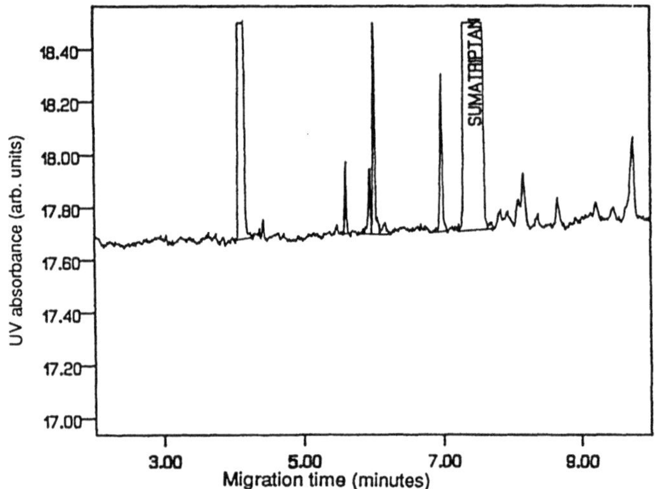

Figure 2.2 Separation of sumatriptan an internal standard and a range of impurities. Data reproduced with permission from (13)

An internal standard (isoxsuprine) was employed to improve precision to 0.8% RSD. A short (14 cm) capillary was employed to give an analysis time of less than 2 minutes. Tablet assay results generated by the CE method were in line with the label claim. Similarly (8) levels of water-insoluble dothiepin were determined in tablets using 50 mM NaH_2PO_4 electrolyte (pH 4.7) containing 10% propan-1-ol and 10 mM β-cyclodextrin. The samples were prepared in 60:40 acetonitrile : water. A reproducibility result of 0.77 % RSD was obtained for assay of

Table 2.4 Table showing quantitative analyses of formulations using CE and HPLC

Sample	HPLC	CE
Salbutamol		
Tablets (4 mg/tablet)	4.05, 4.00	3.94, 3.72
Infusions (1 mg/mL)	1.04, 1.02	0.99, 1.00
Syrups (0.4 mg/mL)	0.40, 0.40	0.41, 0.39
Terbutaline		
Tablets (5 mg/tablet)	4.92, 4.89	4.68, 4.67
Ampoules (0.5 mg/mL)	0.51, 0.51	0.52, 0.50
Syrups (0.3 mg/mL)	0.31, 0.30	0.30, 0.30
Fenoterol		
Tablets (2 mg/tablet)	1.84, 1.86	1.88, 1.97
Respirator	4.98, 4.92	4.89, 4.91

Reproduced with permission from (2)

six tablets using the method. The assay was applied (8) on six separate days giving an overall RSD of 1.87 %.

Alternatively (1) samples of water-insoluble basic drugs can be prepared in acidic solutions in which they are protonated. Often a 1:10 dilution of the separation electrolyte produces a solution with a sufficiently low pH to solubilise water-insoluble compounds. Alternatively if the compound also possesses an acidic function, it may be possible to dissolve the compound in dilute NaOH and analyse using a low pH electrolyte (11). Obviously appropriate consideration to analyte stability under these conditions would be required when selecting an appropriate dissolving solvent.

Levels of various water soluble non-benzodiazepinic anxiolytic drugs were determined (9) using a tris-phosphate buffer. RSD values for migration times were 0.4 % ($n = 30$) and 0.7 % ($n = 20$) for intraday and intraday respectively. Replicate testing of samples ($n = 6$) gave RSD values of ±1.2 % for zalospirone, ±1.3 % for gepirone and ±1.0 % for ipsapirone. Data was comparable to HPLC data.

Samples of water insoluble opiod alkaloids such as morphine, thebaine and papaverine were dissolved in water–DMSO–1M acetic acid (95:0.5:4.5 v/v). 6-amino caproic acid (50 mM) adjusted to pH 4.0 with glacial acetic acid was used (10) as background electrolyte with the addition of 30 mM dimethyl-β-cyclodextrin. Linear regression coefficients were in all cases >0.999. Injection reproducibility was reported as 1.1 % RSD. Recovery values for morphine spiked into a cough syrup preparation were 99.9–100.6 %. Limits of detection were typically 0.08–0.3 mg/L. Levels of morphine in various formulations were determined by both the CE method and a validated HPLC method (Table 2.5) and were shown to be in good agreement.

Table 2.5 Analysis of opiod formulations

	Pectyl Cough mixture	Pectyl Strong Cough mixture	Opium Tincture
HPLC method	0.20 mg/ml	0.29 mg/ml	11.6 mg/ml
CE method	0.20 mg/ml	0.31 mg/ml	11.4 mg/ml

Reproduced with permission from (10)

2.2.2 High pH

2.2.2.1 Water soluble acidic drugs

Simple phosphate, borate or phosphate-borate mixtures, with or without pH adjustments, are employed (14–16, 18,19) in the range pH 6–10. A common electrolyte used (14) to separate a wide range of acidic drugs is 10 mM borate which has a natural pH of 9.5. These electrolytes give stable electro-osmotic flow (EOF) conditions and allow fast, and efficient, separations of water-soluble acidic compounds

The use of a simple 10 mM borate buffer (natural pH ~9.5) has been extensively validated (14) as a general method for the analysis of a range of acidic drugs. Operation with this simple buffer gave good precision for peak area ratios (routinely less than 1 %) and a very stable EOF (RSD for relative migration time 0.1 %). Validation showed acceptable linearity, reagent stability (shelf-life of 3 months) and linearity. Borate has minimal UV background which permits use of low UV wavelengths to achieve acceptable sensitivities in the order of single figure mg/L values. Aminobenzoic acid and β-naphthoxy acetic acid were used (14) as internal standards to give good precision and accuracy (Table 2.6). Figure 2.3 shows a typical separation of an acidic drug (warfarin) and the two internal standards. Water-insoluble acidic compounds were dissolved in 10 mM NaOH. However, due attention should be paid to sample stability using this approach as it was observed, for example, that acetylsalicylic acid rapidly decomposed under these conditions. The method has been repeated in several laboratories world-wide.

A borate buffer has also been used (18) for the determination of enalapril maleate which is used for the treatment of hypertension. The method was shown to give good precision (RSD = 0.62 %, n = 10) with no internal standard. The method was stability indicating and gave linearity correlation coefficients greater than 0.999. Samples were dissolved in distilled water to maximise stacking effects. Ten Vasotec tablets containing enalapril were assayed (18) for content uniformity on two different CE instruments and in both cases gave an equivalent average result to that obtained by HPLC assay.

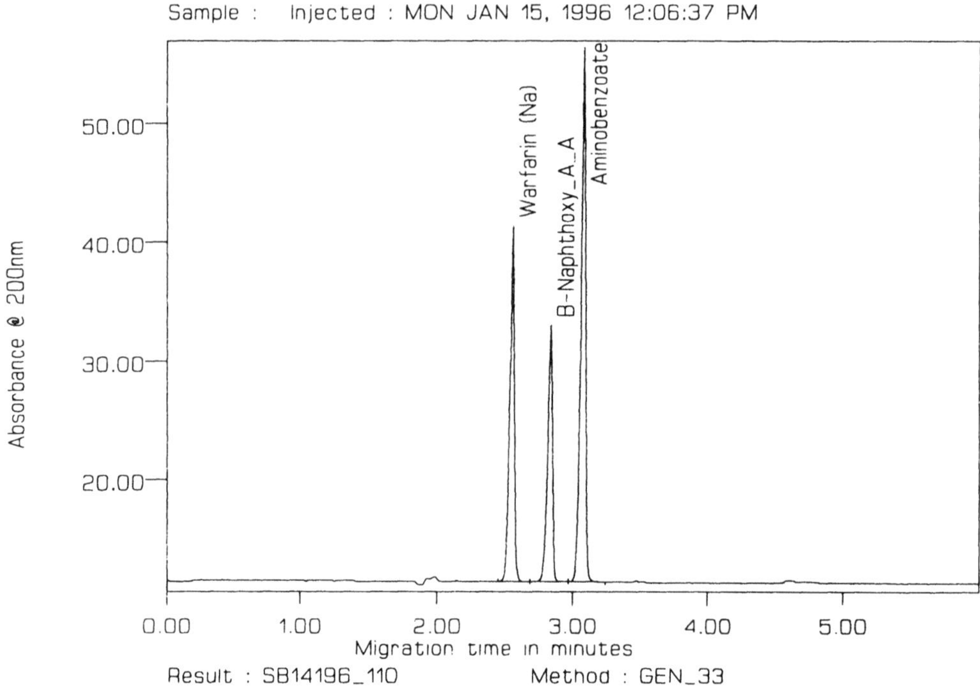

Figure 2.3 Separation of warfarin and two internal standards using a borate buffer.Separation conditions: 10 mM borate, +6.5 kV, 27 cm (total length) × 75 μm capillary, 200 nm, 30°C, 0.1 mg ml^{-1} in water.

Table 2.6a Precision of injection for a range of acidic drugs.

Solute	RMT	PAR
GW1 (calibration)	0.23	0.34
GW1 (sample)	0.19	0.56
GW2 (calibration) Beckman	0.21	0.76
GW2 (calibration) Hewlett-Packard	0.13	1.31
GW2 (sample)	0.34	0.89
Levothyroxine	0.32	0.58
Omperazole	0.31	0.89

PAR = peak area ratio
RMT = relative migration time

Table 2.6b Precision of injection %RSD ($n = 10$)

Tablet	Label claim (mg/tablet)	CE result (mg)	% label claim
GW1	50	50.2	100.4%
GW2	200	198	99%
GW2	400	402	100.5%
Levothyroxine	0.1	0.103	103%
Omeprazole	20	20.7	103%

Reproduced with permission from (14)

A systematic study on the effect of the electrolyte components used (20) for the high pH separation of non-steriodal anti-inflammatory acidic drugs. The optimum electrolyte was found to be 75 mM glycine adjusted to pH 9.1 with triethanolamine. This gave RSD values of around 0.1 and 0.5 % for within-day reproducibility of migration times and peak areas. Samples were prepared in 7.5 mM glycine pH adjusted to pH 9.1 with triethanolamine to maximise sample solubility and stacking effects. Linearity correlations of 0.9999 were obtained over the range 2–100 mg/L. Average accuracy results were 102 % recovery.

The analysis of aminoglycoside antibiotics, such as gentamicin sulphate, is difficult by conventional techniques as they have poor UV absorption coefficients and require derivatisation prior to HPLC analysis. However, use of high concentration (150 mM) borate buffers enabled (19) direct analysis by CE with detection at 195 nm. The borate is able to complex on-column with the hydroxyl groups present on the gentamycin to give an effective negative charge. Gentamycin content was determined (19) by CE in a range of injectable solutions, with a simple dilution being the only sample pre-treatment step. The CE results obtained were in line with those obtained by microbial assay and the label claim. Repeatability of the method was shown by preparing different capillaries and buffer solutions on four different occasions.

Levels of the antibiotic cefotaxime were determined (17) in drug substance and tablets using a pH 8 electrolyte. Careful control of rinsing and operating procedures enabled RSD values of 0.2 and 0.3 % for migration time and peak area precision to be obtained.

2.2.2.2 *Water insoluble acidic drugs*

Water-insoluble acidic compounds can be prepared in organic solvents or high pH aqueous solutions. For instance (7) levels of the acidic drug loxiglumide were determined in injection solutions and tablets using a pH 6 Tris electrolyte. Samples were prepared in a 60:40 water : acetonitrile mixture. An internal standard was used to give good injection precision and good linearity >0.999.

2.2.3 MECC

MECC conditions are employed when selectivity requirements for an separation exceed the simple mobility differences obtainable in FSCE. MECC is also used for neutral compounds or when analysing mixtures of neutral and charged solutes. A typical electrolyte composition (28, 34–36) for MECC may be 10 mM borate containing 40 mM SDS. Analysis of very water-insoluble drugs can be performed by addition of organic solvents to the MECC electrolyte. Levels of organic solvents such as methanol or acetonitrile may be used up to 30 % v/v to improve solute solubility. Alternatively (43, 44) lower levels such as 5–10 % of solvents such as propanol, acetone, DMSO or butan-1-ol can be employed.

Other approaches to analysis of water-insoluble compounds using SDS based MECC, include use of SDS in combination with cyclodextrins, or with high levels of urea (45) which significantly improves solute solubility. An alternative MECC approach (46) is to use bile salts, such as sodium cholate, to generate the micelles instead of SDS. Micellar solutions containing bile salts are very effective (40) in solubilising lipophilic compounds such as corticosteroids and benzothiazepin analogues.

To-date, the most extensive validation and evaluation of the analytical performance of an MECC method has been conducted by Thomas et al (34) who have used MECC for analysis of hydrochlorothiazide and chlorothiazide. The method involved use of a 20 mM borate buffer containing 30 mM SDS. A 100 micron capillary was employed to give large peak areas to minimise integration-related errors. Careful control of the method settings such as injection time, temperature and sample concentration enabled sub 1% RSD values to be routinely obtained with no internal standard. Recovery values ranged from 99.5–100.6 for hydrochlorothiazide over the range 50–150 % of target concentration. Detector linearities were greater than 0.998. Analyses were successfully repeated on different days by different analysts using different capillaries to demonstrate robustness. The method was successfully validated (34) according to USP guidelines.

A MECC method has been validated to United States Pharmacopeia (USP) guidelines for determination of cephradine for routine quality control (31). Recovery experiments were performed with results covering the range 99.8–100.2. Linearity data covering 50–150 % of the nominal concentration gave a correlation coefficient of 0.9996. Several batches of product were analysed by both MECC and HPLC and the results showed no significant difference. The electrolyte contained both SDS and Brij 35 surfactants which stabilised the separation and gave an average RSD value of 0.37 % for the area of repeated injections, an RSD of 0.40 % was obtained for the migration time obtained (31) in 23 repeated injections.

An SDS-based MECC method for the quantitation of paracetamol in capsule formulations has been evaluated in a cross-company method transfer exercise (38) involving seven independent pharmaceutical companies. The separation was successfully repeated by each company by careful control of the capillary conditioning procedure. An internal standard (4-hydroxy acetophenone) was employed to give good precision values for response factors of the calibration solutions (average precision for the seven companies was 1.5 % RSD). Use of rinse cycles of 0.1 M NaOH and running electrolyte between injection, in conjunction with temperature control gave an average RSD of 0.52 % for the relative migration time of para-cetamol (compared to the internal standard). Each of the seven companies analysed four capsules in duplicate. The overall mean MECC result of 298 mg/capsule (99.4 % of label claim) compared well (38) with the HPLC data of 302 mg/capsule (100.8 % of label claim). Figure 2.4 shows separations obtained by 2 of the companies – the capsules also contained caffeine and this, and the internal standard, were repeatedly separated from the paracetamol.

Ten benzodiazepines were simultaneously resolved (27) using an MECC electrolyte containing 25 mM SDS, 20 % v/v methanol in 75mM glycine – 250mM triethanolamine buffer. RSD values for migration times in repeated injections was 0.3–0.5 %. Linearity data over the range 1–50 mg/L gave correlation coefficients of $\geq$ 0.9993 (n = 6). Good between day migration time repeatability was obtained.

A range of xanthine derivatives have been determined (42) in the quality control of pharmaceutical formulations. The nine xanthines, including caffeine, theobromine and theophylline, were quantified in 30 mM borate solution containing 60 mM SDS. The validation data showed precision values of <1 % for migration times and 1–2 % for peak areas. Recoveries of drug were on the range of 98.8–101.4 with RSD values of 0.27–1.12 % for reproducibility of recovery. Linearity data ranged from 0.99903 to 0.99998.

Various β-lactam antibiotics have been assayed (25) using a phosphate-borate pH8 electrolyte containing 50 mM SDS. The method gave precisions of 0.67–0.78 for migration time precision and 1.29–1.97 for peak area precision (no internal standard). Compounds analysed include amoxicillin, ampicillin and penicillin V potassium. Assay results were in line with declared potency results.

Neutral or ionisable water-insoluble compounds are invariably analysed by MECC. Samples are generally prepared in organic solvent–water mixtures or in MECC electrolyte which is quite solubilizing. For very water-insoluble compounds such as steroids, samples are prepared in 100 % organic solvents such as methanol. For example (36) methanolic solutions of various non-steroidal antiflammatory (NSAID's) drugs including naproxen were injected with RSD values of around 1 %. Assay figures were in line with label claim. Donato et al (37) have performed a more extensive investigation into the MECC of NSAID's. Samples were dissolved in running electrolyte and an internal standard (flurbiprofen) was employed. Separation was performed using a 50 mM borate solution containing 40 mM SDS. Various pharmaceutical formulations were analysed (37) including suppositories, capsules and injection solutions.

Various anti-histamines have been quantified (26) by MECC using a phosphate-borate electrolyte containing SDS, β-cyclodextrin and tetrabutylammonium sulphate. Samples from tablet formulations were dissolved in methanol and directly analysed. Assay results were in agreement with label claim.

Levels of the particularly water-insoluble compound hydroquinone, present in skin-toning cream, were determined (35) using SDS-based electrolyte containing 10 % methanol. Caffeine was used as an internal standard for the hydroquinone analysis. Methanolic solutions of Topsym cream (which contains 0.5 % of the steroid, fluocinonide) were directly analysed (39) using a pH 9 phosphate-borate buffer containing 100 mM sodium cholate. The assay figures confirmed the label claim. In the same study (39) levels of diltiazem in tablets were determined using the same electrolyte composition. Good linearity (>0.998) with recoveries of approximately 100 % were reported.

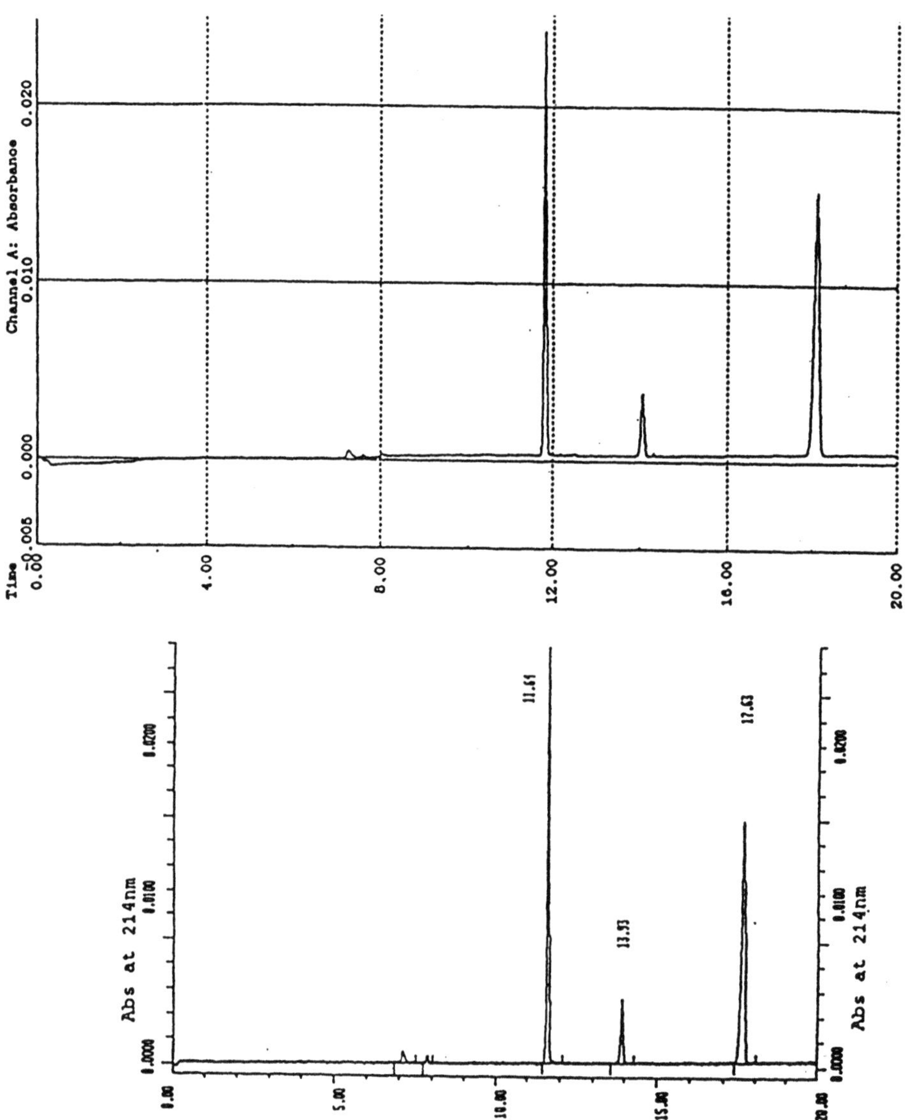

Figure 2.4 MECC of paracetamol and caffeine.
Migration order 1. paracetamol, 2. caffeine and 3. 4-hydroxyacetophenone (int.std) Instrument specific settings. Conditions: electrolyte: 40 mM disodium tetraborate, 125 mM sodium lauryl sulphate; detection at 210 or 214 nm; capillary: fused silica 72 cm × 50 μm (1, 3) 57 × 50 μm (2); temperature: 40°C.

2.4 Identity Confirmation Testing

Product strength and identity confirmation testing forms an important part of the workload of many pharmaceutical analysis laboratories. The testing can involve use of a separative method to confirm that the product contains the required active species and that the active drug is present at the required level. This is especially useful in the areas of quality assurance and clinical supplies testing where many samples types (and sample strengths) may be analysed. Often for example in clinical trials samples covering several drugs at different strengths (together-with placebos) will be supplied to clinicians. CE is particularily effective in this area as it is both time and cost effective.

2.5 General Considerations in Quantitative Analysis

2.5.1 Precision

Hydrodynamic sample introduction is generally used in CE in which the sampling end of the capillary is immersed in the sample solution and a pressure difference applied (positive pressure or vacuum). The volume of sample solution injected onto the column is related (47) to the pressure difference (mbar) across the capillary, the capillary diameter and length, the sample solution viscosity and the injection time (seconds).

Injection precisions of 1 % RSD or less are typically obtained using standard commercial HPLC instrumentation. Typical CE RSD values of 1–3 % RSD were reported in early (16, 29, 36) CE reports on home-made equipment. This increased variability may not be unexpected given that the typical volumes injected for HPLC and CE are 10 µl and 10 nl respectively. However, a number of reports have also described (1, 13, 34, 42) precisions of 1% RSD or better for CE on automated commercial equipment when carefully controlling factors such as sample concentration, use of an internal standard, constant temperature, sample injection time and sample solubility. The influence of these factors (and other subtle factors) on precision have been extensively studied (48, 49). Table 2.7 list some of the major and minor parameters affecting injection precision and a number of these points are discussed below.

2.5.1.1 Sample concentration

Exceptional separation efficiencies of many thousands of theoretical plates may be obtained using relatively dilute sample solutions. However, this results in small peaks which, as in HPLC, amplifies integration errors. Several reports (50, 51) have confirmed the highly significant improvement in precision obtained using higher sample concentrations at the expense of peak efficiency and shape. The use of a high concentration is often mandatory when attempting to simultaneously determine the main component and related impurities of pharmaceuticals. An additional approach to minimising integrator-related imprecision is to employ (52) a wider bore capillary which produces larger peaks, higher peak area counts and reduced integration errors.

Table 2.7 Listing of factors affecting precision

	Comments
Major considerations	
Internal standard	Use whenever possible
Constant temperature	Always use
High sample concentration	Solubility considerations
Injection time	Avoid 1–2 seconds injection times
Preconditioning injections	Perform 2 blank injections
Sample viscosity	Match samples/standards
Dissolving solvent	Optimise during development
Electrolyte composition	Reduce depletion effects
Minor considerations	
Buffer fluid levels	Visually level
Sample fluid levels	Visually level
Injection end of capillary	Immerse in electrolyte vial
Electrolyte co-injection	Assess experimentally
Voltage ramp	Always use
Delays following injection	Minimise delays
Constant current	Assess experimentally

Reproduced with permission from ref. 48

2.5.1.2 Migration time precision

The area of a peak obtained in CE is related to both its corresponding height and width. The peak width is related to both concentration and the speed at which the peak migrates through the detector. Therefore, the peak area is directly related (53) to migration time. For instance, halving the operating voltage approximately doubles both migration time and also doubles the area of the corresponding peaks. Therefore, any variability in migration time will be reflected in variable peak areas leading to an apparently poorer precision.

Control of migration time is possible to less than 1 % RSD (1,14,27) using appropriate buffering electrolytes, pre-separation conditioning rinses and constant temperatures. The most significant factor to control is pH as this governs both solute charge and the level of EOF. Application of a voltage causes a net movement of ions within the system (electrolysis) which can result in a change of electrolyte composition in the electrolyte vials at different ends of the capillary (Figure 2.5). This electrolysis effect is known (54) as "buffer depletion".

Therefore unless the electrolyte employed has an adequate buffering capacity then buffer depletion can result in significant pH differences in the two electrolyte reservoirs. For example (55) prolonged voltage applied across reservoirs containing a phosphate electrolyte, at a pH out of its buffering pH range, resulted in significant differences of pH in the earth

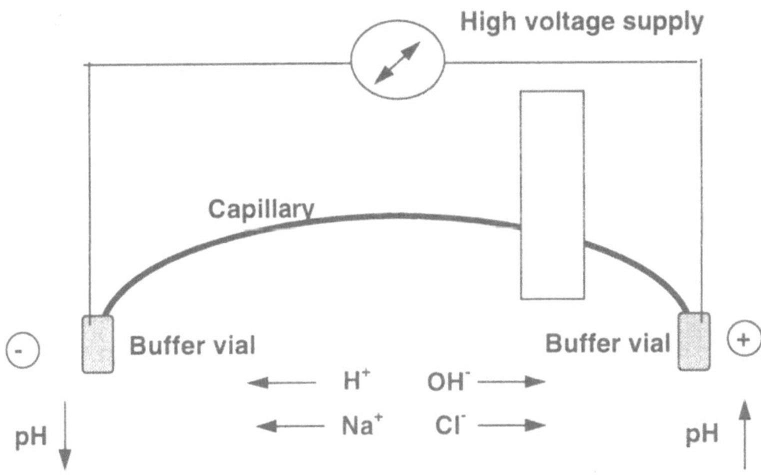

Figure 2.5 Principles of buffer depletion. Reproduced with permission from (38)

electrode and ground electrode reservoirs. The principal approaches to the reduction of buffer depletion effects are (56) are to employ a high electrolyte concentration, narrow bore capillary, rapid analysis time, and replenishment of buffer vial contents.

The condition of the capillary wall also affects the level of EOF and must be controlled. Sample constituents may become adsorbed onto the wall altering the surface. Flushing of the capillary between injections can lead to maintenance of a constant capillary wall condition over multiple injection sequences. Often a simple 1–2 minute flush with fresh electrolyte is sufficient to maintain a consistent EOF. However, harsher rinsing regimes using caustic or acid solutions may be required dependent upon the separation requirements. The rinsing regime should be optimised during method development.

Consistent migration times can only be achieved when operating at a consistent temperature as migration times are inversely related to temperature. In addition, the partitioning of solutes into micelles or complexation agents, such as cyclodextrins, is highly regulated by temperature. Therefore it is important to employ a constant temperature to maintain consistent selectivity.

2.5.1.3 Internal standard

The majority of imprecision is related to variability in the volume of sample solution injected into the capillary due to the technical difficulties of nanoscale injections. If an internal standard is incorporated into the sample solution then this can be used (57) to compensate for this injection volume variability. The variability largely arises from changes in the injection pressure or time, changes in the solution viscosity due to temperature changes, or due to subtle factors such as solution levels in the sample vials or evaporation losses. All of these potential problems can be effectively eliminated by use of an appropriate internal standard. The internal standard should be selected such that it is well-separated from the main component and related substances, it is readily soluble in both the separation electrolyte and sample dissolving solvent, and is suitably stable with an adequate UV response. The internal standard concentration should be relatively high to minimise integrator errors. Table 2.1 shows that a significant number of reported assay procedures employed an internal standard.

The use of internal standards allows calculation of relative migration time or mobilities has been widely adopted in CE to improve solute peak identification. The use of a reference peak in the separation reduces much of the errors associated with the migration time precision. These factors include variations in electrolyte temperature/viscosity, EOF rates, and sample matrix or joule heating effect. Table 2.8 shows the improvement in data obtained when using a marker peak as the reference point for calculation of migration time ratios (58). The calculation of migration time ratios also improves inter-day precision as it reduced the impacts of other factors such as variation in capillary-to-capillary and electrolyte preparation. For instance the RSD for the mobility of nitrophenol was 3.7 % when measured over the course of 12 days. This data was improved (58) to 1.0 % RSD when measured using an internal standard to calculate migration time ratios.

The improvement in precision is greatest when the migration time of the solute is near that of the reference peak (59) as the factors affecting the migration time variability may vary during the course of an individual analysis. For example in the analysis of a wide range of drug of forensic interest, flunitrazepam (migration time 29.36 min) was used as an internal standard. The RSD for migration time and relative migration time for morphine (migration time 11.81 min) were 3.9 and 3.2 % RSD respectively. However there was a considerable improvement in precision for peaks migrating closer to the internal standard peak for example thebaine (migration time 34.47 precision for migration time was 6.6 % RSD which was improved to 0.4 % RSD when calculating migration time ratios. The precision of relative migration time data can be further improved by use of several marker peaks as reference points within each separation as their relative position to each other compensates for changes in the factors affecting migration times that may occur during the course of a separation. The use of 2 or 3 marker peaks to calculate relative migration times can improve the precision data from around 1 % RSD to less than 0.1 % RSD (60).

2.5.1.4 Constant temperature

The volume injected in CE is highly related (47) to the sample solution viscosity. At higher temperatures viscosity is lower and more solution is introduced into the capillary than at

Table 2.8 Variation in migration time and migration time ratios for 60 replicate injections of an amino acid mixture

Solute	Migration time			Migration time ratio		
	Range	*Average*	*%RSD*	*Range*	*Average*	*%RSD*
Mesityl oxide	3.9–4.2	4.0 ± 0.2	5.0	n/a	1.00	n/a
Tryptophan	6.0–6.4	6.3 ± 0.2	3.2	1.49–1.52	1.50 ± 0.01	0.8
Phenylanaline	6.2–6.7	6.5 ± 0.2	4.6	1.60–1.65	1.63 ± 0.02	1.1
Lysine	6.4–6.9	6.7 ± 0.2	3.0	1.67–1.69	1.68 ± 0.01	0.6
Alanine	7.7–8.2	8.0 ± 0.3	3.8	1.99–2.02	2.00 ± 0.02	0.8

Reproduced with permission from reference 58

lower temperatures for the same injection time. Therefore, if an internal standard is not used, it is essential to employ a constant operating temperature. Some autosamplers also allow temperature control of the vials. It may be advantageous to fill the autosampler prior to analyse to allow time for all solutions to reach a constant temperature. A suitable pause prior to the initial analysis may be 30 minutes (61) or to perform a number of system suitability analyses prior to commencing an analytical sequence.

2.5.1.5 Sample injection time

Most sample injection devices are based on pressure differential processes in which the capillary is dipped into the sample solution and a valve briefly opened. When the valve is opened a pressure differential is set-up across the capillary which causes sample solution to be forced into the capillary. The volume of solution entering the capillary is related to both the duration and level of the pressure differential. The length of time can be accurately controlled with a high level of precision. However the pressure differential cannot be regulated to such high precision. Therefore, many commercial injection systems have a feedback mechanism in which the actual pressure differential is measured and the injection time is automatically adjusted to compensate for pressure fluctuations. Therefore, when using a positive pressure injection system of 5 psi, if the actual pressure is only 4.8 psi then a 5.0 second injection time will automatically be extended to 5.2 seconds. The ability to successfully compensate for pressure fluctuations is dramatically reduced when using short (1-2 second) injection times, therefore it is advisable to employ longer times.

2.5.1.6 Sample (internal standard) solubility

The sample should be sufficiently soluble in both the sample dissolving solvent and separation electrolyte to prevent problems of on-capillary precipitation or adsorption. This can often be assessed experimentally by addition of a few drops of sample solution into a vial containing the electrolyte. It is important to note that on-capillary sample concentration (referred to as "stacking") will occur (62) if the sample is dissolved in a lower conductivity solution than the run electrolyte. This stacking may result in up to a ten-fold increase in on-capillary sample concentration which may then cause precipitation problems. Precipitation problems may lead to extremely poor peak shapes with resulting difficulties in accurate integration. The sample and internal standard concentration, dissolving solvent and electrolyte composition should all be appropriately optimised during method development.

2.5.1.7 Other factors affecting precision

Several minor factors can give rise (48) to injection volume variability. These include losses of sample solution due to liquid expansion (63) following initial application of the separation voltage, contamination of the outside of the capillary (64) and variable droplet formation at the detector end of the capillary. For instance (61) it is advisable to perform two injections prior to initiation of a sequence to allow the system to settle and the capillary to equilibrate.

Carryover effects can be appreciable (49, 64, 65) in CE if analysing high concentration samples and precautions should be adopted to prevent this. Contamination of the run electrolyte with sample solution on the outside of the capillary can occur and this can lead to

a)   b)

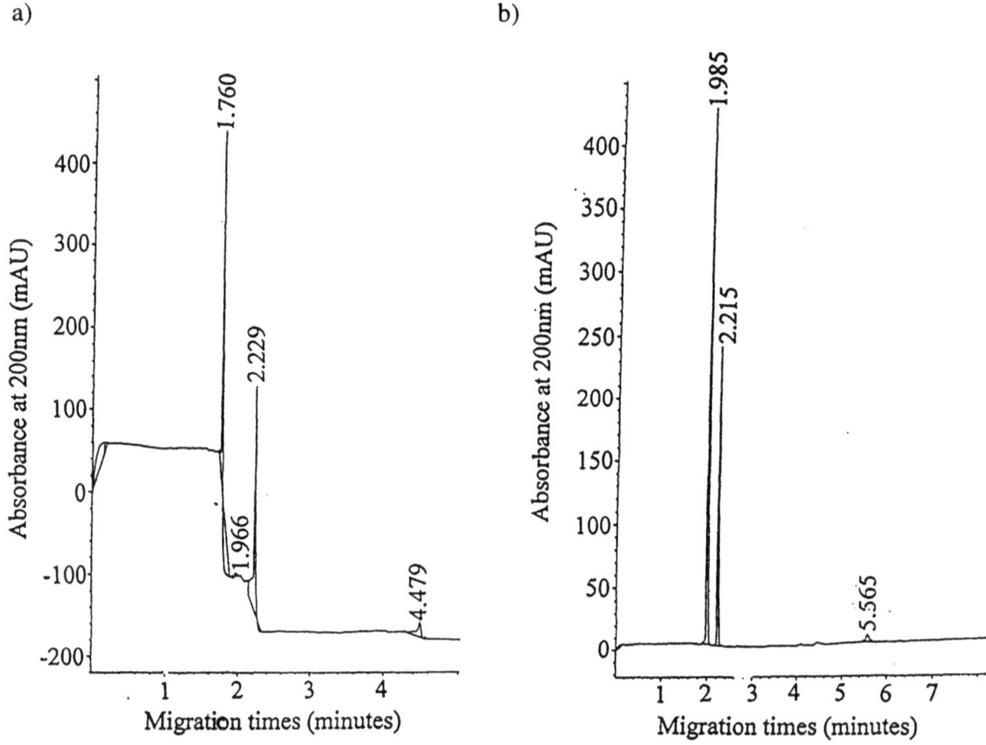

Figure 2.6 Analysis of a 3 component sample solution by MECC (without [Figure 2.6a] and with [Figure 2.6b] an additional capillary dip step in the method). Separation conditions: 20 mM LiDS/12.5 mM Na_2HPO_4/10 % v/v IPA, detection at 200 nm, 34 cm × 50 µm, 8 kV.

very poor replicate separations and steps on the baseline (Figure 2.6a).To prevent this the sampling end of the capillary can be dipped into a separate vial of buffer before being inserted into the buffer vial used for separation purposes. This additional step prevents contamination of the electrolyte solution used for the analysis and produces an improved separation (Figure 2.6b).

2.5.2 Accuracy

The injection volume in CE is highly related (47) to the sample solution viscosity. Lower injection volumes occur for more viscous samples which is different to HPLC where a constant volume is applied onto the column and viscosity effects are negligible. Many pharmaceutical formulations contain viscosity-increasing excipients such as cellulose, sugars, cyclodextrins and lactose. Therefore lower injection volumes will be obtained (67) for the sample solutions when compared to equal length injections times of standard solutions prepared in dissolving solvent. This lower injection volume would give apparently low assay results. This problem can be overcome (67) by the use of internal standards.

2.5.3 Linearity

Peak areas are preferred in quantitative CE determinations as the linear range of peak areas is considerably better than for peak height (67).

Correlation coefficients for detector response (peak area) with sample concentration of greater than 0.999 are routinely expected in HPLC. Intercept peak area values within the range lower than ±2 % of the peak area obtained for analysis of a solution at the correct assay concentration are also routinely obtained. Use of an internal standard has been shown (67) to improve linearity data as the errors of measurements are reduced which significantly reduces the scatter of the points on the calibration line. For example the data for detector linearity using peak areas was 0.9985 which was improved (67) to 0.9999 when the same data was plotted using peak area ratios

2.6 Comparison of CE and HPLC for Drug Assay

The major advantage of CE over HPLC is its simple, inexpensive routine operation. Capillary and reagent costs are minimal compared to HPLC. CE electrolyte solutions are generally aqueous-based with daily requirements of 10–20 ml. This reduced solvent consumption is a key feature in comparison to HPLC where considerable expense is incurred for organic solvent purchase and disposal. The ability to quantify a range of sample types using a single set of CE conditions (1, 14, 20, 23, 26–29) is another strong feature as this can have considerable savings in analysis and system set-up times.

The major disadvantage of CE is that fixed volume injectors are not available and therefore the precision of injection may be poorer than HPLC unless an internal standard and high sample concentrations are employed. In addition sample matrix effects can have a higher influence in CE compared to HPLC.

Overall, it can be summarised that the use of quantitative CE for the analysis of drug substance and pharmaceutical formulations is entirely possible. Optimised methods can give analytical performance data equivalent to HPLC methods. Control of a number of different factors such as temperature, high sample concentration and sample solution viscosity are important in obtaining good injection precision. The use of appropriate internal standards is recommended wherever possible. Consideration to the solubility of both the sample and internal standard should be given to optimise peak shape and precision. The accuracy of CE methods has been widely demonstrated. However it is noted that consideration should be given to sample viscosity effects, principally through use of internal standards. Acceptable linearity data can be achieved for peak area with sample concentration. Shorter linear ranges are obtained when using peak heights. Linearity correlation coefficients and intercepts are improved through use of internal standards. Sample matrix effects such as high salt contents, can cause alterations in the separations achieved.

References

1. Altria KD, Frake P, Gill I, Hadgett T, Kelly M, and Rudd D R, Validated capillary electrophoresis method for the assay of a range of basic drugs and excipients, *J. Pharm. Biomed. Anal.*, 13 (**1995**) 951-957.

2. Ackermans MT, Beckers JL, Everaerts FM, and Seelen IGJA, Comparison of isotachophoresis, capillary zone electrophoresis and high-performance liquid chromatography for the determination of salbutamol, terbutaline sulphate and fenoterol hydrobromide in pharmaceutical dosage forms, *J. Chromatogr.*, 590 (**1992**) 341-353.

3. Pluym A, Van Ael W, and De Smet M, Capillary electrophoresis in chemical/pharmaceutical quality control, *TRAC*, 11 (**1992**) 27-32.

4. Clark BJ, Barker P, and Large T, The determination of the geometric isomers and related impurities of dothiepin in a pharmaceutical preparation by capillary electrophoresis, *J. Pharm. Biomed. Anal.*, 10 (**1992**) 723-726.

5. Birrell HC, Brightwell MD, and Camilleri P, A capillary electrophoresis assay for eminase, *Chromatographia*, 39 (**1994**) 325-328.

6. Fanali S, Cristalli M, Nardi A, Ossicini L, and Shukla SK, Capillary zone electrophoresis in pharmaceutical analysis, *Il Farmaco*, 45 (**1990**) 693-702.

7. Fanali S, Nardi A, Quaglia MG, and Farina P, Determination of Loxiglumide in pharmaceutical preparations by using capillary zone electrophoresis, *Il Farmaco*, 45 (**1990**) 703-706.

8. Fanali S, Cristalli M, Nardi A, and Quaglia MG, Capillary zone electrophoresis for drug analysis - rapid determination of minoxidil in pharmaceutical formulations, *Il Farmaco*, 450 (**1992**) 711-722.

9. Quaglia MG, Farina P, Bossu E, and Dell'Aquila C, Analysis of non-benzodiapenic anxiolytic agents by capillary zone electrophoresis, *J. Pharm. Biomed. Anal.*, 13 (**1995**) 505-509.

10. Bjornstdottir I and Hansen SH, Determination of opium alkaloids in opium by capillary electrophoresis, *J. Pharm. Biomed. Anal.*, 13 (**1995**) 687-693.

11. Altria KD and Chanter YL, Validation of a capillary electrophoresis method for the determination of a quinolone antibiotic and its related impurities, *J. Chromatogr.*, 652 (**1993**) 459-463.

12. Altria KD and Rogan MM, Reduction in sample pretreatment requirements by using high-performance capillary electrokinetic separation methods, *J. Pharm. Biomed. Anal.*, 8 (**1990**) 1005-1008.

13. Altria KD and Filbey SD, Quantitative analysis of sumatriptan by capillary electrophoresis, *J. Liq. Chromatogr.*, 16 (**1993**) 2281-2292.

14. Altria KD, Bryant S, and Hadgett T, Validated capillary electrophoresis method for the assay of a range of acidic drugs and excipients, *J. Pharm. Biomed. Anal.*, 15 (**1997**) 1091-1101.

15. Hoyt AM and Sepaniak MJ, Determination of benzylpenicillin in pharmaceuticals by capillary zone electrophoresis, *Anal. Letters*, 22 (**1989**) 861-873.

16. Arrowood S and Hoyt AM, Determination of cimetidine in pharmaceutical preparations by capillary zone electrophoresis, *J. Chromatogr.*, 586 (**1991**) 177-180.

17. Fabre H and Castanela Penalvo G, Capillary electrophoresis method as an alternative method for the determination of cefotaxime, *J. Liq. Chromatogr.*, 18 (**1995**) 3877-3887.

18. Qin X-Z, Ip DP, and Tsai EW, Determination and rotamer separation of enalapril maleate by capillary electrophoresis, *J. Chromatogr.*, 626 (**1992**) 251-258.

19. Flurer CL and Wolnik KA, Quantitation of gentamicin sulfate in injectable solutions by capillary electrophoresis, *J. Chromatogr.*, 663 (**1993**) 259-263.

20. Bechet I, Fillet M, Hubert Ph, and Crommen J, Quantitative analyses of non-steroidal anti-inflammatory drugs by capillary zone electrophoresis, *J. Pharm. Biomed. Anal.*, 13 (**1995**) 497-503.

21. Ackermans MT, Everaerts FM, and Beckers JL, Determination of aminogycloside antibiotics in pharmaceuticals by capillary zone electrophoresis with indirect UV detection coupled with micellar electrokinetic capillary chromatography, *J. Chromatogr.*, 606 (**1992**) 229-235.

22. Boonkerd S, Lauwers M, Detaevernier MR, and Michotte Y, Separation and simultaneous determination of the components in an analgesic tablet formulation by micellar electrokinetic chromatography, *J. Chromatogr.*, 695 (**1995**) 97-102.

23. Fujiwara S and Honda S, Determination of ingredients of antipyretic analgesic preparations by micellar electrokinetic capillary chromatography, *Anal. Chem.*, 59 (**1995**) 2773-2776.

24. Nishi H, Fukuyama T, Matsuo M, and Terabe S, Effect of surfactant structures on the separation of cold medicine ingredients by micellar electrokinetic chromatography, *J. Pharm. Sci.*, 79 (**1990**) 519-523.

25. Wainright A, Capillary electrophoresis applied to the analysis of pharmaceutical compounds, *J. Micro. Sep.*, 2 (**1990**) 166-175.

26. Ong CP, Ng CL, Lee HK, and Li SFY, Determination of antihistamines in pharmaceuticals by capillary electrophoresis, *J. Chromatogr.*, 588 (**1991**) 335-339.

27. Bechet I, Fillet M, Hubert P, and Crommen J, Determination of benzodiazepines by micellar electrokinetic chromatography, *Electrophoresis*, 15 (**1994**) 1316-1321.

28. Sun P, Mariano GJ, Barker G, and Hartwick RA, Comparison of micellar electrokinetic capillary chromatography and high-performance liquid chromatography on the separation and determination of caffeine and its analogues in pharmaceutical tablets, *Anal. Lett.*, 27 (**1994**) 927-937.

29. Nishi H, Fukayama T, Matsuo M, and Terabe S, Separation and determination of the ingredients of a cold medicine by micellar electrokinetic chromatography with bile salts, *J. Chromatogr.*, 498 (**1990**) 313-323.

30. Fabre H and Castaneda Penalvo G, Capillary electrophoresis as an alternative method for the determination of cefotaxime, *J. Liq. Chromatogr.*, 18 (**1995**) 3877-3887.

31. Emaldi P, Fapanni S, and Baldini A, Validation of a capillary electrophoresis method for the determination of cephradine and its related impurities, *J. Chromatogr. A*, 711 (**1995**) 339-346.

32. Nishi H, Fukuyama T Matsuo M, and Terabe S, Separation and determination of lipophilic corticosteroids and benzothiazepin analogues by micellar electrokinetic chromatography using bile salts, *J. Chromatogr.* , 513 (**1990**) 279-295.

33. Thomas BR and Ghodbane S, Evaluation of a mixed micellar electrokinetic capillary electrophoresis method for validated pharmaceutical quality control, *J. Liq. Chromatogr.*, 16 (**1993**) 1983-2006.

34. Thomas BR, Fang XG, Chen X , Tyrell RJ, and Ghodbane S, Validated micellar electrokinetic capillary chromatography method for the quality control of the drug substances hydrochlorothiazide and chlorothiazide, *J. Chromatogr.*, 657 (**1994**) 383-394.

35. Sakodinskaya I K, Desiderio C, Nardi A, and Fanali S, Micellar electrokinetic chromatographic study of hydroquinone and some esters, *J. Chromatogr.*, 596 (**1992**) 95-100.

36. Weinberger R and Albin M, Quantitative micellar electrokinetic capillary chromatography: linear dynamic range, *J. Liq. Chromatogr.*, 14 (**1991**) 953-972.

37. Donato MG, Van-den-Eeckhout, Van-den-Bossche W and Sandra P, Capillary zone electrophoresis and micellar electrokinetic capillary chromatography of some non-steroidal anti-inflammatory drugs (NSAIDs), *J. Pharm. Biomed. Anal.*, 11 (**1993**) 197-201.

38. Altria KD, Clayton NG, Harden RC, Hart M, Hevizi J, Makwana J, and Portsmouth MJ, An inter-company cross-validation exercise on capillary electrophoresis testing of dose uniformity of paracetamol content in formulations, *Chromatographia*, 39 (**1994**) 180-184.

39. Nishi H and Terabe S, Application of micellar electrokinetic chromatography to pharmaceutical analysis, *Electrophoresis*, 11 (**1990**) 691-701.

40. Nishi H, Fukuyama T Matsuo M, and Terabe S, Separation and determination of lipophilic corticosteroids and benzothiazepin analogues by micellar electrokinetic chromatography using bile salts, *J. Chromatogr.*, 513 (**1990**) 279-295.

41. Dang Q-X, Yan L-X, Sun Z-P, and Ling D-K, Separation and simultaneous determination of the active ingredients in theophylline tablets by micellar electrokinetic capillary chromatography, *J. Chromatogr.*, 630 (**1993**) 363-369.

42. Korman M, Vindevogel J, and Sandra P, Application of micellar electrokinetic chromatography to the quality control of pharmaceutical formulations: the analysis of xanthine derivatives, *Electrophoresis*, 15 (**1994**) 1304-1309.

43. Otsuka K, Higashimori M, Koike R, Karuhaka K, Okada Y, and Terabe S, Separation of lipophilic compounds by micellar electrokinetic chromatography with organic modifiers, *Electrophoresis*, 15 (**1994**) 1280-1284.

44. Bretnall AE and Clarke GS, Investigation and optimisation of the use of micellar electrokinetic chromatography for the analysis of six cardiovascular drugs, *J. Chromatogr.*, 700 (**1995**) 173-178.

45. Terabe S, Ishihama Y, Nishi H, Fukuyama T, and Otsuka K, Effect of urea addition in micellar electrokinetic chromatography, *J. Chromatogr.*, 545 (**1994**) 359-368.

46. Nishi H, Fukuyama T Matsuo M, and Terabe S, Separation and determination of lipophilic corticosteroids and benzothiazepin analogues by micellar electrokinetic chromatography using bile salts, *J. Chromatogr.*, 513 (**1990**) 279-295.

47. Rose DJ and Jorgenson JW, Characterisation and automation of sample introduction methods for capillary electrophoresis, *Anal.Chem.*, 60 (**1998**) 642-648.

48. Altria KD and Fabre H, Approaches to optimisation of precision in capillary electrophoresis, *Chromatographia*, 40 (**1995**) 313-320.

49. Lux JA, Yin H-F, and Schomburg G, Construction, evaluation and analytical operation of a modular capillary electrophoresis instrument, *Chromatographia*, 30 (**1990**) 7-15.

50. Altria KD, Goodall DM, and Rogan MM, Quantitative determination of drug counter-ion stoichiometry by capillary electrophoresis, *Chromatographia* , 38 (**1994**) 637-642.

51. Ryder DS, Determination of sodium vinyl sulphonate in water-soluble polymers using capillary zone electrophoresis, *J. Chromatogr.*, 605 (**1992**) 143-147.

52. Altria KD, Wood T, Kitscha R, and Roberts-McIntosh A, Validation of a capillary electrophoresis method for the determination of potassium counter-ion levels in an acidic drug salt, *J. Pharm. Biomed. Anal.*, 13 (**1995**) 33-38.

53. Altria KD, Essential peak area normalisation in capillary electrophoresis, *Chromatographia*, 35 (**1993**) 177-182.

54. Shafaati A and Clark BJ, Development of a capillary zone electrophoresis method for atenolol and its related impurities in a tablet preparation, *Anal. Proceedings*, 30 (**1993**) 481-483.

55. Zhu T, Sun Y-L, Zhang C-X, Ling D-K, and Sun Z-P, Variation of pH of the background electrolyte as a result of electrolysis in CE, *JHRCC*, 17 (**1994**) 563-564.

56. Kelly MA, Altria KD, and Clark BJ, Approaches used in the reduction of buffer electrolysis effects in routine capillary electrophoresis procedures, *J. Chromatogr. A*, 768 (**1997**) 73-80.

57. Dose EV and Guiochon GA, Internal standardisation technique for capillary zone electrophoresis, *Anal. Chem.*, 63 (**1991**) 1154-1158.

58. Yang J, Bose S, and Hage DS, Improved reproducibility in capillary electrophoresis through the use of mobility and migration time ratios, *J. Chromatogr. A*, 735 (**1996**) 209-220.

59. Tagliaro F, Smith FP, Turrina S, Equisetto V, and Marigo M, Complementary use of capillary zone electrophoresis and micellar electrokinetic capillary chromatography for mutual confirmation of results in forensic drug analysis, *J. Chromatogr. A*, 735 (**1996**) 227-235.

60. Riekkola M-J and Jumpannen JH, Capillary electrophoresis of diuretics, *J. Chromatogr. A*, 735 (**1996**) 151-164.

61. Altria KD, Gill I, Howells J, Luscombe CN, and Williams RZ, Trace analysis of detergent residues by capillary electrophoresis, *Chromatographia*, 40 (**1995**) 527-531.

62. Chien RL and Burghi DS, On-column sample concentration using field amplification in CZE, *Anal. Chem.*, 64 (**1992**) 489A-496A.

63. Knox JH and McCormack KA, Volume expansion and loss of sample due to initial self-heating in capillary electrophoresis (CES) systems, *Chromatographia*, 38 (**1994**) 279-282.

64. Lalljie SPD and Sandra P, Practical and quantitative aspects in the analysis of FITC and DTAF amino acid derivatives by capillary electrophoresis and LIF detection, *Chromatographia*, 40 (**1995**) 519-526.

65. Altria KD and Williams RZ in preparation.

66. H Watzig, Appropriate calibration functions for capillary electrophoresis I. Precision and sensitivity using peak areas and heights, *J. Chromatogr. A*, 700 (**1995**) 1-7.

67. Altria KD and J Bestford, Main component assay of pharmaceuticals by capillary electrophoresis - considerations regarding precision, accuracy and linearity data, *J. Cap. Elec.*, 3 (**1996**) 13-23.

3 Determination of Drug Related Impurities

3.1 Introduction

It is necessary to demonstrate that the purity of pharmaceuticals and formulated products is of sufficient and consistent quality. This testing requires the use of analytical methods that are both selective and sensitive. The method must be selective such that it can resolve all the likely and actual impurities from each other and from the main drug peak. The permitted maximum levels of the total impurity content and each named impurity for a particular sample is specified in submission documents. The method should be selective for likely synthetic by-products and residual intermediate compounds – this can be a difficult task as the by-products will be structurally similar to the drug and the intermediates may have very different polarity and solubility. The specification levels of the impurities may be as low as 0.1 % w/w of the main drug – therefore suitable sensitive and quantitative methods are required. Sensitivity requirements as low as 0.01 % w/w may be specified for particular impurities if they are known to have a detrimental effect on the safety or quality of the product if present at higher levels.

Samples of drug substance and formulated products are stored in order to obtain data on the shelf-life of samples and products. A variety of storage conditions are used including extremes of temperature, humidity and light intensity. Stability-indicating analytical test methods are therefore required to test the degraded samples. The selectivity and sensitivity requirements for a stability-indicating method are as stringent as for a purity test method. Stability testing involves analysis of a large volume of samples as a wide range of various storage conditions and sample strengths are tested. Generally HPLC is the most commonly technique used to determine related impurity content as this offers the selectivity and sensitivity required as well as being fully automated to handle the large numbers of samples. TLC is also used to determine related impurities but this does not generally offer the same degree of selectivity, sensitivity or automation possibilities.

CE is increasingly being viewed as an alternative and complement to HPLC for the determination of drug related impurities. These determinations are probably the principal role of CE within pharmaceutical analysis and represents a challenge to both selectivity and sensitivity capabilities of the technique. The main component and structurally related impurities often have very similar chemical properties which places great requirements on the selectivity necessary. An advantage CE has over its chromatographic counterparts is that high separation efficiencies obtained in CE can often translate a small degree of selectivity into an acceptable resolution. Detection limits of 0.1% area/area are widely accepted as a minimum requirement for a related impurities determination method and this is possible by CE.

This chapter subdivides the reports of drug impurity determinations into low pH, high pH and MECC applications. Table 3.1 shows that CE has been used to determine related impu-

Table 3.1 Reported application of CE to separation of drug impurities

Compound(s)	Electrolyte composition	Sample	Comments	Ref.
Low pH				
Alkaloids	FSCE, pH 4.5	Test mixtures	0.1% impurities detected by CE-MS	1
Amoxycillin & impurities	MECC, SDS	Drug substance	Deuterated and non-deuterated solvents compared	2
Anti-thromboxane agent and impurities	MECC, SDS with MeOH	Test mixture	High temperature improved separation	3
Anti-viral and impurities	FSCE, pH 2.5	Drug substance	Experimental design for robustness studies	4
Atenolol and impurities	FSCE, pH 9.7	Tablets	Impurity determination in tablet formulataion	5
Benzylpenicillin and degradation impurities	FSCE, pH 9	Gastric juices	Degradation measured with time	6
Codeine and by-products	FSCE, various pH's	Drug formulations	Impurities detected at 0.01% level	7
Domperidone and related substances	FSCE, pH 4	Drug substance	Unknown impurity detected	8
Dothiepin	FSCE, pH 4.8, IPA and CD	Tablets	Experimental design used in method development	9
Fluparoxan related impurities	FSCE, pH 2.5	Drug substance	Fraction collection of CE impurity peak	10
Fluparoxan and impurities	FSCE, pH 2.5	Crude drug substance	High speed separation using short capillary	11
Levothyroxine and impurities	FSCE, pH 2.5	Test mixtures	Various synthetic and degradative impurites separated	12
Mitoguazone	FSCE, pH 3	Stressed solutions	Cross-validation with HPLC	13
Minoxidil and impurities	FSCE, pH 3	Formulations	Quantitative determination	14
Non-benzodiazepinic anxiolytic agents	FSCE, pH 3	Drug substance	Quantitative determination	15
Opiod alkaloids and impurities	FSCE, Cds, pH 4	Crude drug / formulations	Quantitative determination	16
Pyridine-4-carboxylic acid and impurities	FSCE, pH 3	Drug substance	Optimised separation of acidic compounds at low pH	17
Quinolone & impurities	FSCE, pH 2	Drug substance	Method validation performed	18

Table 3.1 continued

Compound(s)	Electrolyte composition	Sample	Comments	Ref.
Ranitidine	FSCE, pH 2.5	Syrup formulation	Sample pretreatment reductions shown	19
Ranitidine and impurities	FSCE, low pH	Drug substance	Method validation	20
Ranitidine & impurities	FSCE, low pH	Drug substance	Peak area normalisation study	21
Ranitidine impurity	FSCE, low pH	Solution stability	Solution stability study with time	22
Remoxipride and impurities (chiral and achiral)	pH 3, TAA, OH-P-β-CD	Drug substance	Optimisation of CD and ion-pair reagents	23
Remoxipride and related compounds	FSCE, pH 3, CD's and ion-pair	Drug substance	Impurities monitored at 0.1% level	24
Salbutamol impurities	FSCE, pH 2	Drug substance	Cross-correlation between CE and HPLC	25
Salbutamol impurities	FSCE, pH 2	Drug substance	Peak identification by co-injection procedure	26
Sumatriptan and impurites	FSCE, pH 2	Injection solutions	Range of degradation and synthetic impurities	27
Tetracycline and degradtion products	FSCE, pH 3.9, and EDTA	Test mixtures	Unknown impurities detected	28
High pH				
Acetylcysteine and impurities	FSCE, pH 8.6, 5% PEG	Drug substance	Impurities detected at < 0.05 %	29
Antitumour peptide antagonist	pH 12.7	Drug substance	Profiliing by CE and LC-MS	30
Cephalosporin	pH 8.6	Liquid formulations	Solution stability monitored	31
Doxycycline	pH 10 EDTA	Drug substance	LOD of 0.2 %	32
Gentamycin and related impurities	FSCE, pH 9.4	Injection solutions	Additional impurities detected	33
Penciclovir	pH 7	Drug substance	Faster than HPLC with the same LOD	34
Oxytetracycline	pH 11 EDTA	Drug substance	Purity control and stability studies	35
Pilocarpine and its *trans* epimer	pH 8 and CD	Ophthalmic solution	Quantitative determination	36

Table 3.1 continued

Compound(s)	Electrolyte composition	Sample	Comments	Ref.
Riboflavin-5-phosphate and imps	FSCE, pH 9	Test mixtures	Qualitative determination	37
Various drugs	FSCE, pH 7, and CD	Drug substance	Batches of drug substance analysed	38
MECC				
Basic drugs	pH 4, Tween	Drug substance	LOD's of less than 0.1 %	39
Cefuroxime axetil and degradation imps	MECC, SDS	Formulations	Sample pretreatment reductions shown	19
Cefotaxime	MECC, SDS	Drug substance	Cross-validation with HPLC	40
Cephalexin and degradation im-purities	MECC, SDS and ion-pair	Capsules	Stability indicating method	41
Cephradine and its related impuri-ties	MECC, SDS	Various	Validation to USP, inclu-ding comparison with HPLC	42
Cholesterol lowe-ring agent	MECC, SDS	Drugs substance and capsules	Method submitted to FDA	43
Dilitiazem & impurities	MECC, SDC	Tablets	Range of impurities monito-red at 0.2–0.4 %	44
Enalapril maleate and rotamer	MECC, SDS	Tablets	Method validation study	45
Fluparoxan and 5 impurities	MECC, SDS and IPA	Test mixtures	Method development details	46
Fluticasone and impurities	MECC, SDS and MeOH	Test mixture	Additional impurity detected	46
Heroin impurities	MECC, SDS with DMSO or ACN	Drug seizures	Impurity profiles used for identification	47
Heroin impurities	MECC, SDS	Test mixtures	Seizures analysis	48
Heroin impurities	MECC, SDS, CD	Seizure	Impurity profiling of drug seizures	49
Hydrochlorothiazi de and chloro-thiazide	MECC,SDS	Drug substance	Succesfully validated to USP guidelines	50
Pilocarpine and its degradation products	MECC,SDS	Test mixtures	Method development details	51
Salicylamide	MECC, SDS	Drug substance	Impurities determined at <0.1%	52

Table 3.1 continued

Compound(s)	Electrolyte composition	Sample	Comments	Ref.
Sulfixoxazole, phenazopyridine, and their related impurities	MECC,SDS	Tablets	LOD, linearity and specificity	53
Various antibiotics and imps	FSCE and MECC	Drug substances	Suppliers identified by impurity profiling	54
Vasotec (antihypertensive)	MECC with Brij 35	Test mixtures and formulations	Stability indicating method	55
Xanthine and derivatives	MECC, SDS	Tablets	Impurities at 0.1% level and below	56

where :

Brij 35 = non ionic surfactant

CD = cyclodextrin

DMSO = dimethylsulphoxide

EDTA = ethylenediaminetetraacetic acid

FSCE = free solution capillary electrophoresis

IPA = isopropanol

MECC = micellar electrokinetic capillary chromatography

PEG = polyethylene glycol

SDS = sodium dodecyl sulphate (surfactant)

SDC = sodium deoxycholate (surfactant)

TAA = tetra-alkylammonium salts (ion-pairing reagent)

Tween = non-ionic surfactant

rities in a wide range of drugs. Examples from each of the three main separation mechanisms employed are discussed. A section covering method performance and validation clearly shows that CE methods are capable of validation in this area and can often give equivalent routine performance to HPLC methods.

Currently the vast majority of drug-related impurity determinations are performed by HPLC which can offer the desired sensitivity for trace level determinations and offers a high degree of automation. A wide variety of stationary phases and operating modes makes HPLC applicable to all drug classes. The typical detection limits for drug-related impurities by HPLC are 0.1 % or low as 0.01 % and this can be routinely met in the majority of circumstances using conventional HPLC UV detectors. It is common practise to employ a secondary support analytical technique to verify HPLC impurity data. In the past this secondary testing has been largely performed by TLC. However, the high degree of automation is not readily available except in highly sophisticated TLC systems. Modern CE instruments are capable of the required sensitivity, automation and precision to approach that of HPLC and CE has now become established as an efficient and complementary alternative to HPLC.

The ability of CE to give a different selectivity to HPLC and/or TLC provides a further means to characterise the impurity content and profiles in drugs. For example levels of domperidone impurities were quantified by TLC, HPLC and CE in drug substance (8). Table 3.2 shows the result from HPLC and CE for selected impurities determined in 3 drug substance

Table 3.2 Levels of domperidone related impuriites as determined by CE and HPLC

Batch number	R45571 content		R48557 content		Unknown impurity	
	CE	HPLC	CE	HPLC	CE	HPLC
1	0.24	0.26	0.15	0.35	0.17	–
2	0.22	0.23	0.15	0.34	0.24	–
3	0.26	0.27	0.15	0.30	0.18	–

Reproduced with permission from reference 8

batches. Good agreement between the 3 techniques was obtained for total impurity levels. However, CE resolved 2 components which both co-eluted in the HPLC and TLC methods.

CE also has an additional advantage over TLC in that with commercial autosamplers and on-line detection simultaneous assay of the main component and related impurities by CE is possible.

One of the attractive features of CE compared to TLC is that peaks are generated which can be integrated and the areas used to calculate the impurity content as a directly as area% of the main peak. This makes comparisons with HPLC data much more simple than with TLC determinations. There is however an additional requirement in handling CE data to calculating impurity levels in that it is necessary to divide the observed area of each peak by its migration time (21). This normalisation is necessary since faster migrating peaks move through the detector at a greater speed than their slower counterparts. Therefore, faster moving peaks have smaller peak widths and correspondingly smaller peak areas as the peak area obtained is related to both the peak height and peak width. The sum of these "normalised peak areas" is used to calculate impurities as % area/area. When impurity determinations are to be expressed as %w/w through the use of external standards, providing the precision of migration time is acceptable this normalisation process is not required. This was experimentally demonstrated (21) using a solution of ranitidine spiked with a known amount of an impurity. The impurity migrated before the main peak and was spiked at 9.1%, the non-normalised data indicated a result of 6.0 % whilst the normalised data confirmed the expected 9.1%.

3.2 Separations Using Low pH Electrolytes

Analysis of basic drugs by HPLC can represent (57) problems due to peak tailing. This problem does not occur so frequently in CE and impurities of basic drugs are generally well separated at pH 2–4. The selectivity obtained under these conditions is based on differences between the charge/mass ratios for the drug and its related impurities. The majority of degradation impurities and synthetic intermediates tend to be smaller than the drug compound and therefore migrate before the main peak. In addition dimeric impurities may be doubly charged compared to the parent compounds (25, 27) at low pH and migrate before the main peak. Conversely in HPLC dimeric impurities are generally strongly retained on column and often require use of a gradient. Often a simple NaH_2PO_4 buffer pH adjusted to pH 2.5 with concentrated H_3PO_4 gives a useful initial separation for the separation of the related impurities of basic drugs as these are generally water soluble and fully ionised as cations at this low pH. An additional advantage of this simple electrolyte systems is that they have low background UV absorbance and operation (11, 25) at low UV wavelengths such as 190–210 nm is possible where many drug compounds have significantly enhanced UV absorbance coefficients. Indeed many small synthetic intermediates such as imidazole are very difficult to analyse by HPLC but present few problems (58) at low pH with detection at 200 nm. It may therefore be possible to detect impurities with very limited chromophores using low wavelength detection in CE that would be undetectable by HPLC which has typical UV cut-off ranges of 205–210 nm depending on the solvent used.

If mobility differences alone do not provide sufficient selectivity then various complexing agents may be added to the electrolyte to suitably alter the migration speed (mobility) of the drug and its impurities. The most frequently employed approach is the addition of millimolar quantities of cyclodextrin (9, 16, 23) into the electrolyte. The migration times obtained then reflect both the solutes electrophoretic mobility and its partitioning with the cyclodextrin. EDTA has also been employed (28) to adjust the separation order of various tetracycline's by on-capillary derivatisation with the EDTA. The anionic complexes were resolved and directly quantified (28) by UV absorbance detection.

Other factors that can be optimised in method development include the addition of ion-pair reagents (24) and the ionic strength (24) of the electrolyte. Both of these influence the shape of the main peak and appropriate optimisation may allow resolution of a closely resolved impurity. For instance, various types and concentrations of ion-pair reagent were employed (24) in conjunction with addition of cyclodextrin in the optimisation of the separation of remoxipride impurities. For example, Figure 3.1 shows separation of remoxipride spiked with a range of synthetic and degradative impurities at the 0.1 % level. The peak shape and resolution was effectively manipulated by the choice and concentration of these additives. The concentration of both cyclodextrin and ion-pair reagent were simultaneously optimised to adjust the peak shape and selectivity in the separation of remoxipride and 8 impurities. The final method conditions used a mixture of tetrapropylammonium (8 mM) and tetrabutylammonium (12 mM) ions and methyl-beta-cyclodextrin (20 mM) which allowed all impurities to be resolved and detected at 0.05 % in the presence of the main component.

Resolution can also be adjusted by the addition of organic solvents (9) such as iso-propanol and methanol to the electrolyte as this effects the pKa values of the basic drugs. Use of non-aqueous electrolytes has also proved useful for the resolution of impurities (see Chapter 13).

Adjustment of the peak shape to allow resolution of closely related migrating peaks can be achieved by varying the electrolyte constituents. For example to obtain an efficient peak shape for later migrating cationic drugs it is appropriate to use an electrolyte containing triethanolamine as the cationic constituent of the buffer. Conversely if the solute is a small and/or highly charged compound then sodium is as an appropriate cationic constituent. Figure 3.2 shows the efficient separation of a basic drug and related impurities using an electrolyte containing triethanolamine adjusted to pH 2.5 with phosphoric acid. When this analysis was performed using sodium phosphate adjusted to pH 2.5 with phosphoric acid the peak shape obtained (66) was very asymmetric and the impurity peaks were poorly resolved from each

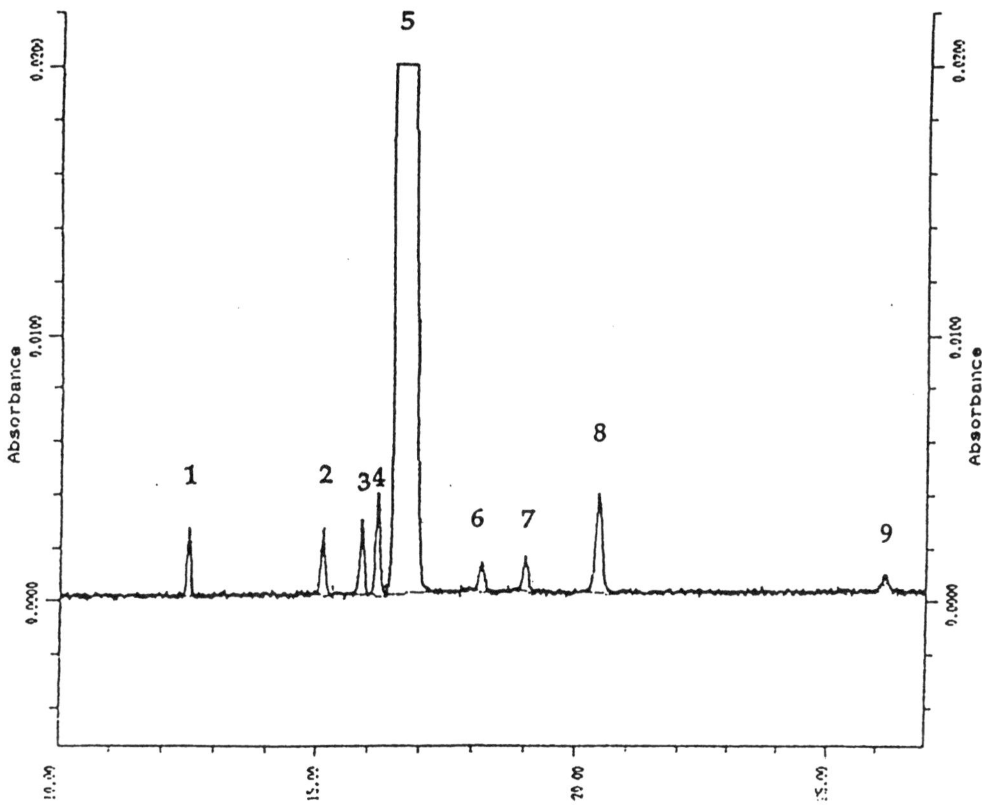

Figure 3.1 Separation of remoxipride spiked with a range of synthetic and degradative at the 0.1% level. (Reproduced with permission from reference 24)

other and the main peak. The triethanolamine ion can also become adsorbed onto the capillary surface resulting in a slightly positively charged capillary surface which serves to prevent adsorption of positively charged basic drug ions.

The choice of the anionic component of the buffer can also influence the selectivity obtained. For instance (20) in the method development for the separation of ranitidine and its related impurities a wide range of buffer types were assessed including phosphate, TRIS, acetate, and citrate. Acceptable resolution was only obtained using the citrate buffer. A similar result was obtained in the separation of a range of beta-blockers where citrate gave a significantly better resolution than phosphate (59).

In many circumstances there is not a need for a full characterisation profile of the test sample and a fast analysis time is more essential. This may be the case for example in the testing of a batch during production where in-process testing is performed. In-process testing involves monitoring the levels of the reactants and products during a synthesis until the batch contains less than a specified level of precursor(s) and/or % purity of the product. In this type of testing speed of analysis is of major importance and the analytical methods used do not generally fully characterise the impurity content of the product. Use of short capillaries

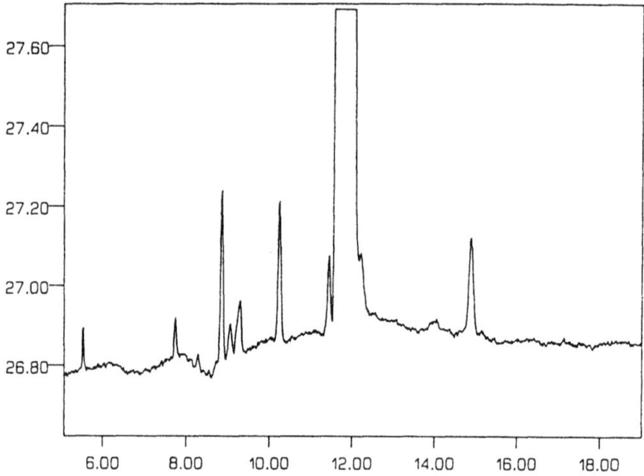

Figure 3.2 Separation of a basic drug from a range of its impurities using a pH 2.5 triethanolamine-phosphoric electrolyte. Conditions: 50 mM phosphoric acid adjusted to pH 2.5 with triethanolamine, sample concentration 0.5 mg/ml in water, 230 nm, + 10 kV, 30°C, 37 cm × 50 μm capillary. (Reproduced with permission from reference 66)

coupled with high voltages can allow extremely short analysis times to be attained. For example Figure 3.3a shows the separation (11) of fluparoxan impurities within 2 minutes using a 27 cm capillary. Figure 3.3b shows the same separation using a 57 cm capillary. Although some degree of resolution is sacrificed a good indication of the purity of the test substance is rapidly obtained with the short capillary.

When attempting separation of acidic compounds optimal resolution is often achieved at a pH close to the pKa value of the various acids. Therefore it may often be appropriate to separate closely related acidic species at low pH to exploit potential differences in pKa values between the acids. For example, resolution of impurities of the drug intermediate pyridine-4-carboxylic acid was maximised (28) at pH 3.

Water insoluble basic compounds may often be dissolved in a 1 to 10 dilution of the buffer or pH adjusted water. Amphoteric compounds may (29) be beneficially dissolved in a high pH solution, but analysed using a low pH electrolyte. This approach has been used in the determination of impurities in a water insoluble amphoteric quinolone antibiotic (29). The compound was only soluble at pH extremes of less than 2 and greater than 10. The sample was dissolved in NaOH solution and analysed with a pH 1.5 electrolyte. Generally samples should be dissolved in pure water or dilute buffer if possible as this maximises resolution and peak efficiencies (11, 20, 23).

3.3 High pH

At high pH the migration direction of acidic components is against the EOF flow which maximises mobility differences. Operation with standard electrolytes (31, 33) such as phosphate (pH 7) or borate (pH 9.5) often lead to useful initial separations for acidic compounds. As in low pH separations, selectivity can be altered by addition of cyclodextrins, ion-pair reagents and organic solvents. Addition of organic solvents leads to a decrease in EOF and reduced ionisation of acids which may be beneficial. Alternatively, EOF can be reduced by increasing the viscosity of the electrolyte by the addition (1) of polymeric substances such as cellulose or polyvinyl alcohol.

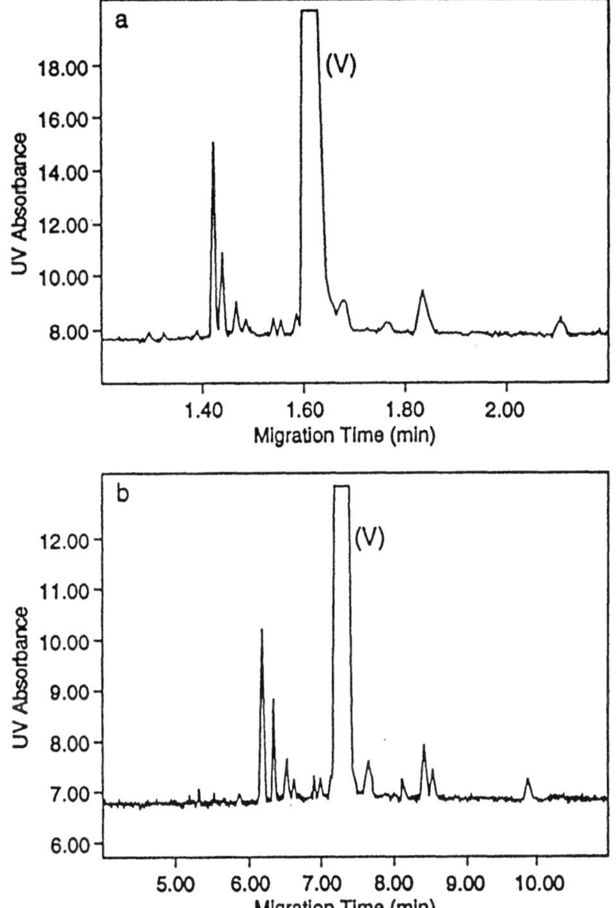

Figure 3.3 Rapid separation of fluparoxan and related impurites; a) Separation of fluparoxan impurities using a 27 cm capillary; b) Separation using a 57 cm capillary. (Reproduced with permission from reference 11)

A worker from the FDA demonstrated (33) that higher levels of gentamycin impurities were detected using CE compared with that achieved by the USP registered HPLC method. Gentamycin has a poor chromophore and therefore needs to be derivatised prior to HPLC analysis. However, not all impurities appear to be derivatised to an equal extent. CE allowed the direct analysis of gentamycin and impurities with low UV wavelength detection. High concentrations of borate were used in the buffer as borate complexes with the aminoglycosides which results the formation of negative charged species which are well separated at high pH. The complexed forms also have significantly increased UV activity. Further work from the same laboratory has been performed (60) on the analysis of aminoglycoside antibiotics by capillary electrophoresis. High concentration borate buffers were used (60) to separate aminoglycoside impurities with detection at 195 nm. For example, 0.4 % levels of streptomycin was determined in dihydrostrephtomycin. Potency values were also determined by CE. TTAB was added to the buffer to reverse the peak migration order to allow trace level determinations of streptomycin.

CE can also be applied to the purity testing of intermediate compounds. For example the impurity content of diethylenetriaminepentaacetic acid dianhydride has been determined by CE. These compounds have no appreciable UV activity and are detected by their pre-separation derivatisation with Zn^{2+} ions which was added into the sample solutions. (61). A pH 10 boric acid buffer with detection at 200nm were used for the determinations and a limit of detection of 0.02% was found for the major degradation products.

A stability indicating method for enalapril has been reported (62) which gave detection limits of 0.2 % for the monitored impurities. Separation and quantitation of a range of tetracycline impurities was achieved (63) by on-capillary derivatisation with EDTA. The anionic complexes were resolved and directly quantified.

3.4 MECC

This approach would be adopted when dealing with uncharged solutes or mixtures of charged and neutral species. This approach may also be considered when simple mobility differences prove insufficient in free solution CE. The selectivity can be manipulated in a similar fashion to those parameters employed in reversed phase HPLC and these include addition of cyclodextrin (49), ion-pair reagents (41), organic solvents (46, 47). Additional selectivity manipulation can be achieved by varying (52) the type and concentration of surfactant.

Water insoluble compounds are generally analysed using MECC. Samples can be prepared in 100 % organic solvents but this can produce problems of out-gassing when employing extended injection times. To minimise the potential difficulties it is best to prepare the sample in a solvent containing the minimum % of organic required to solubilise the sample. The sample should be soluble in the electrolyte used or on-capillary precipitation can occur. For example (39) levels of impramine N-oxide hydrochloride impurities were determined at the 0.01% level using a low pH electrolyte containing levels of the neutral surfactant Tween.

Figure 3.4 shows separation (52) of a range of acidic or neutral salicylamide impurities using a MECC method employing SDS. A detection wavelength of 214 nm allowed detection of trace impurities at the 0.01 % level. Repeated analysis produced an RSD of 9.3 % of the area of a peak relating to an impurity spiked at the 0.1 % level.

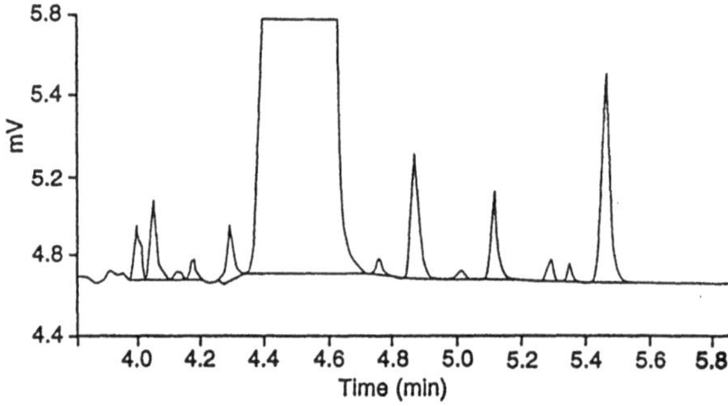

Figure 3.4 MECC separation of a range of acidic or neutral salicylamide impurities. (Reproduced with permission from reference 52)

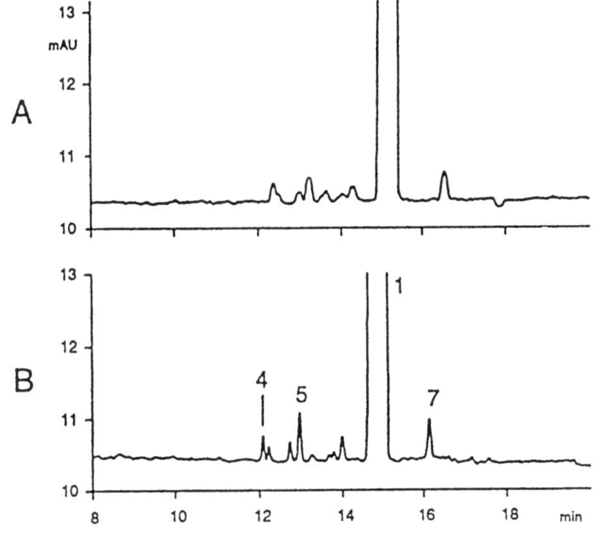

Figure 3.5 Influence of sample solvent composition of the detection and resolution of neutral (peaks 4 and 7) and anionic (peak 5) impurities present at low levels in an acidic drug; a) Sample dissolved in separation; and b) dissolved in optimised diluent. (Reproduced with permission from reference 64)

The use of MECC to separate the charged and neutral impurities in an acidic drug (SB 209247) has been shown (64). The impurities were separated using an electrolyte containing 50 mM borate:ACN (65:35 %v/v) containing 50 mM SDS. Figure 3.5 shows that the choice of sample dissolving solvent has a pronounced effect on the peak efficiency and resolution obtained. The optimal resolution being when the sample solvent contained a lower ionic strength than the buffer to permit focusing (stacking).

3.5 Comparison with HPLC

The data generated on impurity profiling is often compared to that obtained by chromatographic methods (typically HPLC). The principles of separation in CE are entirely different to HPLC and therefore a good agreement between the two techniques strongly supports the integrity of the data. This technique combination is now established in many laboratories and has become a suitable replacement for the conventional use of TLC and HPLC in combination. Apart from routine investigations, this combined use is of particular importance during method validation.

The differences in selectivity between CE and HPLC can result in discrepancies in results with one technique showing an under-estimation in impurity levels. This occurrence signifies that further method optimisation is required. Critical events in the development of a formulation such as synthetic route or process changes or at key drug stability timepoints are times when this may occur. Literature examples include the discovery of additional impurities in tetracycline (63) domperidone (8) and fluticasone (46).

3.5.1 Sensitivity

The principal disadvantages of the use of CE compared to HPLC for determining related impurities are the possible requirements for higher sample concentrations. When operating HPLC and CE at the same wavelength it may be necessary to use 2–5 times more concentrated samples for CE to obtain an equivalent limit of detection. This may represent problems for poorly soluble drugs. This lower sensitivity is principally due to the use of the capillary for on-capillary detection. The section of the capillary used for detection purposes may for example only be 75 micron wide (for a 75 micron capillary) and 200 micron long. This compares very disfavourably with a 1 cm^3 HPLC flowcell.

Aqueous based electrolytes are often employed in CE which have low UV absorbance coefficients allowing detection wavelengths such as 200 nm to be routinely employed. This can compensate for sensitivity problems. Other approaches include the use of wider bore capillaries, and modified capillaries (z-cells and bubble cell modifications). Many impurities or small intermediates have poor chromophores making their quantitation at traditional HPLC

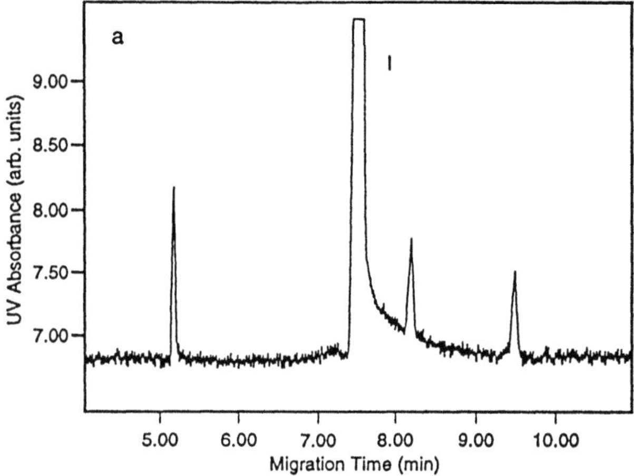

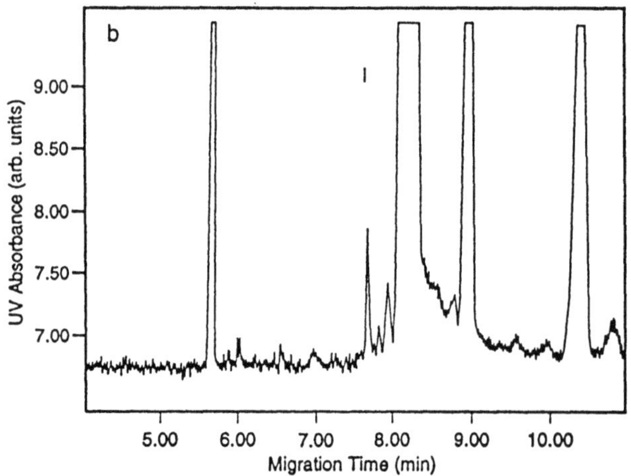

Figure 3.6 High-low quantitation, a) 1 second injection; b) 10 second injection. (Reproduced with permission from reference 65)

wavelengths difficult or impossible. This may be of particular importance for degradative processes where reactions may lead to the loss of the functionality providing the chromophore.

The use of low UV wavelengths may compensate for the inherent poor sensitivity in CE. For example (25) when operating at 200 nm there is a 10 fold increase in signal for salbutamol and its impurities compared with that at 276 nm which is the HPLC wavelength.

The linear dynamic range of CE detectors are generally more limited than for HPLC detectors. This can be of concern when attempting to monitor very low level impurity contents by CE as signal for the highly concentrated main peak may exceed the dynamic range of the detector. This problem can also occur in HPLC and an approach to the problem has been developed. A small injection is performed in which the main peak is on-scale and then a larger injection is performed in which the main peak is off the scale of the detector. The trace level impurities can be measured in the off-scale separation and the expected peak area for the off-scale main peak can be calculated by multiply the on-scale peak area by a suitable factor. This is termed high-low chromatography and has also been used in CE (65). A short injection time (i.e. 1 second) was performed to produce a separation with the main peak on-scale (Figure 3.6a), and a larger injection time (i.e. 10 seconds) was then used to produce a separation with the main peak off-scale (Figure 3.6b). The longer injection time gave considerably enhanced sensitivity for the minor components. The peak area of the off-scale peak is then calculated by multiplying the peak area of the on-scale injection by the ratio of the injection times (i.e. 10:1). Impurities were quantified as % area/area of the calculated area for the off-scale peak. This is increased the detection limit for fluparoxan impurities (65) from 0.1 % to 0.01 %.

3.6 Peak identity confirmation

For a complex impurity separation where there are many peaks with very close migration times confirmation of peak identity by its migration times may not always sufficient for positive identification. If this is the case then the sample solution is usually spiked with authentic impurity standard and re-analysed to confirm the peak identity. This spiking can be achieved automatically using standard CE instrumentation (26). An additional sampling step can be programmed into the separation method to achieve the spiking. Figure 3.7a shows the separation of a salbutamol sample solution (5 second injection time) which also contains two dimeric impurities. Figure 3.7b shows the separation obtained from a 5 second injection of the salbutamol sample solution followed immediately by a 5 second injection of a solution of the impurity of interest (bis ether). Following both injections the voltage was applied and the separation given in Figure 3.7b was produced. The stacking effects that occur during the immediate portion of the separation focus and mix the two sample injection zone together. The impurity identity is clearly confirmed and no loss in resolution is observed from this dual-injection procedure. This approach is especially useful if spiking is needed for several peaks during method development.

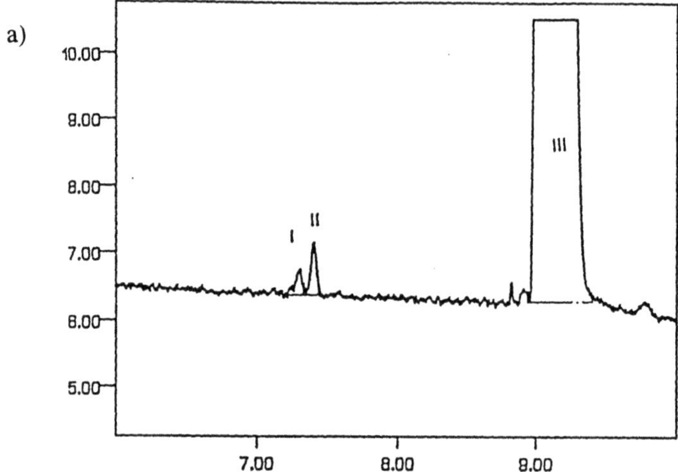

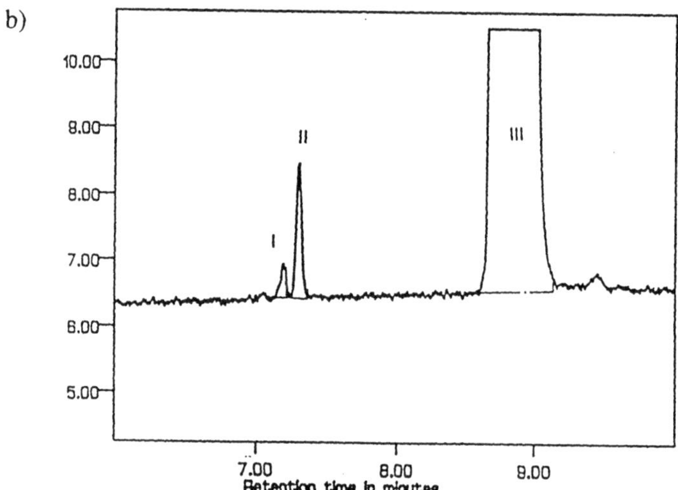

Figure 3.7 Peak identification by automatic spiking using co-injection; a) Separation of a salbutamol sample solution (5 second injection time); b) Separation obtained following a 5 second injection of the salbutamol sample solution followed immediately by a 5 second injection of a impurity standard solution. (Reproduced with permission from reference 26)

The recent advent of UV diode array detector (DAD) technology greatly assists in method development and peak assignment confirmation. Figure 3.8 shows the use of this detector to measure the spectra of a 0.1% impurity (66).

The hyphenation of CE-MS also allows the possibility of determining peak purity and obtaining additional characterisation of the separated peaks. For example impurity levels of 0.1 % in alkaloid and peptide samples were monitored by CE-MS (1). The impurity levels

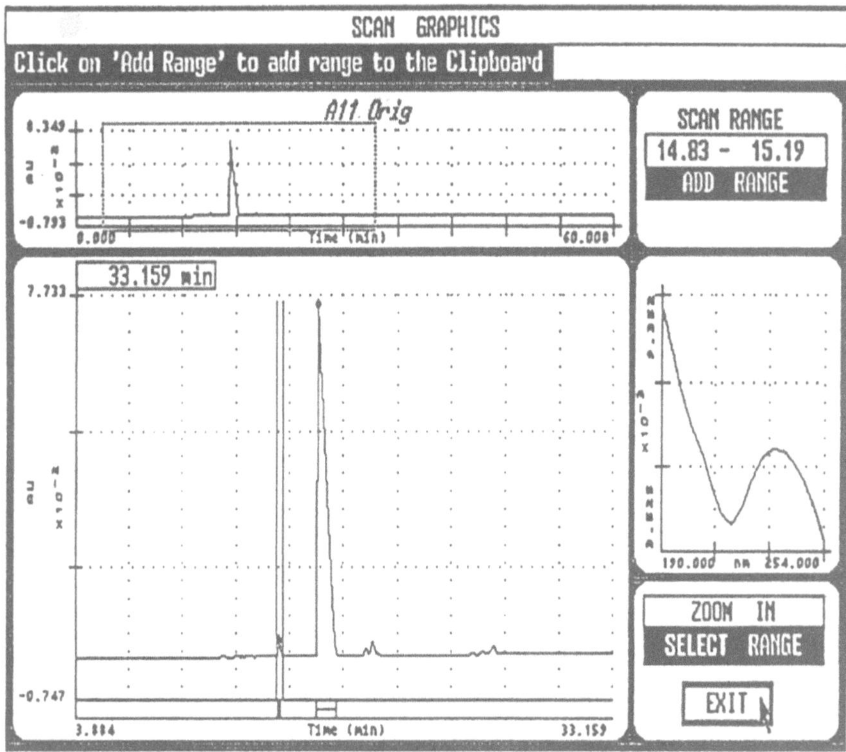

Figure 3.8   Spectra of an 0.1 % impurity present in a drug substance material. (Reproduced with permission from reference 66)

were also verified by in-line UV absorbance detection. Figure 3.9 shows (67) the separation of a drug substance material containing 3 impurities present at 0.1 % of the main component loading. The separation capillary initially passes through the UV detector on the CE instrument used and the UV profile is observed (Figure 3.9a), the peaks exit the capillary and enter the MS much later and there is a better resolution of the compounds (Figure 3.9b). The mass ions of the separated peaks can be clearly determined at this 0.1% level.

Due to the small sample volumes employed in CE, the possibility of conducting micro-preparative scale separations are very limited. However a number of reports have been published in the literature (10, 69).

3.7 Applications

There are number of application types including chemical purity testing of manufactured batches of drug substance and formulated product, analysis of samples following storage under predetermined conditions and profiling of impurities to characterise the origin of the sample.

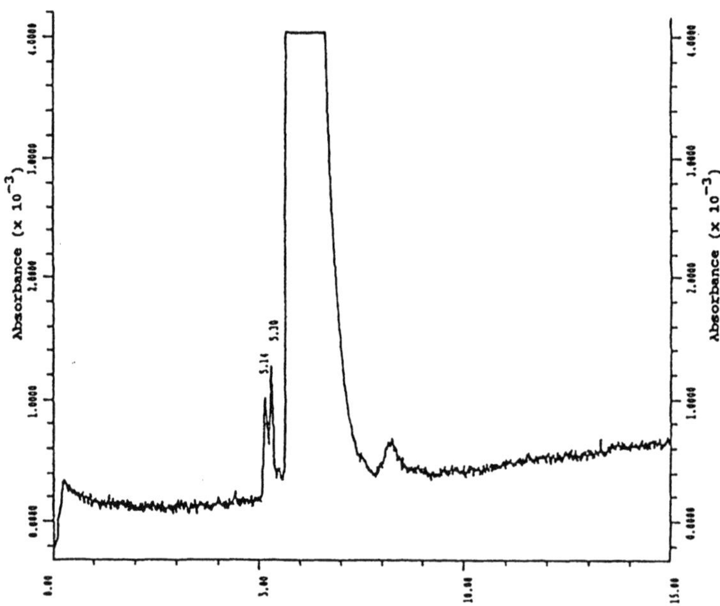

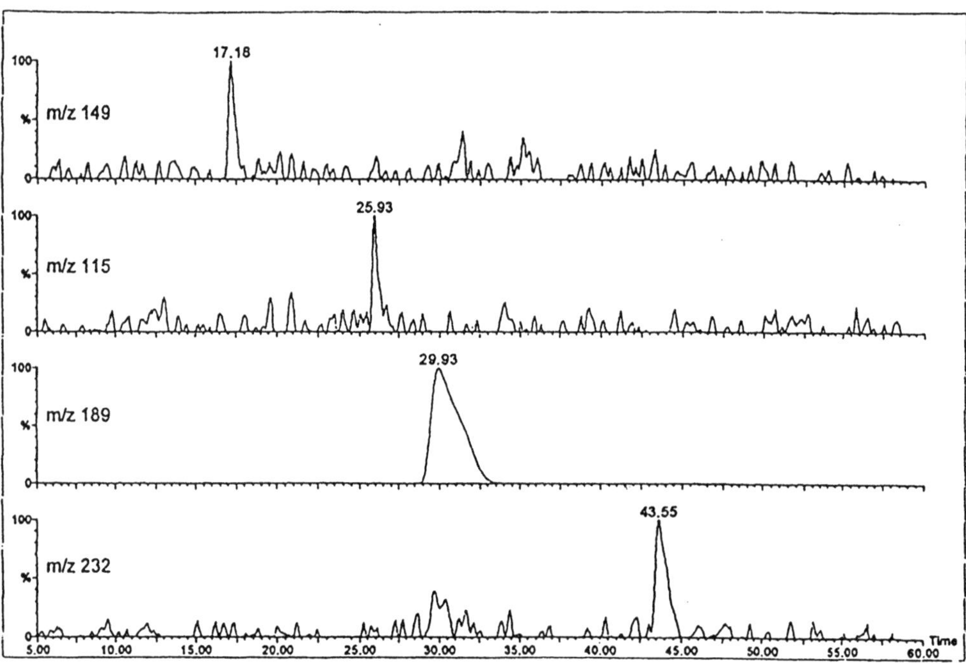

Figure 3.9 Detection of impurities by UV and MS measurements. (Reproduced with permission from reference 67)

3.7.1 Chemical purity testing of drug substance

The complementary nature of HPLC and CE was highlighted in a study (25) of the determination of two dimeric salbutamol impurities. These relatively large impurities were strongly adsorbed onto the HPLC column and therefore had lengthy retention times. However, in CE the dimeric impurities had a charge $Z = +2$ and therefore migrated before the salbutamol ($Z = +1$). Drug substance batches were tested by CE and HPLC using external standards of the impurities for quantitation. Table 3.3 shows the good correlation between the two techniques.

A pH 8 boric acid-tricine-EDTA buffer containing 33 % v/v acetonitrile has been used (69) to evaluate the purity of a metal chelator pharmaceutical (Sanofi-Winthrop TMT-CNS). Migration time repeatability for the drug and 5 related impurities was 0.1 % RSD ($n = 6$). Linearity over the range 0.002 to 1 mg/ml gave a correlation coefficient of 0.9996. A limit of detection of less than 0.1 % was reported. Six purity determinations on a single drug batch gave a RSD of 0.25 % for drug purity. Specificity was shown by injection of collected HPLC impurity peaks onto the CE and by use of peak purity options on the CE DAD detector.

Table 3.3 Comparison of impurity levels in batches of salbutamol drug substance by CE and HPLC

Batch	Bis ether (%w/w)		Dimer (%w/w)	
	CE	HPLC	CE	HPLC
1	0.14	0.16	0.08	0.08
	0.14	0.16	0.08	0.08
2	0.10	0.11	0.06	0.07
	0.10	0.11	0.07	0.06
3	0.20	0.19	0.13	0.11
	0.20	0.19	0.14	0.10
4	0.12	0.13	0.07	0.06
	0.15	0.14	0.08	0.05
5	0.13	0.14	0.08	0.06
	0.12	0.13	0.07	0.05
6	0.31	0.28	0.18	0.17
	0.31	0.26	0.19	0.15
7	0.07	0.09	0.05	0.04
	0.08	0.10	0.06	0.03
8	0.38	0.38	0.20	0.18
	0.44	0.38	0.22	0.19
9	0.37	0.38	0.19	0.18
	0.37	0.35	0.19	0.19
10	0.75	0.66	0.39	0.33
	0.77	0.67	0.40	0.35

Reproduced with permission from reference 25

Impurities have been determined in a water insoluble quinolone antibiotic (18). The compound was only soluble at pH extremes of less than 2 and greater than 10. The sample was dissolved in NaOH solution and analysed with a pH 1.5 electrolyte. A detection limits of 0.1 % were demonstrated during validation. Linearity was measured in two exercises (1–150 % and 20–150 % of target concentration) correlation coefficients of 0.9990 and 0.9997 were obtained respectively. A single sample was injected 10 times and precision values of 0.4 and 0.6 % RSD were obtained for migration time and peak area respectively.

Indirect UV detection is widely used for the determination of inorganic drug counter-ions (Chapter 5). Similarly indirect UV detection can be used for the determination of impurities that have no chromophores. An electrolyte containing benzyltrimethylammoniun bromide as the UV absorber has been used to determine levels of a non-UV absorbing impurity in clidinium bromide drug substance (70). A limit of detection of 0.008 % was obtained for the impurity using indirect UV detection. Results for batch testing obtained by CE and TLC agreed well, for instance a result of 0.44% was obtained by CE which compared with a result of 0.4-0.45% by TLC for the same sample.

A MECC method has been used (40) to monitor the chemical purity of cefotaxime drug substance. Figure 3.10 shows both the separation of a test mixture containing a range of likely impurities and analysis of a drug substance batch. MECC data was shown to be statistically equivalent to HPLC impurity data. Sample degradation during the analysis was limited by use of a CE instrument with a cooled autosampler tray facility. Linearity was demonstrated (correlation coefficient 0.9993) over the range 0.05 to 1.50 g/L. A limit of detection of 0.02 % was obtained for the MECC method.

3.7.2 Chemical purity testing of formulated product

Impurity levels in tablets containing sulfisoxazole have been determined (53) by MECC. Linearity for concentration was demonstrated over the range 0.3–1000 mg/L with a correlation coefficient greater than 0.9999. Limits of detection of 0.03 % were reported for specific impurities. Results were in a agreement with TLC data for impurity contents, the MECC method having a better LOD. Three separate analytical methods are used to profile tablets and the MECC method offers the possibility of a single method replacing all three with considerable cost and time savings .

Extensive validation has been performed on a CE method for the determination of impurities in mirtazapine drug substance present in Remeron tablets (71). An experimental design was used to optimise the separation conditions in terms of the pH, buffer concentration and % methanol content. An experimental design was also used to assess the optimal sample dissolving solvent. It was found that a diluted form of the buffer gave the best performance in terms of repeatability for migration times, consistent current throughout the separation and improved peak area precision. Figure 3.11 shows the variation in current monitored throughout separations where various sample diluents were used. Diluent 4 was used as the optimal solvent (20 mM phosphoric acid in ethanol:water 35:65 v/v). Selectivity was demonstrated by preparing stressed samples by exposure to high levels of temperatures, humidities and light intensity. Detector response linearity (correlation greater than 0.995) for each of the im-

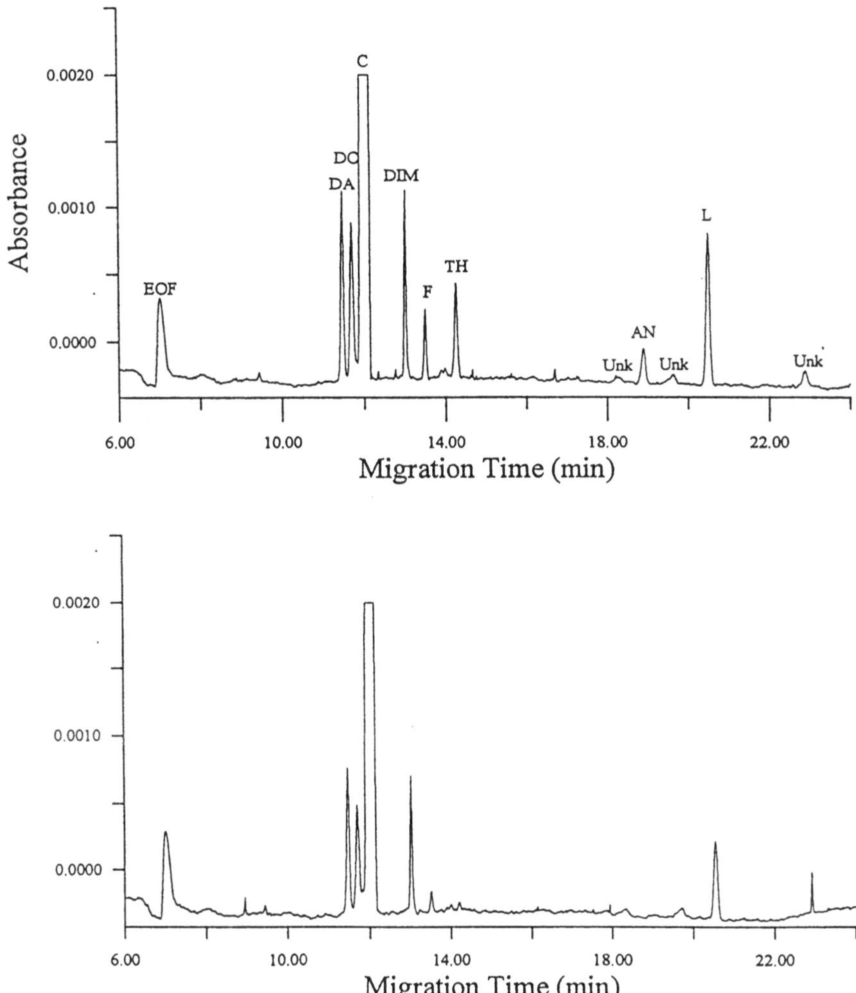

Figure 3.10 Separation of a test mixture of cefotaxime and related impurities and profiling a batch of drug substance. (Reproduced with permission from reference 40)

purities was demonstrated over the range 0.25–2.0 % of the expected mirtazarpine range. Recovery was demonstrated by spiking placebo tablet formulations (recovery data was well within the range of 90–110 %). Use of an internal standard allowed 0.5 % RSD values to be obtained for peak area ratios. Limits of detection of 0.02–0.04 % were obtained for the individual impurities. A correlation coefficient of 0.9997 was obtained between the CE results and those generated using the established HPLC method.

An MECC method has been validated (42) to USP guidelines for the simultaneous assay of the antibiotic cephradine and determination of related impurities content in an aqueous formulation containing sodium carbonate. An RSD of 0.88 % was recorded for recovery

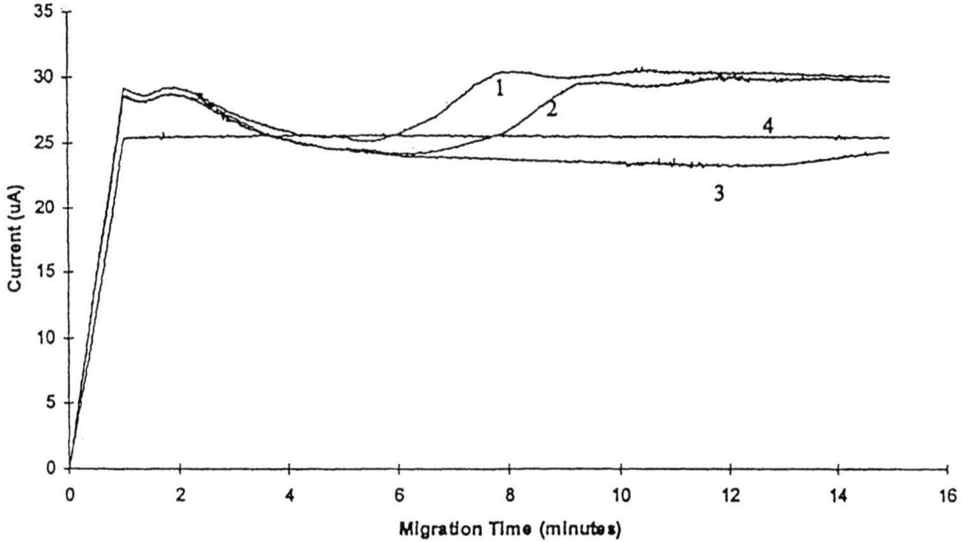

Figure 3.11 Monitored current for various sample diluents. (Reproduced with permission from reference 71)

experiments with average recoveries varying between 99.8–100.2. Correlation coefficients of greater than 0.9993 were obtained. Table 3.4 shows HPLC and MECC assay results for 9 sample batches. The assay results showed no significant difference of the 95 % confidence limits.

Table 3.4 Comparison of HPLC and MECC data for the analysis of cephradine content in an aqueous formulation

Batch No.	HPLC	MECC
057/04	73.5	73.5
064/04	74.2	72.3
065/04	73.2	72.1
066/04	73.6	72.0
067/04	72.5	73.0
055/04	75.3	73.8
058/04	71.8	71.2
059/04	74.0	72.9
060/04	70.4	69.8
Average	*73.2*	*72.3*

Reproduced with permission from reference 42

3.7.3 Stability testing

A CE method has been used (31) to monitor the solution stability of a cephalosporin in solution. Validation included specificity for known degradants and synthetic impurities, linearity over the required range and precision. Analysis was repeated by different analysts on different days. Migration time precision was less than 1 % RSD. Injections were performed from sample solutions every 30 minutes to monitor the solution stability on-line. The cephalosporin (Roche compound RO 23-9424) was found to be twice as stable in a L-arginine-sodium benzoate-saline solution than when prepared in water. The CE data obtained (31) allowed structural elucidation of an unknown degradant formed by hydrolysis.

A combination of both HPLC and CE were used (34) in studying the degradation of penciclovir. CE gave a faster analysis, 7 minutes compared to 22, and required no organic solvents. Impurities were identified by their diode array profiles obtained in both CE and HPLC. The peak time precision was better in CE (0.42 % RSD) compared to HPLC (0.84 %) and both methods gave equivalent sensitivity at 0.5 mg/L.

In-situ solution stability has been performed (22) in which a sample solution of a ranitidine related impurity was placed on the autosampler tray and repeatedly analysed. The initial sample solution was analysed by both CE and HPLC (8.2 and 8.7 % area/area impurity content respectively). The rate of solution degradation was followed for a specific impurity of ranitidine. The sample solution was re-injected several times over the course of 9.25 hours. The reaction was followed (Figure 3.12) in an unattended injection sequence over 9 hours until only 2 % of the original component remained as the main component had degraded to two earlier migrating components.

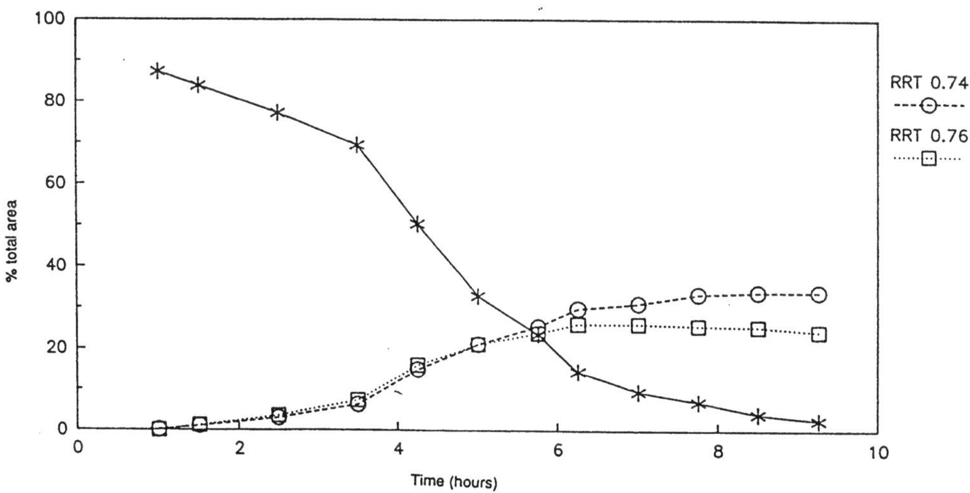

Figure 3.12 Time dependent plot of the degradation of a ranitidine related impurity and formation of two hydrolysis products. (Reproduced with permission from reference 22)

3.7.4 Impurity profiling

CE has been used (60) by FDA investigators to chemically profile bulk pharmaceuticals from various manufacturers. Batches of various drug substances such as antibiotics were obtained from different pharmaceutical suppliers and analysed by CE (60). The impurity profiles in terms of the number of impurities and their respective levels were found to be indicative of the suppliers.

MECC has also been used to profile the constituents of heroin seizures (47, 57) with good cross-correlation with HPLC. Levels of individual heroin impurities were shown (47) to be indicative of the synthetic route and country of origin of the heroin material.

References

1. Hsieh FYL, Cai JC and Henion J, Determination of trace impurities of peptides and alkaloids by capillary electrophoresis-ion spray mass spectrometry, *J. Chromatogr.*, 679 (**1994**) 206-211.

2. Okafo GN and Camilleri P, Micellar electrokinetic capillary chromatography of amoxycillin and related molecules, *Analyst*, 117 (**1992**) 1421-1424.

3. Altria KD , Rogan MM and Finlay G, Electrokinetic separation techniques – a review, *Chrom. and Analysis*, Aug (**1990**) 9-14.

4. Altria KD and Filbey SD, The application of experimental design to the robustness testing of a method for the determination of drug related impurities by capillary electrophoresis, *Chromatographia*, 39 (**1994**) 306-310.

5. Shafaati A and Clark BJ, Development of a capillary zone electrophoresis method for atenolol and its related impurities in a tablet preparation, *Anal. Proceedings*, 30 (**1993**) 481-483.

6. Arrowood S, Hoyt AM Jr. and Sepaniak MJ, Monitoring of benzylpenicillin decomposition in gastric contents by capillary zone electrophoresis, *J. Chromatogr.*, 583 (**1992**) 105-110.

7. Korman M, Vindevogel J and Sandra P, Separation of codeine and its by-products by capillary zone electrophoresis as quality control tool in the pharmaceutical industry, *J. Chromatogr.*, 645 (**1993**) 366-370.

8. Pluym A, Van Ael W and De Smet M, Capillary electrophoresis in chemical/pharmaceutical quality control, *TRAC*, 11 (**1992**) 27-32.

9. Clark BJ, Barker P and Large T, The determination of the geometric isomers and related impurities of dothiepin in a pharmaceutical preparation by capillary electrophoresis, *J Pharm. Biomed. Anal.*, 10 (**1992**) 723-726.

10. Altria KD and Dave K, Peak homogeniety determination and micro-preparative fraction collection by capillary electrophoresis in pharmaceutical analysis, *J. Chromatogr.*, 633 (**1993**) 221-225.

11. Altria KD, Application of high speed capillary electrophoresis to the analysis of pharmaceuticals employing commercial instrumentation, *J. Chromatogr.*, 636 (**1993**) 125-132.

12. Dalal PS, Albuquerque P and Bhagat HR, Development and standardisation of levothryoxine analysis by high-performance capillary electrophoresis, *Anal. Biochem.*, 2111 (**1994**) 34-36.

13. Thompson CE, Gray MR and Baxter MP, The use of capillary electrophoresis as part of a specificity testing strategy for mitoguazone dihydrochloride HPLC methods, *J. Pharm. Biomed. Anal.*, 15 (**1997**) 1103-1111.

14. Fanali S, Cristalli M, Nardi A and Quaglia MG, Capillary zone electrophoresis for drug analysis - rapid determination of minoxidil in pharmaceutical formulations, *Il Farmaco*, 450 (**1992**) 711-722.

15. Quaglia M. G., Farina S., Bossu E. and Dell'Aquila C., Analysis of non-benzodiapenic anxiolytic agents by capillary zone electrophoresis, *J. Pharm. Biomed. Anal.*, 13 (**1995**) 505-509.

16. Bjornsdottir I and Hansen SH, Determination of opium alkaloids in crude opium using non-aqueous capillary electrophoresis, *J. Pharm. Biomed. Anal.*, 13 (**1995**) 1473-1481.

17. Dette C, Watzig H and Uhl H, Characterisation of the important pharmaceutical intermediate pyridine-2, 4-dicarboxylic acid using capillary electrophoresis (CE), *Pharmazie*, 48 (**1993**) 276-280.

18. Altria KD and Chanter YL, Validation of a capillary electrophoresis method for the determination of a quinolone antibiotic and its related impurities, *J. Chromatogr.*, 652 (**1993**) 459-463.

19. Altria KD and Rogan MM, Reductions in sample pretreatment requirements by using high performance capillary electrokinetic separation techniques, *J. Pharm. Biomed. Anal.*, 8 (**1990**) 1005-1008.

20. Kelly MA, Altria KD, Grace C, and Clark BJ, Optimisation, validation and application of a capillary electrophoresis method for the determination of ranitidine hydrochloride and related substances, *J. Chromatogr. A*, in press.

21. Altria KD, Application of High-Low CE for improved quantitative determination of drug related impurities *Chromatographia*, 35 (**1993**) 493-496.

22. Altria KD and Connolly PC, On-line solution stability determination of pharmaceuticals by capillary electrophoresis, *Chromatographia*, 37 (**1993**) 176-178.

23. Stalberg O, Brotell H, and Westerlund D, Capillary electrophoretic separation of basic drugs using surface-modified capillaries and derivatised cyclodextrins as structural/chiral selectors, *Chromatographia*, 40 (**1995**) 697-704.

24. O. Stalberg, Westerlund D, U-B Rodby and S. Schidmt, Determination of impurities in remopxipride by capillary electrophoresis using UV detection and LIF detection; principles to handle sample overloading, *Chromatographia*, 41 (**1995**) 287-294.

25. Altria KD, Quantitative analysis of salbutamol related impurities by capillary electrophoresis, *J. Chromatogr.* , 634 (**1993**) 323-328.

26. Altria KD and Luscombe DCM, Application of capillary electrophoresis as a quantitative identity test for pharmaceuticals employing on-column standard addition, *J. Pharm. Biomed. Anal.*, 11 (**1993**) 415-420.

27. Altria KD and Filbey SD, Quantitative analysis of sumatriptan by capillary electrophoresis, *J.Liq. Chromatogr.*, 16 (**1993**) 2281-2292.

28. Zhang CX, Sun ZP, Ling DK and Zhang YJ, Separation of tetracycline and its degradation products by capillary zone electrophoresis, *J. Chromatogr.*, 627 (**1992**) 281-286.

29. Watzig H, Dette C, Aigner A and Wilschowitz L, Analysis of acetylcysteine by capillary electrophoresis (CE). 2. Determination of side components, *Pharmazie*, 49 (**1994**) 249-252.

30. Reubsaet JE, Beijnen JH, Bult A, Hop ED, Vermaas R, Kellekule Y, Kettenes-van den Bosch JJ, and Underberg WJM, Structural Identification of the Degradation Products of the Antitumor Peptide Antagonist [Arg, D-Trp, MePhe] Substance P, *Anal. Chem.* (**1995**) 67 4431-4436.

31. Nickerson B , Cunningham B and Scypinski S , The use of capillary electrophoresis to monitor the stability of a dual-action cephalosporin in solution, *J. Pharm. Biomed. Anal.*, 14 (**1995**) 73-83.

32 Van Schepdael A, Kibaya R, Roets E, Hoogmartens J, Analysis of doxycycline by capillary electrophoresis, *Chromatographia,* 41 (**1995**) 367-370.

33. Flurer CL and Wolnik KA, Quantitation of gentamicin sulfate in injectable solutions by capillary electrophoresis, *J. Chromatogr.*, 663 (**1993**) 259-263.

34. Hsu LC , Constable DJC, Orvos DR and Hannah RE, Comparison of high-performance liquid chromatography and capillary zone electrophoresis in penciclovir biodegradation kinetic studies , *J. Chromatogr. B*, 669 (**1995**) 85-92.

35. Van Schepdael A, Van den Bergh I, Roets E, Hoogmartens J, Purity control of oxytetracycline by capillary electrophoresis, *J. Chromatogr. A*, 730 (**1996**) 305-311.

36. Baeyens W, Weiss G, Van Der Weken G, Van Den Bossche W and Dewaele C, Analysis of pilocarpine and its *trans* epimer, isopilocarpine, by capillary electrophoresis, *J. Chromatogr.*, 638 (**1993**) 319-326.

37. Kenndler E, Schwer C and Kaniansky, Capillary zone electrophoresis, *J. Chromatogr.*, 508 (**1991**) 203-207.

38. Ng Cl, Ong CP, Lee HK and Li SFY, Determination of pharmaceuticals and related impurities by capillary electrophoresis, *J. Chromatogr.*, 680 (**1994**) 579-586.

39. Hansen SH, Bjornsdottir I and Tjornelund J, Separation of basic drug substances of very similar structure using micellar electrokinetic chromatography, *J. Pharm. Biomed. Anal.*, 13 (**1995**) 489-495.

40. Castaneda, Penalvo G, Julien E and Fabre H, Cross validation of capillary electrophoresis and high-performance liquid chromatgraphy for cefotaxime and related impurities *Chromatographia.* 42 (**1996**) 159-164.

41. Sciacchitano CJ, Mopper B, and Specchio JJ, Identification and separation of five cephalosporins by micellar electrokinetic capillary chromatography, *J. Chromatogr.*, 657 (**1994**) 395-399.

42. Emaldi P, Fapanni S, and Baldini A, Validation of a capillary electrophoresis method for the determination of cephradine and its related impurities, *J. Chromatogr. A*, 711 (**1995**) 339-346.

43. Bretnall AE, Hodgkinson MM and Clarke GS, Micellar electrokinetic chromatography stability indicating assay and content uniformity determination for a cholesterol-lowering drug product, *J. Pharm. Biomed. Anal.*, 15 (**1997**) 1071-1075.

44. Nishi H, Fukuyama T Matsuo M and Terabe S, Separation and determination of lipophilic corticosteroids and benzothiazepin analogues by micellar electrokinetic chromatography using bile salts, *J. Chromatogr.*, 513 (**1990**) 279-295.

45. Qin XZ, Ip DP and Tsai EW, Determination and rotamer separation of enalapril maleate by capillary electrophoresis, *J. Chromatogr.*, 626 (**1992**) 251-258.

46. Smith NW and Evans MB, Capillary zone electrophoresis in pharmaceutical and biomedical analysis, *J. Pharm. Biomed. Anal.*, 12 (**1994**) 579-611.

47. Trennery VG, Wells RJ and Robertson J, The analysis of illicit heroin seizures by capillary zone electrophoresis, *J. Chrom. Sci.*, 32 (**1994**) 1-6.

48. Wienberger R and Lurie IS, Micellar electrokinetic capillary chromatography of illicit drug substances, *Anal. Chem.*, 63 (**1991**) 823-827.

49. Lurie IS, Chan KC, Spratley TK, Casale JF and Issaq HJ, Separation and detection of acidic and neutral impurities in illicit heroin via capillary electrophoresis, *J. Chromatogr. B*, 669 (**1995**) 3-13.

50. Thomas BR, Fang XG, Chen X , Tyrell RJ and Ghodbane S, Validated micellar electrokinetic capillary chromatography method for the quality control of the drug substances hydrochlorothiazide and chlorothiazide, *J. Chromatogr.*, 657 (**1994**) 383-394.

51. Charman W N, Humberstone A J and Charman S A, Analysis of pilocarpine and its degradation products by micellar electrokinetic capillary chromatography, *Pharm. Res.*, 9 (**1992**) 1219-1223.

52. Swartz M, Method development and selectivity control for small molecule pharmaceutical separations capillary electrophoresis, *J. Liq. Chromatogr.*, 14 (**1991**) 923-938.

53. Nickerson B, Scypinski S, Sokoloff H and Sahota S, Separation of sulfixoxazole, phenazopyridine, and their related impurities by micellar electrokinetic chromatography, *J. Liq. Chromatogr.*, 18 (**1995**) 3847-3875.

54. Flurer Cl and Wolnik KA, Chemical profiling of pharmaceuticals by capillary electrophoresis in the determination of drug origin, *J. Chromatogr.*, 674 (**1994**) 153-163.

55. Thomas BR and Ghodbane S, Evaluation of a mixed micellar electrokinetic capillary electrophoresis method for validated pharmaceutical quality control, *J. Liq. Chromatogr.*, 16 (**1993**) 1983-2006.

56. Korman M, Vindevogel J and Sandra P, Application of micellar electrokinetic chromatography to the quality control of pharmaceutical formulations: the analysis of xanthine derivatives, *Electrophoresis*, 15 (**1994**) 1304-1309.

57. Weinberger R and Lurie I S, Micellar electrokinetic capillary chromatography of illicit drug substances, *Anal. Chem.*, 63 (**1991**) 823-827.

58. Ong CP Ng CL, Lee HK and Li SFY, Separation of imidazole and its derivatives by capillary electrophoresis, *J. Chromatogr.*, 686 (**1994**) 319-324.

59. Lin C-E, Chang C-C, Lin W-C, and Lin EC, Capillary zone electrophoretic separation of b-blockers using citrate at low pH , *J. Chromatogr. A*, 753 (**1996**) 133-138.

60. Flurer CL, The analysis of aminoglycoside antibiotics by capillary electrophoresis, *J. Pharm. Biomed. Anal.*, 13 (**1995**) 809-816.

61. Bullock J, Assay for the dinhydride of diethylenetriaminepentaacetic acid and its major degradation products by capillary electrophoresis, *J. Chromatogr. B*, 669 (**1995**) 149-155.

62. Qin X-Z, Ip DP and Tsai EW. Determination and rotamer separation of enalapril maleate by capillary electrophoresis. *J. Chromatogr.*, 626 (**1992**) 251-258.

63. Zhang CX, Sun ZP, Ling DK and Zhang YJ, Separation of tetracycline and its degradation products by capillary zone electrophoresis, *J. Chromatogr.*, 627 (**1992**) 281-286.

64. Giles M, Determination of impurities in an acidic drug substance by micellar electrokinetic chromatography, *Chromatographia*, 44 (**1997**) 191-196.

65. Altria KD, Application of high-low CE for improved quantitative determination of drug related impurities, *Chromatographia*, 35 (**1993**) 493-496.

66. Altria KD, Separation and quantitation of drug related impurities by capillary electrophoresis, *J. Chromatogr.*, 735 (**1996**) 43-56.

67. Unpublished data.

68. Camilleri P, Okafo GN, Southan C and Brown R, Analytical and micropreparative capillary electrophoresis of the peptides from calcitonin, *Anal. Biochem.*, 198 (**1991**) 36-42.

69. Bullock J, Capillary electrophoretic purity method for the novel metal chelator TMT-NCS, *J. Pharm. Biomed. Anal.*, 14 (**1996**) 845-854.

70. Nickerson B, The determination of a degradation product in clidinium bromide drug substance by capillary electrophoresis with indirect UV detection, *J. Pharm. Biomed. Anal.*, 15 (**1997**) 965-971.

71. Wynia GS, Windhorst G, Post PC and Mars FA, Development and validation of a capillary electrophoresis method within a pharmaceutical quality control environment and comparison with high-performance liquid chromatography, *J. Chromatogr. A*, 773 (**1997**) 339-350.

4 Separation and Quantitation of Enantiomers

4.1 Introduction

The chiral separation of pharmaceuticals is one of the most active (1–96) and important areas of CE research and there are over 400 publications in the area of separation of enantiomers by CE. A number of reviews have appeared on the subject. (91–98). Given this considerable interest a relatively large chapter of this book is devoted to this subject. A recent survey (99) showed that 21 % of those surveyed used CE to obtain chiral separations. A survey of the use of CE within pharmaceutical companies confirmed (100) that 22 % of companies use CE for chiral separations.

Generally the chiral separation of ionic solutes are achieved in Free Solution Capillary Electrophoresis (FSCE) by the addition of chiral selectors such as cyclodextrins (CDs) into the electrolyte. Alternatively chirally selective electrolyte additives include proteins, carbohydrates, crown ethers and antibiotics are employed. Chiral separations of neutral solutes are generally achieved through use of MECC with chirally selective micelles such as bile salts or using mixtures of surfactants and chiral selectors such as cyclodextrins. Alternatively uncharged solutes can be chirally resolved by use of charged ionic cyclodextrins. The range of CE chiral selectors employed reflects a similar range of chiral stationary phases used in HPLC. Table 4.1 gives a number of examples of the use of the various CE chiral separation options.

Chiral CE methods are capable of validation and can give performance levels similar to those obtained by chiral HPLC methods. Table 4.1 includes many quantitative CE applications with selected details of performance and application.

Undoubtedly the most frequently used selectors in FSCE are cyclodextrins (CD's). Other possibilities (Table 4.1) include crown ethers, carbohydrates, proteins and chiral antibiotics. The majority of pharmaceutical applications have involved the use of CD's. The most widely used approaches in MECC are mixture of SDS and CD's or bile salts which are naturally occurring chiral surfactants.

4.2 Cyclodextrins

Table 4.1 shows that the use of CD's in FSCE mode is by far the most popular means of obtaining chiral separation in CE. Cyclodextrins are naturally occurring carbohydrates which have a toriod (bucket-like) shape (Figure 4.1 gives the structure of beta-cyclodextrin). Analyte molecules can become included into the core of the cyclodextrin through their com-

Table 4.1 Range of drugs chirally resolved by CE

Compounds	Separation Conditions	Ref.	Comments
Low pH, CD's			
7 Acids and bases (racemates)	Low or high pH with cyclodextrins	1	Systematic study of effect of operating parameters
9 Amines (racemates)	Low pH, CD, TAAs	2	Up to 10 mM TAA used to improve resolution
20 Amines (racemates)	Low pH, CD's	3	Optimisation study, effect of counter-ion, triethanolamine best
5 Amines (racemates) including ephedrine, and isoproterenol	Low pH, CD's	4	Coated capillary used to suppress EOF
5 Amines (racemates) including bupivacaine	Low pH with CD, HTAB, HEC	5	< 0.1 % bupivacaine enantiomer quantified
6 Amino alcohols including bambuterol	Various pHs and CDs	7	Optimisation study – effects of CD type and concentration
4 Amino alcohols	pH 3, TAA, OH-P-β-CD	7	Optimisation of CD and ion-pair reagents
Aminoglutethimide	pH 3 various CD's including γ	8	Mixtures of charged and neutral CD's
5 Amphetamines	pH 6 with OHP-β-CD	9	Determination in urine sample, phenylanaline used as internal standard
Anaesthetics	pH 3 DM-β-CD	10	Validated routine method better than 0.1 % LOD
42 Assorted compounds	pH 2.5 with various β-CD's	11	Study on selectivity
10 Basic drugs including ephedrine (racemates)	Low pH, cyclodextrins	12	Limit of detection < 0.5 % trace enantiomer
10 Basic drugs (racemates)	Low pH, CD, HPMC	13	Up to 200 mM γ-CD used
23 Basic drugs (racemates)	pH 3 + various CD	14	All compounds resolved using different cyclodextrins
22 Basic compounds	pH 2.5, β-CD's, TAA's		TAA to reverse EOF
5 Basic drugs including terbutaline	Various pHs and CDs	16	Systematic study – effect of pH and CD
Basic compounds	pH 2.5, CD's TAA	17	Short-chain TAA's to optimise chiral separations
BCH 189	Low pH CD	18	Quantitative monitoring of enzymatic biotransformation , LOD of 0.5 %
BCH 189 and others	Low pH CD's	19	LOD of 0.1 % for BCH 189, cross-correlation with HPLC data

Table 4.1   continued

Compounds	Separation Conditions	Ref.	Comments
9 Beta blockers	pH 2.5, β -CD with TAA	20	EOF modified by addition of TAA's
Bupivacaine (and other basic drugs)	pH 3, CD's with additives	21	Assay in plasma, internal standard, good linearities
7 Chiral drugs	pH 2.5 γ CD	22	7 out of 50 racemates resolved
Clenbuterol and picumeterol	Low pH with CD	22	Optimisation studies on CD, pH and temperature
Clenbuterol	Low pH, OHP- β -CD	23	Method transfer exercise between 7 drug companies
Clenbuterol	Low pH CD	24	Optimisation study using experimental design
Denopamine, trimetoquinol, and timepidium	Low pH, β-CD polymer	25	0.1 % LOD, Spiking experiments - recoveries 95–104 %
Dihydropyridine and racemic amines	Low or high pH with CDs	26	Various neutral and ionisable CDs used for separations, syrup samples analysed
Epinephrine	Low pH, cyclodextrin	27	Pseudoephedrine internal standard, formulation analysis
Epinephrine (adrenaline)	Low pH, cyclodextrin	28	Commercial adrenaline solution analysed
Ergot alkaloids	Low pH, cyclodextrins	29	< 0.5 % trace enantiomer coated capillary
Fenfluramine	Low pH, TM-β-CD	30	Cross-validation with HPLC
Fluparoxan	Low pH, β-CD, urea, IPA	31	Full method validation
Imidazoles (racemates)	pH 3, various CD's, MeOH	32	S-BE β-CD found to be most efficient CD
Hydroxy acids	pH 5 + amino CD	33	L-isomer always eluted later
Lilly drug LY248686	Low pH cyclodextrin	34	Full validation, 0.1 % LOD
Metaproterenol and isoproterenol	low pH DM-β-CD	35	40–50 seconds analysis
Naproxen	pH 5, CD + TBAH	36	Validation including 0.1 % LOD, repeatability
Neutral compounds	pH 4.5 and CM-β-CD	37	Method optimisation
Neutral compounds	Mono(6-amino-6-deoxy)-beta-cyclodextrin	38	Drug- CD binding constants measured
Phenylalanine enantiomers	Crown ether + CD agarose gel	39	Beneficial effect of CD and crown ether
Picumeterol	Low pH CD	40	Study on the effects of peak area normalisation

Table 4.1 continued

Compounds	Separation Conditions	Ref.	Comments
Pinacidil	pH 2, OHP- β -CD, 0.05 % HPC	41	Range of CD's evaluated
Propanolol	pH 3 CM-β-CD	42	Validation LOD – 0.03%
Propanolol	pH 3, various β-CD's, MeOH, ACN	43	Increasing solvent content decreased resolution
Quinagolide, Sandoz ENX 792	Low pH, cyclodextrin	44	0.2 % inactive ENX 792 detected
Salbutamol	Low pH, DM-β-CD	45	Range of chiral and non-chiral impurities resolved simultaneously
Tiaprofenic	pH 4 various CD's	46	Determination in tablets with internal standard
Terbutaline and propanolol	Low pH cyclodextrins	47	Up to 30 % methanol used
Terbutaline	Low pH, DM-β-CD	48	Preparative collection of a single enantiomer from a racemate
Terbutaline and ephedrine	pH 3, DM-β-CD	49	1 % ephedrine, clinical analysis using SIM-MS
Trogers base, tryptophan and others	low pH, cyclodextrin and crown ethers	50	Addition of both crown ether and CD improved tryptophan resolution
Verapamil and norverapamil	Low pH, cyclodextrins	51	Assay in plasma, acceptable validation data
Low pH, crown ether			
25 Amines (racemates)	Low pH, crown ether	52	90 % chirally resolved, 60 % to baseline
5,6-dihydroxy-2-aminotetralin	Low pH crown ether	53	0.3 % LOD for trace enantiomer, validation data
Methoxamine and octopamine	pH 2, crown ether	54	0.5 % inactive methoxamine detected
Di- and tri-peptides	pH 2.5, crown ether	55	Separation optimisation
High pH, CD's			
10 Acids and bases (racemates)	pH 6–10, SBE-β-CD	56	Optimisation studies
Barbital and phenobarbital	High pH, CD + ion-pair	57	Sodium-d-camphor-10-sulphonate additive
Dihydropyridine and 5 racemic amines	High, or low pH with CDs	58	Various neutral and ionisable CDs used for separations
Ephedrine and related compounds	Various pH's and CD's	59	Low pH with neutral CD or high pH with SBE-CD
Lilly development compound	High (and low pH) with SBE-β-CD	60	Migration order switched at high or low pH

Table 4.1 continued

Compounds	Separation Conditions	Ref.	Comments
n-acetyl cysteine and related compounds	Borate, PEG, no CD	61	0.4 % LOD, separation from impurities
Organic acids	pH 7, CD	62	Chiral separation related to degree of substitution
Phenylalanine (dansylated)	pH 9 with CD	63	25 μm capillary, analysis in nutritional formulation 70 seconds
Sandoz drug SDZEAA 494s (and amino acids)	pH 11, CD	64	Compound dansylated pre-column, LOD of 0.1 %
Various acids, bases and neutrals	Sulphated -β CD, pH 7	65	Range of concentrations and MeOH additions investigated
Warfarin	pH 8, CD, MeOH	66	Enantioselective metabolism studies conducted, internal standard used for improved precision

SDS, MECC, CD's

Compounds	Separation Conditions	Ref.	Comments
Bristol-Myers Squibb drug BMS-180431-09	SDS MECC with CD	67	Validation including LOD of 0.06 %, recoveries, robustness
Nalbuphine and other diastereoismers	SDS, MECC, with THF or ACN	68	LOD or 0.05 %, 0.005 % with z-cell
9 NSAID's and barbiturates (racemates)	MECC, SDS, cyclodextrin	69	Low pH with cyclodextrins used, analysis in plasma
Propanolol, atenolol, alprenolol	SDS, CD, pH 9.3	70	Reproducibility study - migration times
Valine (OPA derivative)	SDS, pH 3	71	0.13 % LOD, automated derivatisation

Carbohydrates

Compounds	Separation Conditions	Ref.	Comments
7 Acidic drugs including ibuprofen and warfarin	pH7, dextrins	72	Analysis of commercial ibuprofen samples
Anti-coagulants, and NSAID's	Maltodextrins, pH 7	73	1 % LOD's found, quantitation study
Chlorpheniramine and chloroquine	Heparin or dextran sulphate, pH 6	74	Heparin best, Rs of up to 17 achieved for chlorpheniramine
Dilitiazem and trimetoquinol	Chondroitin and heparin, various pH's	75	0.2 % LOD for Dilitiazem and trimetoquinol
Oxamnique	Heparin and CD's	76	0.23 % LOD, RSD for 10 injections < 1%
Trimetoquinol and related compounds	Dextran sulfate, pH 5.5	77	Separations better using CD's or bile salts

Proteins

Compounds	Separation Conditions	Ref.	Comments
5 Basic drugs (racemates)	pH 5 with CBH 1	78	Plug of protein solution in capillary used for separation

Table 4.1 continued

Compounds	Separation Conditions	Ref.	Comments
Kynurenine	pH 8 + HSA	79	Separation order identical to HPLC elution order
Leucovorin	pH 7, BSA	80	PEG coated capillary to reverse order
Warfarin, promethazine and pindolol	pH 7 + various proteins	81	Isopropanol improved resolution at low % v/v levels
Bile salt MECC			
22 assorted compounds	STDC, CD's	82	SDS and CD also used in some resolutions
Dilitazem and trimetoquinol	Bile salt MECC	83	1 % inactive trimetoquinol detected
Serotonin agonist and derivatives	Bile salt MECC + CD	84	GITC derivatives separated, < 1 % LOD
Various including amphetamine	MECC, SDS methanol	85	Samples derivatised with GITC pre-separation
Antibiotics			
Assorted neutral racemic compounds	Vancomycin with Cu^{2+}	86	Improvement in chiral separation by the addition of Cu^{2+} ions.
Binaphthyl compounds and synthetic intermediates of diltiazem	Aminoglycosidic antibiotics	87	Range of compounds separated with high selectivity values
100 Racemates including NSAID's	pH 7 Vancomycin	88	Resolution factors of up to 20 obtained – all 100 compounds resolved
120 Racemates	Ristocetin A, pH 6	89	Resolution factors up to Rs 40
Isoproterenol	Rifamycin B and Rifamycin SV	90	Indirect UV detection at 350 nm, 0.1 % LOD
Reviews	FSCE and MECC	91	58 references
	FSCE and MECC	92	109 references
	FSCE, MECC	93	111 refs
	Theory	94	8 refs
	General overview	95	146 refs
	FSCE and MECC	96	82 refs.

where :

ACN = acetonitrile	MECC = micellar electrokinetic capillary
Brij 35 = neutral surfactant	chromatography
BSA = bovine serum albumin	MeOH = methanol
CBI = cellobiohydrolase I	NSAIDS = non steroidal anti-imflammatories
CD = cyclodextrin	OHP-β-CD = hydroxypropyl-beta-cyclodextrin
CM-β-CD = carboxymethyl-beta-cyclodextrin	PEG = polyethyleneglycol
DM-β-CD = dimethyl-beta-cyclodextrin	SBE- β-CD = sulfobutyl-beta-cyclodextrin
EOF = electro-osmotic flow	SDS = sodium dodecyl sulphate (surfactant)
FSCE = free solution capillary electrophoresis	SDC = sodium deoxycholate (surfactant)
HEC = hydroxyethyl cellulose	SIM-MS = single-ion monitoring mass spectro-
HPMC = hydroxypropyl methyl cellulose	scopy
HSA = human serum albumin	STDC = sodium taurodeoxycholate (surfactant)
HTAB = hexyltrimethyammonium bromide	THF = tetrahydrofuran
IPA = isopropanol	TM-β-CD = trimethyl-beta-cyclodextrin
LOD = limit of detection	TAA = tetra-alkylammonium salts (ion-pairing
	reagent)

plexation into the CD cavity. The rim of the cyclodextrin contains a number of chiral hydroxyl groups which can enantioselectively interact with a chiral analyte during its migration along the capillary. Drug enantiomers can fit inside the cavity of the CD and have individual binding constants. If the enantiomers of a compound have different binding constants then it is possible to chirally resolve them using cyclodextrin additions into the CE electrolyte.

There are a wide range of both native and derivatised CDs commercially available. The native CDs, α,β,γ, possess different numbers of glucose subunits, being six, seven and eight respectively. The γ CD has the largest cavity and can therefore be used to separate larger compounds. These surface hydroxyl groups can be chemically replaced with groups such as hydroxypropyl- and dimethyl- groups. These derivatised CD's have different enantioselectivities than the native CD's and are generally more water soluble than the naturally occurring CD's.

Ionic, chargeable CD's offer the possibility of separation of neutral drug enantiomers or enhanced separation of ionic drugs. Several CE specific derivatived CD's are commercially available including those with amino, sulphate or carboxylic groups . Table 4.1 gives a number of examples of the use of these derivatised CD's.

For example (101) separations of neutral and basic racemates were performed using five different anionic cyclodextrin derivatives as chiral selectors; carboxymethylated β-CD, β-CD phosphate sodium salt, sulfobutylether (SBE) β-CD sodium salt, carboxymethylated γ-CD, and γ-CD phosphate sodium salt. 40 basic racemates were successfully separated with a limit of detection 1% for the enantiomeric impurity.

Positively charged CD's are also available and methylamino-beta-CD and hepta-methyl-amino-β-CD have been used as chiral selectors for the enantiomeric separation of a number of acidic (arylpropionic acids, warfarin and acenocoumarol) and basic compounds (terbutaline, ketamine, chlorpheniramine and isoproterenol).

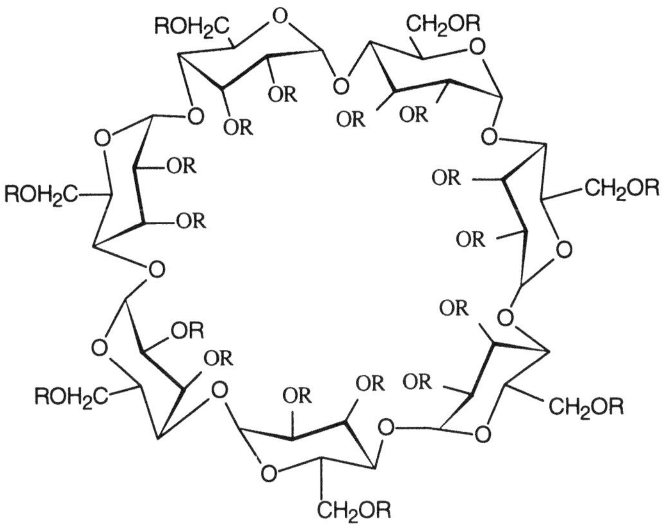

Figure 4.1 Beta-cyclodextrin structure.
R = H on native CD's but can be replaced by neutral or ionic groups

4.2.1 Low pH electrolytes containing cyclodextrins

Use of low pH electrolytes such as phosphate (pH 2–3) containing cyclodextrins has been widely employed to resolve chiral basic drugs. The drug enantiomers are positively charged and migrate towards the anode. The cyclodextrins are uncharged and retard the movement of the enantiomers when they form complexes. The enantiomer-CD complex can be stabilised through interactions such as solute hydrogen bonding with the hydroxyl groups on the rim of the CD cavity. The cyclodextrin hydroxyl groups have a fixed chirality and therefore each enantiomer will complex to a different extent in the CD. There is little or no EOF at this low pH and therefore the CD effectively acts a stationary phase. The complexation is in an equilibrium which is governed by the binding constant of the enantiomer with the CD. The enantiomer which forms the most stable complex with the CD will therefore be more retained and will be detected last. Figure 4.2 shows the process diagramatically for the resolution of a basic drug at low pH.

4.2.2 Cyclodextrins at high pH

High pH electrolytes can be used to resolve chiral acidic drugs (Table 4.1). Under these conditions the neutral cyclodextrins are swept along the capillary by the EOF while the acidic drugs migrate against the EOF. Complexation of the drug enantiomers with the CD therefore reduces the migration time of the negatively charged drug. Figure 4.3 shows the process

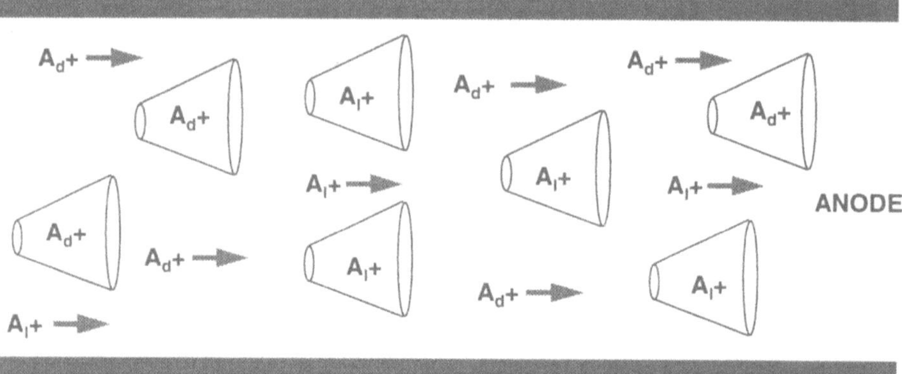

Figure 4.2 Principles of chiral separation for basic compounds using CD additives,
where : A_l^+ and A_d^+ are "l" and "d" enantiomers respectively.

diagramatically. The enantiomer with the strongest interaction with the uncharged CD will is detected first. This migration order is opposite to that obtained for low pH separations where the more strongly bound enantiomer is detected second.

High pH electrolytes in combination with negatively charged CD's can be used to separate neutral compounds (Table 4.1). For example a high pH electrolyte containing SBE-β-CD has been (60) used to chirally resolve a Lilly development drug and to determine trace enantiomer levels below 0.5 %.

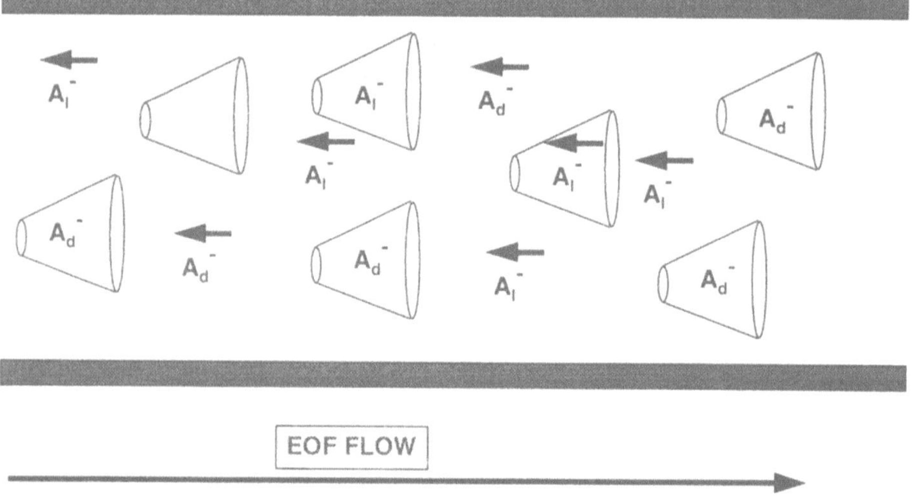

Figure 4.3 Principles of separation for chiral acidic drugs

4.2.3 Cyclodextrins in Non-aqueous CE

Chiral separations of pharmaceutical racemic amines such as trimipramine, mianserin, and thioridazine have been achieved (102) by non-aqueous capillary electrophoresis (NACE). β- and γ- cyclodextrins and various β-CD derivatives were dissolved in three organic solvents, formamide (FA), n-methylformamide (NMF), and n,n-dimethylformamide (DMF). Chiral separations for a range of compounds were achieved. Chiral separation of 12 dansyl amino acids has been achieved (103) by capillary electrophoresis using β-cyclodextrin dissolved in n-methylformamide or formamide.

4.3 Crown ethers

Crown ethers are chiral selectors which are appropriate for the resolution of chiral primary amines in CE. The chiral separation is achieved through enantioselective-chelation of the NH_3^+ group of the solute with the crown ether. The extent of enantiorecognition depends on the secondary interactions of the crown ether substituents with the functional groups on the analyte. Many drugs and intermediates have primary amine functions and crown ethers have also been employed for the resolution of a number of optically active amines of pharmaceutical interest (52-55).

18-crown-6 forms an inclusion complex (Figure 4.4) with the $-NH_3^+$ group, and enantiorecognition depends (54) on the secondary interactions of the crown ether substituents with the functional groups on the analyte. Other background electrolyte components must be chosen with care, as small cations such as Na^+ can competitively bind with the crown ether and cause a loss in resolution.

4.3.1 Crown ethers in Non-aqueous CE

Crown ether has also been used in CE to resolve (104) eight primary amino compounds including aromatic amines, amino acids and amino alcohols. (+)-18-crown-6 tetracarboxylic acid was dissolved in formamide containing tetra-n-butylammonium perchlorate as a supporting electrolyte.

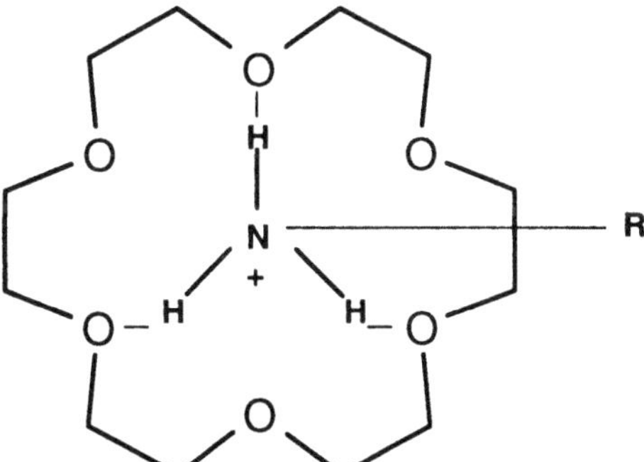

Figure 4.4 Structure of crown ether

4.4 SDS-CD MECC

Cyclodextrins can be added to SDS based MECC electrolyte to achieve (67-71) chiral resolutions. This is especially useful for neutral compounds which cannot readily be separated by FSCE (unless ionic CD's are used). In SDS-CD MECC the solutes are included into either, or both, the CD cavity or the micelle. If the distribution coefficient between these two phases is different for individual enantiomers then a chiral resolution can be obtained. Figure 4.5 shows the process diagramatically as the analyte enantiomers (A_d and A_l) are distributed differently between the SDS, CD and electrolyte phases.

4.5 Carbohydrates

A range of carbohydrates, other than cyclodextrins, are chirally selective and have been employed in CE. For instance maltodextrins (mixtures of linear linked D-glucose polymers) have been used for chiral resolution of a range of acidic drugs (105). Heparins and cyclodextrins were used to chirally resolve oxamnique at pH 3 (76) with migration time stability of better than 1 % RSD and detection limits of 0.23 % for the trace enantiomer.

Dextran has also been used to obtain chiral separations (74, 77, 106) to achieve chiral separations.

4.6 Proteins

At moderate to high pH values proteins are generally negatively charged and migrate against the EOF direction. Drugs enantiomers can selectively bind to proteins. Therefore chiral separations can be achieved (78–81) by the addition of trace levels of proteins to mid-to-high pH electrolytes. Proteins such as bovine serum albumin (BSA) (80, 81, 107) and cellobiohydrolase I (CBH I) (78) has been successful in chiral separation of a range of drugs including β-blockers, warfarin, benzoin and tryptophan.

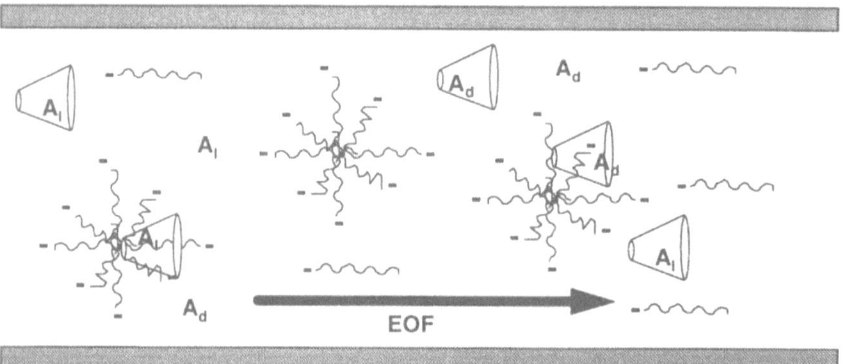

Figure 4.5 CD modified SDS MECC diagram, where: A_l and A_d are "l" and "d" enantiomers respectively.

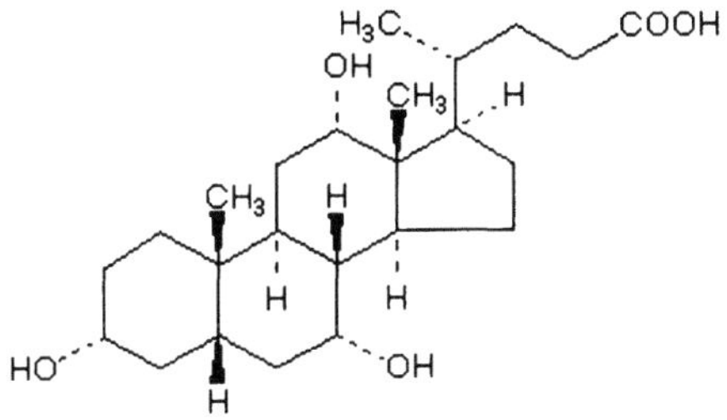

Figure 4.6 Structure of cholic acid

4.7 Bile salt MECC

An alternative approach is to incorporate bile salts into the electrolyte as these are readily available, micelle-forming surfactants which have native chirality within their structures. Enantiomers can differentially bind into the chiral micelle and resolution can be obtained. Bile salts such as sodium cholate (Figure 4.6), deoxycholate, or taurodeoxycholate are added at 25–100 mM levels to neutral and alkaline electrolytes. The bile salts form sheet-like micelles and the drug enantiomers can differentially fit inside the layers of bile salt in the micelle. A number of drugs have been chirally resolved (Table 4.1) using bile salt MECC (82-85). Bile salts are especially using in the chiral resolution of large water insoluble drugs as the interior of the micelle is very hydrophobic.

4. 8 Antibiotics

Chiral analytes can enantioselective bind to glycopeptide antibiotics such as Vancomycin (see Figure 4.7). Addition of these antibiotics to the FSCE electrolyte can often lead to impressive chiral resolutions. For example Vancomycin has been used to resolve over 100 different racemic drugs and amino acids (88).

The use of these antibiotic additives is somewhat limited by their appreciable UV absorbance signals which can make trace level determinations difficult. However it is possible to use these high UV absorbance values for the antibiotics to perform indirect UV detection. For example (90) using indirect UV detection at 350 nm, a UV absorbance maxima for Rifamycin, allowed detection of trace enantiomeric impurities. A limit of detection of 0.1 % of isoproterenol was obtained (Figure 4.8) with this indirect UV detection method.

Vancomycin forms a stable complex with Cu^{2+} in neutral aqueous solutions. Addition of copper ions can improve the enantioselectivity of native Vancomycin in chiral CE separations (86).

Other antibiotics are chiral and can be used to generate chiral separations in CE and three kinds of aminoglycosidic antibiotics; fradiomycin sulfate, kanamycin sulfate and streptomycin sulfate, have been employed (106) in the chiral separation of binaphthyl compounds and synthetic intermediates of diltiazem.

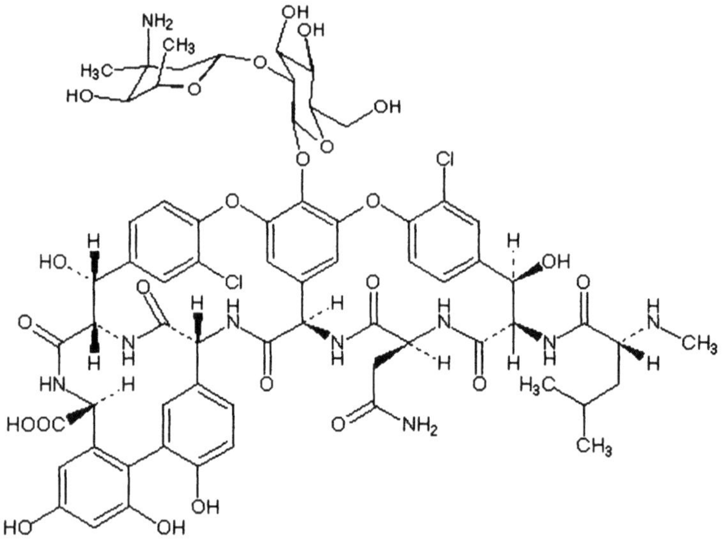

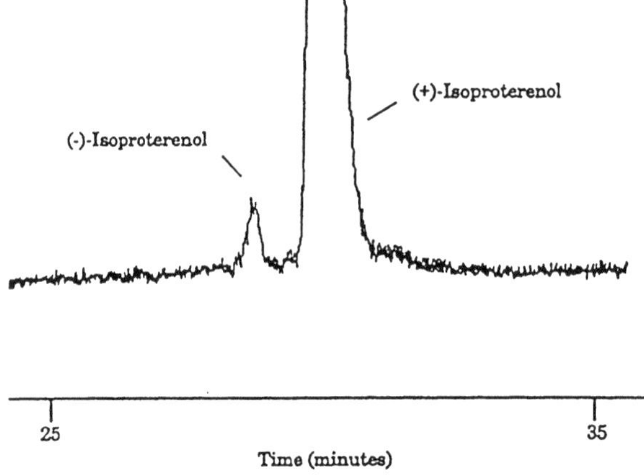

Figure 4.7
Structure of Vanco-
mycin

4.9 Synthetic surfactants

A novel synthetic surfactant with a chiral head group, (r)- or (s)-n-dodecoxycarbonylvaline
(DDCV), was used to achieve (108) enantiomeric separations of twenty basic pharmaceutical compounds by micellar electrokinetic chromatography (MEKC). Most of these compounds were beta-agonists (anti-asthmatic, bronchodilators) or beta-antagonists (anti-hyper-
tension, anti-angina). 100 mM zwitterionic CHES (2-[n-cyclohexylamino] ethanesulfonic
acid) and 10 mM triethylamine the precision for migration times, relative migration times
and selectivities of better than 1 %, 0.1 % and 1 % RSD, respectively.

Figure 4.8 Detection of 0.1%
isoproterenol by indirect UV
detection using Rifamycin as the
chiral selector. Reproduced with
permission from reference 90.

4.10 Method development

4.10.1 Selector type

A number of factors should be optimised during chiral CE method development. The principal one is to determine the most appropriate additive to achieve the desired chiral separation. If the compound is a relatively small water soluble basic drug then use of a low pH electrolyte containing CD would be appropriate. However, if the compound is a water insoluble polyaromatic neutral compound then a bile salt based MECC method may be the most appropriate.

The main consideration during the development of a CD based FSCE chiral separation is optimisation of the selector type and concentration. A range of different chiral selectors to should been explored (34, 96) to establish the CD which provides the optimum chiral resolution. Figure 4.9 shows the separations obtained for the racemic drug clenbuterol using a range of native and derivatised cyclodextrins (alpha, hydroxypropyl-alpha, beta, methyl-beta, dimethyl-beta, hydroxyethyl-beta, hydroxypropyl-beta, gamma and hydroxypropyl-gamma). This shows that the use of α-CD was inappropriate as the clenbuterol was too large to form a stable inclusion complex in the CD cavity. Increasing the size of the α-CD cavity by chemically attaching hydroxypropyl groups expanded the cavity size and allowed a partial chiral resolution. β-CD has a larger cavity and therefore allowed a separation to be achieved. Increasing the β-CD cavity by using derivatised β-CD's improved the resolution with OHP-β-CD providing the best resolution. Use of γ-CD and derivatised γ-CD was ineffective as the cavity was too large and stabilisation of an inclusion complex was impossible and therefore no chiral resolution was obtained.

In a similar exercise a wide range of CD's were evaluated (41) for the chiral separation of pinacidil. The optimal separation was found to be with hydroxypropyl-β-CD. The resolution was improved using 0.05% hydroxypropyl cellulose to increase solution viscosity.

It is also possible to use combinations of CD's to give improved chiral CE separations (109). For example dual systems of cyclodextrins, implementing the cationic mono(6-amino-6-deoxy)-beta-cyclodextrin (β-CD-NH$_2$) and a neutral CD (trimethyl-beta-CD (TM-β-CD) or dimethyl-beta-CD (DM-β-CD) for the separation of neutral enantiomers which were insufficiently resolved using electrolytes containing only one CD.

There is also an optimum CD concentration to maximise chiral resolution increases over this optimum lead to losses of resolution (110). Addition of insufficient CD does not allow a full separation to be achieved as there are only a limited number of CD's present in solution to form inclusion complexes. Conversely, if the CD concentration is too high both enantiomers will spend a high proportion of their time complexed with the CD and there will be a reduced opportunity for differentiation between the enantiomers. If the chiral molecule binds strongly with the CD then only a relatively low CD concentration is required for optimal chiral resolution. However if the compound only weakly binds then a much higher CD concentration is required. The range of CD concentrations used therefore vary significant and vary between 0.1–200 mM. Figure 4.10 shows that different solutes have a range of optimal CD concentrations for maximum chiral resolution.

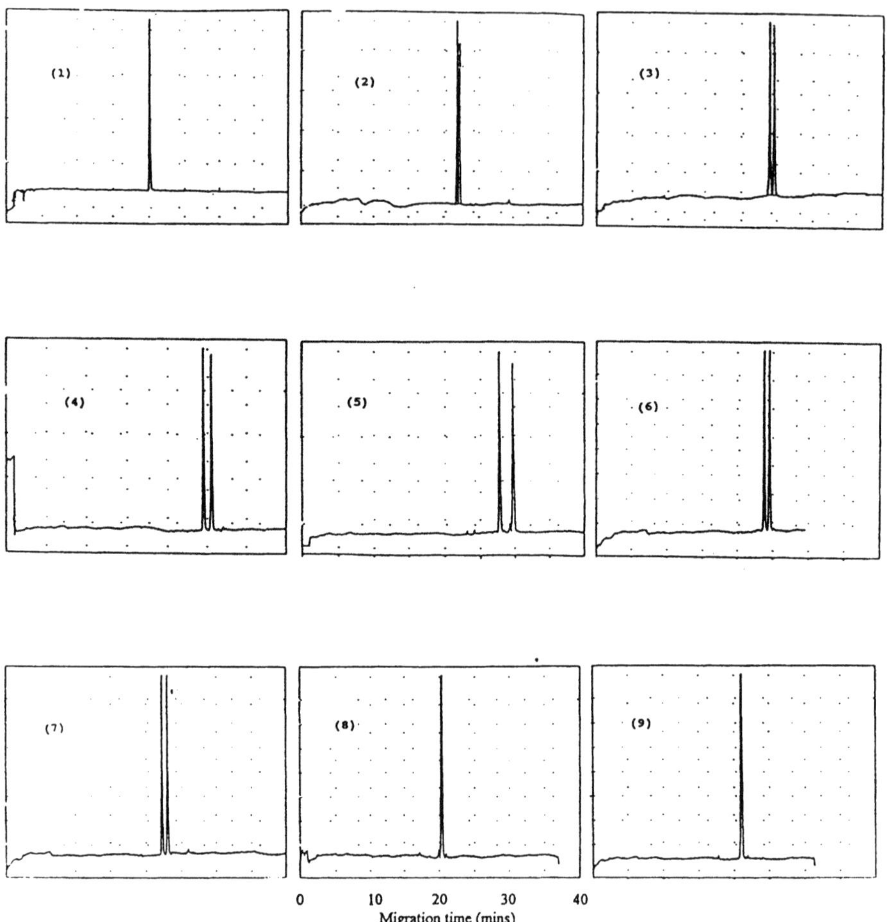

Figure 4.9 Enantiomeric separation of clenbuterol using a range of native and derivatised CD's. Selector: 16 mM of (1) α- (2) hydroxypropyl-α- (3) β- (4) methyl-β- (5) dimethyl-β- (6) hydroxyethyl-β- (7) hydroxypropyl-β- (8) γ- and (9) hydroxypropyl-γ-CDs. BGE: pH 4.0, 0.2 M disodium hydrogenphosphate / 0.1 M citric acid. Capillary 57 cm × 75 μm, voltage 13 kV, ambient temperature. Detection: UV at 214 nm. Injection: 5 s pressure (reproduced with permission from reference 22).

Each drug enantiomer has a specific interaction (binding) with the CD and therefore binding constants can be determined through treatment of CE data. The mobility of the solute in both the absence and presence of CD in the buffer can be measured. This concept has been illustrated (38) through the separations of neutral and anionic enantiomers in the presence of a cationic cyclodextrin, mono(6-amino-6-deoxy)-beta-cyclodextrin.

Similar concentration considerations are necessary when selecting bile salt concentrations in MECC and simultaneous optimisation of both the SDS and CD concentrations in CD based MECC.

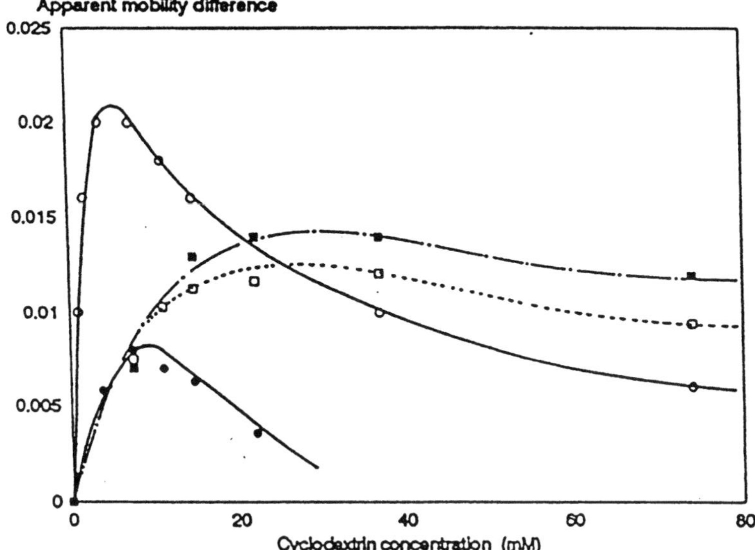

Experimentally determined apparent mobility difference (10^{-4} cm^2 / V s) for the different β-blockers at different MeBCD concentrations.
O = propanolol; ■ = atenolol; ● = metoprolol; □ = oxprenolol.

Figure 4.10 Effect of CD concentration on the chiral separation of a range of basic drugs. Reproduced with permission from reference 110.

4.10.2 Temperature

The use of lower temperatures improves chiral resolution at the expense of longer analysis times (22). The lower temperature reduces the solubility of the solute in the electrolyte and therefore favours solute inclusion into the CD. Electrolyte viscosity is also lower at reduced temperatures which reduces diffusive band broadening.

4.10.3 Electrolyte selection

Higher electrolyte concentrations reduce peak tailing and this enhances separation (22). However excessive concentration can lead to problems of Joule heating and cause reduced performance. The choice of electrolyte species can also have a significant effect on the quality of the resolution achieved. For instance (111) the chiral separation of tryptophan was dramatically improved moving from a phosphate buffer pH 2.5 to triethanolamine-phosphoric pH 2.5. Figure 4.11 shows a similar improvement (10) for the detection of 0.1 % R-ropivacaine using electrolyte with triethanolamine and phosphate. The choice of electrolyte primarily effects the peak shapes obtained.

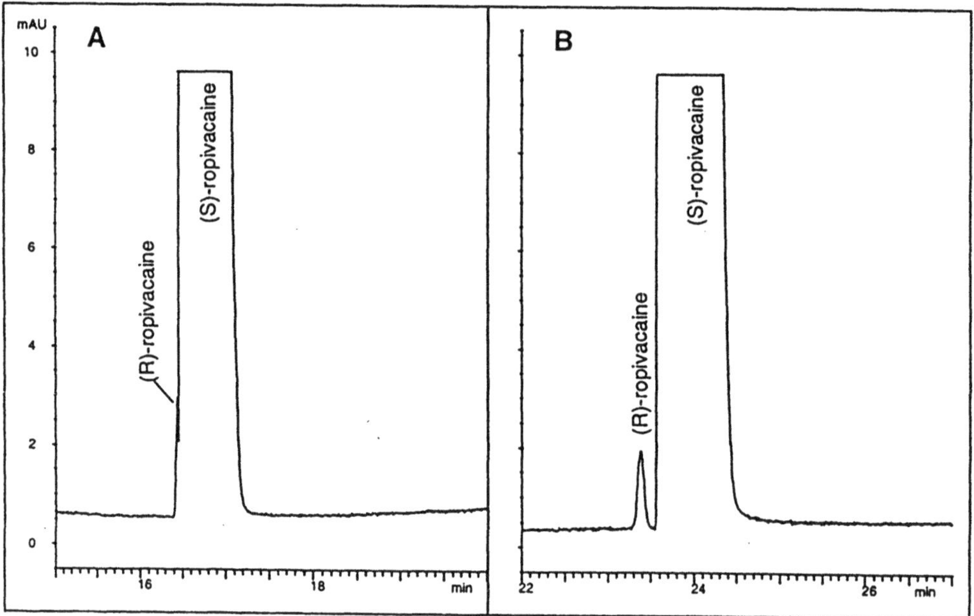

Figure 4.11 Determination of 0.1% R-ropivacaine with either phosphate or triethanolamine buffer. Separation conditions: a) 0.1M H_3PO_4 adj. to pH 3.0 with NaOH or b) 0.1M H_3PO_4 adj. to pH 3.0 with triethanolamine. 80.5cm × 50 micron, 30 kV, 30°C, 5 sec 50 mBar, detection at 206 nm, 133 mg DM-β-CD in 10 ml of buffer. Reproduced with permission from reference 10.

4.10.4 MECC electrolyte optimisation

Chiral surfactants such as bile salts can be used to achieve chiral separations. In these circumstances optimisation involves selection of the most appropriate type and concentration of bile salt. Bile salt separations are also often improved by the addition of organic solvents (such as IPA) or CD's (82). Combinations of SDS and CD's have been used for chiral separations (Table 4.1) and simultaneous optimisation of the SDS concentration and CD type/ concentration is required to obtain the optimal conditions.

4.11 Quantitative Applications

Application areas include enantiomeric purity testing of single enantiomer compounds, reaction rate monitoring, stability testing, and the enantiomeric separation of drugs in clinical samples.

The need to correct peak areas in CE analysis is important (40) for the accurate quantitation of enantiomer levels. Peak areas are related to solute response factor and concentration and the analyte residence time in the detector cell. As the second enantiomer is moving more slowly along the capillary, it spends a longer time within the detector and therefore has a greater peak area than the first enantiomer. This effect is compensated for by division of each peak area by its migration time. Another phenomenon which should be considered in quantitative analysis is the fact that differential binding of the selectands to the selector can result in changes in absorbance values for the individual enantiomers. Fanali and Bocek (28) recognised this phenomenon and proposed a correction factor if this occurs. These changes in absorbance and fluorescence differences will vary depending on the selector/selectand combination. It is therefore recommended that UV response factors for each enantiomer are obtained prior to calculation of quantitative results.

4.11.1 Enantiomeric purity testing

Several applications involving low pH electrolytes containing cyclodextrins have been reported for the enantiomeric purity testing of drugs have been reported (from Table 4.1). For example Nielen (12) demonstrated chiral separation of both norephedrine and ephedrine within 14 minutes. The method was capable of detection of < 1 % of the undesired enantiomers, with detector linearities of > 0.999 and precision's of < 2 % RSD for peak areas. 0.1% L-tryptophan in D-tryptophan has been determined using a low pH buffer containing α-CD (111). Figure 4.11 (above) shows a determination of 0.1 % R-ropivacaine in S-ropivacaine.

Application of a chirally selective MECC methods has been described for the optical purity testing of batches of drug substances (67, 68, 84). Table 4.2 shows recovery data obtained for the enantiomer of a cholesterol-lowering drug using a SDS-CD MECC method. A correlation of 0.9985 was obtained for the main component over the range of 0.15–0.40 mg/ml. A LOD of 0.01 % was obtained for the impurity enantiomer.

Optimisation of the cyclodextrin type and concentration in combination with electrolyte composition (10) enabled detection of 0.1 % R-ropivacaine in the S-enantiomer. Quantitation of 0.5 % level d-nicergoline in the presence of l-nicergoline in the pharmaceutical product Sermion has been shown (29).

Fanali and Bocek (28) reported the enantiomeric resolution of five sympathomimetic drug racemates (ephedrine, norephedrine, epinephrine, nor-epinephrine and isoproterenol) using dimethyl-β-CD. Difference in enantiomeric ratios in two commercial samples of epinephrine were found using this method. An internal standard, isoproterenol, was also used to determine the total amount of epinephrine.

Using optimised conditions, enantiomeric excess data was generated (19) for (R)-picumeterol. The (S)-enantiomer present at 0.2 % of the main component was accurately quantified, and the data shown to compare favourably with those from a previous HPLC method.

Table 4.2 Recovery data for spiking of trace enantiomers in pure drug substance

Spike %w/w	Found	Recovery
0.36	0.37	102.7
0.65	0.65	100.5
0.80	0.84	104.6
1.10	1.11	100.9

4.11.2 Reaction rate monitoring

The robust nature of chiral CE separations makes the monitoring of chirally selective processes possible. A chirally selective CE method was employed (18) to monitor the enzymatic transformation of a racemate to a single enantiomer. The required (+)-enantiomer of the anti-HIV drug 2'-deoxy-3'-thiacytidine (BCH 189) is isolated following the action of cytidine deaminase on the racemic mixture; the enzyme selectively deaminates the (-)-enantiomer to form a uridine analogue and (+)-BCH 189 is then separated by conventional means. Using DM-β-CD as selector, the biotransformation reaction time course was monitored over a 51 hour period (Figure 4.12). Quantitation of the remaining distomer was possible down to 0.5 % level.

4.11.3 Formulation stability testing

Interconversion of enantiomers can occur with drug storage and it is generally necessary to incorporate a chirally specific assay in stability for chiral compounds. The requirement of type of assay is both speed and simplicity as many samples may be analysed. Peterson and Trowbridge (27) monitored the enantiomeric purity of l-epinephrine in a pharmaceutical for-

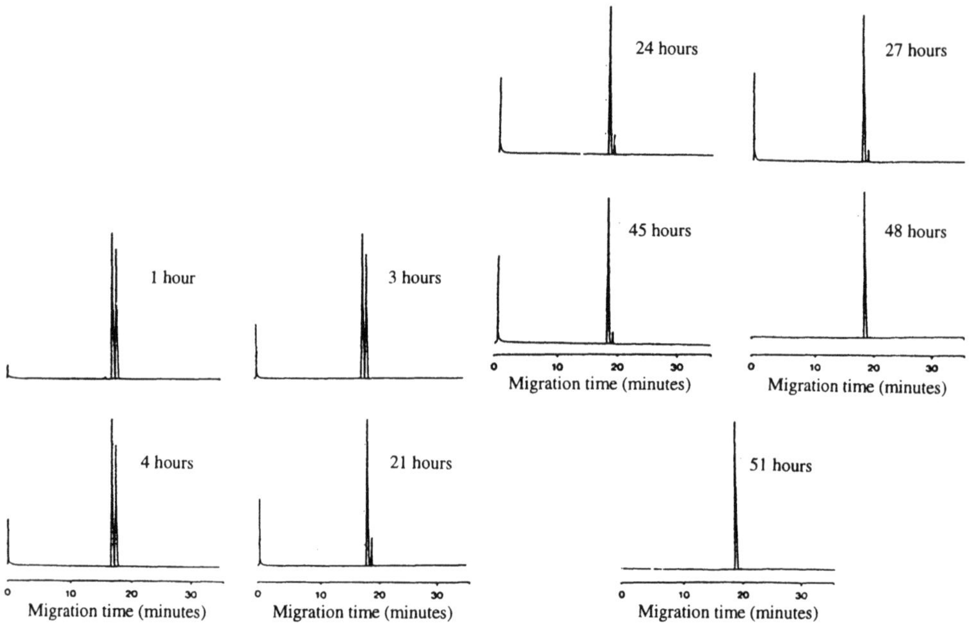

Figure 4.12 Time dependence of percentage of (+) enantiomer remaining in enantioselective enzymatic biotransformation of enzymatic biotransformation of 2'-deoxy-3'-thiacytidine (BCH 189) monitored by CE. Electrolyte: 27 mM DM-β-CD, 0.1M NaH₂PO₄ adjusted to pH 2.3 with H₃PO₄. Capillary: 57 cm × 75 μm, 15 kV, ambient temperature. Detection: UV at 214 nm. Injection: 5s pressure. Reproduced with permission from reference 18.

mulation employing l-pseudoepinephrine as an internal standard. Peak area ratio precision's of 1.8 % RSD with 99 % recoveries and detector linearities of > 0.998 were reported. Stored samples were tested and enantiomeric purity results were found to be within the company's predetermined specifications.

A chiral CE method has also been used (112) to assess both the potency and chiral purity of the Lilly drug LY231514. The method gave an LOD of 0.5 % for the undesired enantiomer and used an internal standard to obtain good injection precision. Results were compared favourably with those generated by HPLC. Dried powder drug formulation material was analysed using the fully validated CE method.

4.11.4 Clinical applications

Often extensive sample work-up of clinical sample solutions is required in chiral HPLC due to the presence of matrix components which may mask the peaks of interest or impair chromatographic performance. However in CE it has been shown that biosamples may often be analysed directly with no sample pre-treatment. A number of chiral bioassay applications have been reported (21, 51, 66, 69). For example Gareil et al (66) have performed studies concerning the enantioselective metabolism of warfarin in patients undergoing warfarin therapy. CE analysis of plasma samples confirmed the preferential metabolism of the (−) enantiomer. A detection limit of 0.2 mg/L of each warfarin enantiomer was reported. Chlorowarfarin was used as an internal standard. The limit of detection for the whole procedure (dichloromethane extraction followed by evaporation to dryness) was of the order of 0.2 mg/L $(6.5 \times 10^{-7} \text{ M})$ of each enantiomer.

A fully validated assay for cicletanine in plasma and urine has been reported by Prunonosa et al (112). The (R)-(−) and (S)-(+) enantiomers were resolved both in standards and biofluids using β-cyclodextrin in the presence of SDS, and the assay applied to a pharmacodynamic study in two volunteers.

4.12 Method Validation

The validation criteria are similar to those employed in the validation of chiral HPLC methods and include limits of detection and quantitation for the undesired enantiomer, linearity of detector response, recovery, precision, freedom from interference and method robustness.

4.12.1 Detection limits

Single enantiomer chiral compounds are often produced to high enantiomeric purities, with less than 1 % of the undesired enantiomers present as an impurity. Therefore the sensitivity of the method for the determination of the levels of trace undesired enantiomers is generally regarded as the most important performance measurement. Detection levels of < 1 % trace enantiomer by CE have been demonstrated by several workers (Table 4.1). Figure 4.13 shows (19) the analysis of a batch of BCH189 containing 0.2 % of the undesired enantiomer, the detection limit was determined as 0.05 %.

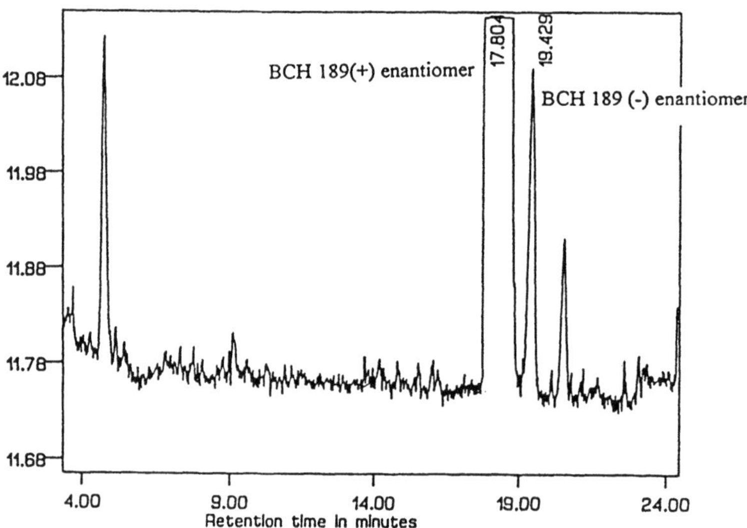

Figure 4.13 Analysis of a BCH189 drug substance batch containing 0.3 % of the (–) enantiomer. Separation conditions : 47 cm × 75 micrometre fused silica capillary, 50 mM dimethyl-β-cyclodextrin in 100 mM NaH_2PO_4 adjusted to pH 2.3 with conc. H_3PO_4, 214 nm, +13 kV. Reproduced with permission from reference 19.

4.12.2 Precision

There is often a need to quantify the total level of enantiomers as well as the ratio of the individual enantiomers. Injection precision in CE is generally poorer that for HPLC. This imprecision is largely attributable to the difficulties involved in reproducibly injecting nanolitre volumes of samples into the capillary. Many quantitative reports have utilised internal standards to alleviate this problem and acceptable precision data has been reported in those cases (27, 51, 111). An internal standard is not required if enantiomeric ratios are being measured as each enantiomer essentially acts as an internal standard for the other and if good precision for peak area obtained. Migration time reproducibility is also required to confirm peak identity for each enantiomer and is generally acceptable with RSD values of less than 2 % obtainable. Relative migration time (RMT) of one enantiomer compared to the other should be considerably better with RSD values of < 1 % being obtained (36, 76).

4.12.3 Linearity

A method may be employed to analyse a wide range of enantiomeric mixture compositions and it is therefore essential to demonstrate suitable linearity of detector response (peak area) for both sample concentration and trace enantiomeric impurity levels. To demonstrate this a single enantiomeric compound is spiked with levels of the undesired enantiomer covering the likely concentration range that may be encountered in routine testing. Adequate and acceptable detector linearity for both main peak concentration and for varying levels of the un-desired enantiomer have been reported with typical linearity data of greater than 0.998 des-cribed (31, 67, 112).

4.12.4 Recovery

To demonstrate method accuracy recovery experiments have been conducted (31, 67, 85, 112) in which single enantiomeric compounds have been spiked with known levels of their stereoisomers. The peak area ratio obtained by CE quantitatively confirmed the spiking level. These experiments also serve to demonstrate the accuracy of the methods under as-sessment. Table 4.2 shows the recovery data obtained during validation (67) of a method for enantio-purity determination of a cholesterol lowering drug.

4.12.5 Cross-validation

Combinations of HPLC and CE are being increasingly applied in method validation studies. Agreement between these two techniques strengthens the validity of the data generated by the final method of choice.

Good agreement between HPLC and CE has been reported in the enantiomeric purity tes-ting of Lilly drug LY 231514 (112) and (84) batches of a serotonin agonist (Table 4.3).

4.12.6 Freedom from interference

Freedom from interference is especially important in the analysis of formulations and clini-cal samples where endogenous peaks may mask the peaks of interest. This has been demon-strated by analysing appropriate samples of dissolving solvents (31, 112) or biosamples (21, 51, 66, 69).

Table 4.3 Comparison of % purity by CE and HPLC for a range of drug substance batches. Repro-duced with permission from reference 84.

Sample ID	HPLC	CE
CS1	38	36
CS2	52	52
CS3	58	58
CS4	72	72
CS5	78	78
CS6	88	87
CS7	92	94
CS8	< 99	< 99

4.12.7 Selectivity

As in other separative techniques co-migration of related impurities with either of the enantiomer peaks is possible. Therefore it is necessary to establish the migration position of all available related substances. This procedure has been performed in a number of studies including the chiral separations of a cholesterol lowering agent (67), salbutamol (45), Lilly drug LY231514 (112). Figure 4.14 shows the selectivity for a chiral CE for the separation of a range of anaesthetics using DM-β-CD.

4.12.8 Robustness evaluation

The effect of small deliberate changes in operating parameters is assessed as part of method validation. Factors investigated include electrolyte concentration, pH, cyclodextrin concentration, voltage, temperature and sample concentration. Many chemically derivatised cyclodextrins have been used (Table 4.1) to achieve separations and variability in the extent of the derivatisation can lead to changes in the degree of selectivity, (34, 62, 112). Therefore it essential to demonstrate that the method can be repeated using cyclodextrin from different manufacturers and lots from the same supplier.

Given the number of factors to be examined it may be appropriate to employ an experimental design such as Plackett-Burman (24) to identify the critical parameters effecting the resolution. A similar Plackett-Burman design has been (114) used to in the optimisation of a chiral separation of dexfenfluramine using dimethyl beta-cyclodextrin. The impact of concentration of CD, concentration of methanol added to the buffer, pH of the background electrolytes, temperature of the capillary and applied voltage were simultaneously evaluated. Optimised conditions were established to monitoring of the levo-rotatory enantiomeric impurity of dexfenfluramine.

4.12.9 Method transfer

The chiral CE method for the separation of the racemic drug clenbuterol has been transferred (23) between seven independent pharmaceutical companies. The separation was obtained using a low pH electrolyte containing OHP-β-CD. All participating companies obtained RSD values of <1% (n = 10) for peak area ratios and <0.05 % RSD for the precision of relative migration times. Linearity correlation coefficients of better than 0.99 were obtained by all the companies. This exercise endorsed the robustness of the method as all companies prepared their own electrolyte independently and used a variety of CE instruments and capillaries from a number of different sources.

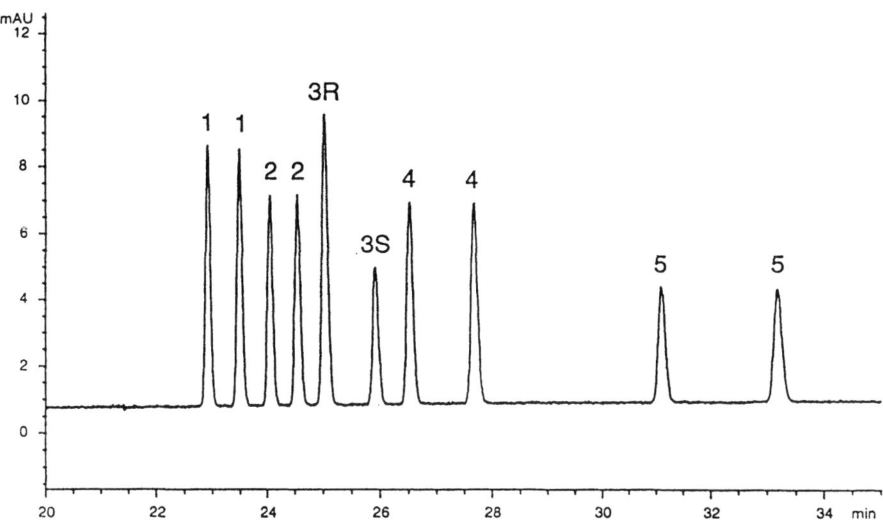

Figure 4.14 Separation of a range of anaesthetics. Separation conditions: 0.1M H_3PO_4 adj. to pH 3.0 with triethanolamine. 80.5 cm x 50 micron, 30 kV, 30°C, 5 sec 50 mBar, detection at 206 nm, 133 mg DM-β-CD in 10 ml of buffer. 1 = metpivacaine, 2 = ethyl–pipecoloxylildide HCl, 3 = ropivacaine, 4 = bupivacaine, 5 = pentyl–pipecoloxylildide HCl. Reproduced with permission from reference 10.

4.13 Conclusions

CE is a useful addition to the separation techniques available for the resolution and quantitation of enantiomers. Methods have been validated and can give similar performance levels to those obtained by HPLC. Successful method transfer has also been demonstrated. Particular features of chirally selective CE methods may include simplicity, ruggedness and low cost when compared to the purchase of expensive chirally selective HPLC columns. Method development in chiral CE can also be highly effective and cost effective compared to HPLC. Application areas of quantitative CE analysis reported to-date include enantiomeric purity testing of drugs and herbicides, reaction rate monitoring, stability testing and the analysis of clinical and forensic samples.

Stop press update

During the final editing stages of this book a number of papers appeared in this rapidly developing area. These are briefly discussed below.

Review articles have appeared on chiral CE (115) and the use of chiral CE in the analysis of biofluids (116).

The quantitative aspects of chiral CE have been further highlighted by a report (117) covering use of a chiral CE method for determining ropivacaine hydrochloride in injection solutions. The method had an LOD of 0.1 % for the enantiomeric impurity. Validation included specificity, linearity, intermediate precision and accuracy. Robustness was tested using an experimental design. A comparison of HPLC and CE data showed no significantly difference.

A variety of new separation approaches have been developed recently. These include the use of mixture of a charged and neutral cyclodextrins to chirally separate nine nonsteroidal anti-inflammatory drugs (118). The chiral separation could be manipulated by varying the type and ratios of the cyclodextrins used.

A combination of a non-chiral crown ether (18-crown-6) and beta-cyclodextrin has been used (119) to achieve enantioselective separations of primary amino compounds. The resolutions achieved where better than those obtained when no crown ether was added. The cationic selector, 2-hydroxypropyltrimethylammonium salt of beta-cyclodextrin, has been used (120) to chirally separate neutral analytes such as hexobarbital and thalidomide.

As discussed earlier in this chapter chiral resolution is improved at lower temperatures. This improvement has been highlighted (121) where the chiral selectivity of separations was increased tenfold when moving from 40°C to –20°C. 5M urea was added to the buffer to prevent freezing at these low temperatures. This approach was used for the chiral separation of a number of basic drugs including octopamine and isoproterenol.

References

1. Peterson T, Separation of drug stereoisomers by capillary electrophoresis with cyclodextrins, *J. Chromatogr.* 630 (**1993**) 353-361.

2. Quang C and Khaledi MG, Direct separation of the enantiomers of β-blockers by cyclodextrin-mediated capillary zone electrophoresis, *JHRCC*, 17, (**1994**), 99-101.

3. Bechet I, Paques P, Fillet M, Hubert P, and Crommen J, Chiral separation of basic drugs by capillary zone electrophoresis with cyclodextrin additives, *Electrophoresis*, 15, (**1994**) 818-823.

4. Fanali S, Separation of optical isomers by capillary zone electrophoresis based on host-guest complexation with cyclodextrins, *J. Chromatogr.*, 474, (**1989**), 441-446.

5. Soini H, Snopek J, and Novotony MV, Chiral separations of basic drugs with modified cyclodextrin-containing buffers by Capillary Electrophoresis, Beckman Application datasheet DS-836, (**1992**) 1-4.

6. Palmarsol S and Edholm LE, Capillary zone electrophoresis for separation of drug enantiomers using cyclodextrins as chiral selectors. Influence of experimental parameters on separation, *J. Chromatogr.*, 666 (**1994**) 337-350.

7. Stalberg O, Brotell H, and Westerlund D, Capillary electrophoretic separation of basic drugs using surface-modified capillaries and derivatised cyclodextrins as structural/chiral selectors, *Chromatographia*, 40 (**1995**) 697-704.

8. Anigbogu VC, Copper CL, and Sepaniak MJ, Separation of the stereoisomers of aminoglutethimide using three capillary electrophoretic techniques, *J. Chromatogr.*, 705 (**1995**) 343-349.

9. Varesio E and Veuthey J-L, Chiral separation of amphetamines by high-performance capillary electrophoresis, *J. Chromatogr. A*, 717 (**1995**) 219-228.

10. Sanger-van de Griend C, Groningsson K, and Westerlund D, Chiral separation of local anaesthetics with capillary electrophoresis. Evaluation of the inclusion complex of the enantiomers with heptakis(2,6-di-O-methyl)-β-cyclodextrin, *Chromatographia*, 42 (**1996**) 263-268.

11. Autamell A, Wells RJ and Wong DYK, Enantiomeric differentation of a wide range of pharmacologically active substances by capillary electrophoresis using modified β-cyclodextrins, *J. Chromatogr.*, 686 (**1994**) 293-307.

12. Nielen MWF, Chiral separation of basic drugs using cyclodextrin-modified capillary zone electrophoresis, *Anal.Chem.*, 65 (**1993**) 885-893.

13. Belder D and Schomburg G, Chiral separations of basic and acidic compounds in modified capillaries using cyclodextrin-modified capillary zone electrophoresis, *J. Chromatogr. A*, 666 (**1994**) 351-365.

14. Heurmann M and Blaschke G, Chiral separation of basic drugs using cyclodextrins as chiral pseudo-stationary phases in capillary electrophoresis, *J. Chromatogr.*, 648 (**1993**) 267-274.

15. Quang C and Khaledi MG, Extending the scope of chiral separation of basic compounds by cyclodextrin-mediated capillary zone electrophoresis, *J. Chromatogr.*, 692 (**1995**) 253-265.

16. St.Pierre LA, and Sentell KB, Cyclodextrins as enantioselective mobile phase modifiers for chiral capillary electrophoresis. Effect of pH and cyclodextrin concentration, *J. Chromatogr.*, 657 (**1994**) 291-300.

17. Quang C and Khaledi MG, Improved chiral separation of basic compounds in capillary electrophoresis using β -cyclodextrin and tetra-alyklammonium reagents, *Anal.Chem.*, 65 (**1993**) 3354-3358.

18. Rogan MM, Drake C, Goodall DM and Altria KD, Enantioselective enzymatic biotransformation of 2'Deoxy-3'-thiacytidine (BCH 189) monitored by capillary electrophoresis, *Anal.Biochem.*, 208 (**1993**) 143-147.

19. Altria KD, Goodall DM and Rogan MM, Quantitative applications and validation of the resolution of enantiomers by capillary electrophoresis, *Electrophoresis*, 15 (**1994**) 824-827.

20. Quang C and Khaledi MG, Direct separation of the enantiomers of β -blockers by cyclodextrin-mediated capillary zone electrophoresis, *JHRCC*, 17 (**1994**) 99-101.

21. Soini H, Riekkola M-L, Novotony MV, Chiral separations of basic drugs and quantitation of bupivacaine enantiomers in serum by capillary electrophoresis with modified cyclodextrin buffers, *J. Chromatogr.*, 608 (**1992**) 265-274.

22. Altria KD, Goodall DM and Rogan MM, Chiral separation of a-amino alcohols by Capillary Electrophoresis using cyclodextrins as buffer additives, I. Effect of varying operating parameters, *Chromatographia*, 34 (**1992**) 19-24.

23. Altria KD, Harden RC, Hart M, Hevizi J, Hailey PA, Makwana J, and Portsmouth MJ, An inter-company cross-validation exercise on capillary electrophoresis. 1. Chiral analysis of clenbuterol, *J. Chromatogr.*, 641 (**1993**) 147-153.

24. Rogan MM, Altria KD, and Goodall DM, Plackett-Burman experimental design in chiral capillary electrophoresis, *Chromatographia*, 38 (**1994**) 723-729.

25. Nishi H, Nakumura k, Nakai H, Sato T, and Terabe S, Enantiomeric separation of trimetoquinol, denopamine, and timepidium by capillary electrophoresis and HPLC and the application of capillary electrophoresis to the optical purity testing of the drugs, *Chromatographia*, 40 (**1995**) 638-644.

26. Schmitt T and Engelhardt H, Optimisation of enantiomeric separations in capillary electrophoresis by reversal of the migration order and using different derivatized cyclodextrins, *J. Chromatogr.*, 697 (**1995**) 561-570.

27. Peterson,T.E.and Trowbridge, D., Quantitation of l-epinephrine and determination of the d-/l-epinephrine enantiomer ratio in a pharmaceutical formulation by capillary electrophoresis, *J. Chromatogr.*, 603 (**1992**) 298-301.

28. Fanali S and Bocek P, Enantiomeric resolution by using capillary zone electrophoresis: resolution of racemic tryptophan and determination of the enantiomer composition of commercial pharmaceutical epinephrine, *Electrophoresis*, 11 (**1990**) 757-760.

29. Fanali S, Flieger M, Steinerova N and Nardi A, Use of cyclodextrins for the enantioselective seperation of ergot alkaloids by capillary zone electrophoresis, *Electrophoresis*, 13 (**1992**) 39-43.

30. Porrà R, Quaglia MG and Fanali S, Determinations of fenfluramine enantiomers in pharmaceutical formulations by capillary zone electrophoresis, *Chromatographia*, 41 (**1995**) 383-388.

31. Altria KD, Walsh AR and Smith NW, Validation of a capillary electrophoresis method for the enantiomeric purity testing of fluparoxan, *J. Chromatogr.*, 645 (**1993**) 193-196.

32. Chankvetadze B, Endresz G and Blaschke G, Enantiomeric resolution of chiral imidazole derivatives using capillary electrophoresis with cyclodextrin-type buffer modifiers, *J. Chromatogr.*, 700 (**1995**) 43-49.

33. Nardi A, Eliseev A, Bocek P and Fanali S, Use of charged and neutral cyclodextrins in capillary zone electrophoresis : enantiomeric separation of some 2-hydroxy acids, *J. Chromatogr.*, 638 (**1993**) 247-253.

34. Rickard EC and Bopp RJ, Optimisation of a capillary electrophoresis method to determine chiral purity of a drug, *J. Chromatogr.*, 680 (**1994**) 609-621.

35. Aumatell A and Guttman A, Ultra-fast chiral separation of basic drugs by capillary electrophoresis, *J. Chromatogr. A*, 717 (**1995**) 229-234.

36. Guttman A and Cooke N, Practical aspects in chiral separation of pharmaceuticals by capillary electrophoresis II. Quantitative separation of naproxen enantiomers, *J. Chromatogr.*, 685 (**1994**) 155-159.

37. Chankvetadze N, Endresz G, and Blaschke G, Capillary electrophoresis enantioseparation of non-charged and anionic chiral compounds using anionic cyclodextrin derivatives as chiral selectors, *J. Cap. Elec.*, 2 (**1995**) 235-245.

38. Lelievre L, Gareil P, and Jardy A, Selectivity in capillary electrophoresis – application to chiral separations with cyclodextrins, *Anal. Chem.*, 69 (**1997**) 385-392.

39. Lin J-M, Nakagama T, and Hobo T, Combined Chiral Crown Ether and β-Cyclodextrin for the Separation of *o*-, *m*-, and *p*-Fluoro-D,L-Phenylalanine by Capillary Gel Electrophoresis, *Chromatographia*, 42 (**1996**) 559-565.

40. Altria KD, Essential peak area normalisation in capillary electrophoresis, *Chromatographia*, 35 (**1993**) 177-182

41. Wang Z, Huang AJ, Sun YL, and Sun ZP, Cyclodextrins as buffer additives for the enantiomeric separation of pinacidil in capillary zone electrophoresis, *J. Chromatogr. A*, 749 (**1996**) 300-303.

42. Fillet M, Bechet I, Chiap P, Hubert Ph, and Crommen J, Enantiomeric purity determination of propanolol by cyclodextrin-modified capillary electrophoresis, *J. Chromatogr. A*, 717 (**1995**) 203-209.

43. Wren SAC and Rowe RC, Thereotical aspects of chiral separations in capillary electrophoresis. II. The role of organic solvent, *J. Chromatogr.*, 603 (**1992**) 235-241.

44. Kuhn R, Stoecklin F, and Erni F, Chiral separations by host-guest complexation with cyclodextrins and crown ethers by capillary zone electrophoresis, *Chromatographia*, 33 (**1992**) 32-36.

45. Rogan MM, Goodall DM, and Altria KD, Enantiomeric separation of salbutamol and related impurities using capillary electrophoresis, *Electrophoresis*, 15 (**1994**) 808-817.

46. Aturki Z, Camera E, La Torre F, and Fanali S, Direct chiral resolution of tiaprofenic acid in pharmaceutical formulations by capillary zone electrophoresis using cyclodextrins as chiral selector, *J. Cap. Elec.*, 2 (**1995**) 213-220.

47. Fanali S, Use of cyclodextrins in capillary electrophoresis: resolution of terbutaline and proprano-lol enantiomers, *J. Chromatogr.*, 545 (**1991**) 437-444.

48. Altria KD, Use of capillary electrophoresis for micro-preparative isolation, *Isolation and Purification*, 2 (**1996**) 113-125.

49. Sheppard RL, Tong X, Cai and Henion JD, Chiral separation and detection of terbutaline and ephedrin by capillary electrophoresis coupled with ion-trap mass spectrometry, *Anal. Chem.*, 67 (**1995**) 2054-2058.

50. Kuhn R, Hofstetter-Kuhn S, Chiral separation by capillary electrophoresis, *Chromatographia*, 34 (**1992**) 505-512.

51. Dethy JM, De Broux S, Lesne M, Longstreth J and Gilbert P, Stereoselective determination of verapamil and norverapamil by capillary electrophoresis, *J. Chromatogr.*, 654 (**1994**) 121-127.

52. Kuhn R, Erni F Bereuter T and Hausler J, Chiral recognition and enantiomeric resolution based on host-guest complexation with crown ethers in capillary zone electrophoresis, *Anal. Chem*, 64 (**1992**) 2815-2820.

53. Castelnovo P and Albanesi P, Determination of the enantiomeric purity of 5,6-dihydroxy-2-aminotetralin by high-performance capillary electrophoresis with crown ether as chiral selector, *J. Chromatogr A*, 715 (**1995**) 143-149.

54. Hohne E, Krauss GJ and Gubitz G, Capillary zone electrophoresis of the enantiomers of amino-alcohols based on host-guest complexation with chiral crown ethers, *J. High Res. Chromatogr.*, 15 (**1992**) 698-700.

55. Kuhn R, Riester D, Fleckenstein B and Weismuller K-H, Evaluation of an optically active crown ether for the chiral separation of di- and tripeptides, *J. Chromatogr. A*, 716 (**1995**) 371-379.

56. Desiderio C and Fanali S, Use of negatively charged sulfobutyl ether-β-cyclodextrin for enantio-meric separation by capillary electrophoresis, *J. Chromatogr. A*, 716 (**1995**) 183-196.

57. Otsuka K and Terabe S, Enantiomeric separation by micellar electrokinetic chromatography, *TRAC*, 12 (**1993**) 125-130.

58. Schmitt T and Engelhardt H, Derivatised cyclodextrins for the separation of enantiomers in capil-lary electrophoresis, *J. High Resol. Chromatogr.*, 16 (**1993**) 525-529.

59. Dette C , Ebel S and Terabe S , Neutral and anionic cyclodextrins in capillary zone electrophoresis : Enantiomeric separation of ephedrine and related compounds, *Electrophoresis*, 15 (**1994**) 799-803.

60. Liu L and Nussbaum MA, Control of enantiomer migration order in capillary electrophoresis using sulfobutyl ether beta-cyclodextrin., *J Pharm Biomed. Analysis*, 14 (**1995**) 65-72.

61. Dette C and Watzig H, Separation of the enantiomers of N-acetylcysteine by capillary electropho-resis after derivatisation with 0-phtaldialdehyde, *Electrophoresis*, 15 (**1994**) 763-768.

62. Valko IE, Billiet HAH, Frank J, and Luyben KCAM, Effect of the degree of substitution of (2-hydroxy)propyl-β-cyclodextrin on the enantioseparation of organic acids by capillary electro-phoresis, *J. Chromatogr.*, 678 (**1994**) 139-144.

63. Sepaniak MJ, Cole RO and Clark BK, Use of native and chemically modified cyclodextrins for the capillary electrophoretic separation of racemates, *J.Liq.Chromatogr.*, 15 (**1992**) 1023-1040.

64. Werner A, Nassauer T, Keichle P and Erni F, Chiral separation by capillary zone electrophoresis of an optically active drug and amino acids by host-guest complexation with cyclodextrins, *J. Chromatogr.*, 666 (**1994**) 375-379.

65. Wu W and Stalcup AM,, Capillary electrophoretic chiral separations using a sulphated β-cyclo-dextrin-containing electrolyte, *J. Liq. Chromatogr.*, 18 (**1995**) 1289-1315.

66. Gareil P, Gramond JP, and Guyon F, Separation and determination of warfarin enantiomers in human plasma samples by capillary zone electrophoresis using a methylated β-cyclodextrin-containing electrolyte, *J. Chromatogr.*, 615 (**1993**) 317-325.

67. Noroski JE, Mayo DJ, and Moran M, Determination of the enantiomer of a cholesterol-lowering drug by cyclodextrin-modified micellar electrokinetic chromatography, *J. Pharm. Biomed. Anal.*, 13 (**1995**) 54-52.

68. Williams RC, Edwards JF, and Ainsworth CR, Analysis of diastereoisomeric impurities in chiral pharmaceutical compounds by capillary electrophoresis, *Chromatographia*, 38 (**1994**) 441-445.

69. Francotte E, Cherkaoul S, and Faupel M, Separation of the enantiomers of some racemic non-steroidal aromatase inhibitors and barbiturates by capillary electrophoresis, *Chirality*, 5 (**1993**) 516-526.

70. Siren H, Jumppanen JH, Manninen K, and Riekkola M-L, Introduction of migration indices for identification ; Chiral separation of some β-blockers by using cyclodextrins in micellar electrokinetic capillary chromatography, *Electrophoresis*, 15 (**1994**) 779-784.

71. Houben RJH, Gielen H, and Van der Wal S, Automated preseparation derivatization on a capillary electrophoresis instrument, *J. Chromatogr.*, 634 (**1993**) 317-322.

72. Quang C and Khaledi MG, Chiral separations of acidic compounds by dextrin-mediated capillary zone electrophoresis, *JHRCC*, 17 (**1994**) 609-612.

73. D'Hulst A and Verbeke N, Quantitation in chiral capillary electrophoresis : Thereotical and practical considerations, *Electrophoresis*, 15 (**1994**) 854-863.

74. Agyei NM, Gahm KH, and Stalcup AM, Chiral separations using heparin and dextran sulphate in capillary zone electrophoresis, *Anal. Chim. Acta*, 307 (**1995**) 185-191.

75. Nishi H, Nakumura K, Nakai H, and Sato T, Enantiomeric separation of drugs by mucopolysaccharide-mediated electrokinetic chromatography, *Anal. Chem.*, 67 (**1995**) 2334-2341.

76. Abushoffa AM and Clark BJ, Resolution of the enantiomers of oxamniquine by capillary electrophoresis and high-performance liquid chromatography with cyclodextrins and heparin as chiral selectors, *J. Chromatogr.*, 700 (**1995**) 51-58.

77. Nishi H, Nakumura k, Nakai H, Sato T, and Terabe S, Enantiomeric separation of drugs by affinity electrokinetic chromatography using dextran sulphate, *Electrophoresis*, 15 (**1994**) 1335-1340.

78. Valtcheva L, Mohammad J, Pettersson G, and Hjerten S, Chiral separation of β-blockers by high-performance capillary electrophoresis based on non-immobilised cellulase as enantioselective protein, *J. Chromatogr.*, 638 (**1993**) 263-267.

79. Vespalec R, Sustacek V and Bocek P, Prospects of dissolved albumin as a chiral selector in capillary zone electrophoresis, *J. Chromatogr.*, 638 (**1993**) 255-261.

80. Barker GE, Russo P and Hartwick RA, Chiral separation of Leucovorin with bovine serum albumin using affinity capillary electrophoresis, *Anal. Chem.*, 64 (**1992**) 3024-3028.

81. Busch S, Kraak JC and Poppe H, Chiral separations by complexation with proteins in capillary zone electrophoresis, *J. Chromatogr.*, 635 (**1993**) 119-125.

82. Autamell A and Wells RJ, Enantiomeric differentation of a wide range of pharmacologically active substances by cyclodextrin-modified micellar electrokinetic capillary chromatography using a bile salt, *J. Chromatogr.*, 688 (**1994**) 329-337.

83. Nishi H, Fukuyama T, Matsuo M and Terabe S, Chiral separation of diltiazem, trimetoquinol and related compounds by micellar electrokinetic capillary chromatography with bile salts, *J. Chromatogr.*, 515 (**1990**) 233-243.

84. Okafo GN, Rana KK and Camilleri P, Improved separation of diastereoisomers in capillary electrophoresis using a mixture of β-cyclodextrin and sodium taurocholate, *Chromatographia*, 39 (**1994**) 627- 630.

85. Lurie IS, Micellar electokinetic capillary chromatography of the enantiomers of amphetamine, methamphetamine and their hydroxyphenethylamine precursors, *J. Chromatogr.*, 605 (**1992**) 269-275.

86. Nair UB, Chang SSC, Armstrong DW, Rawjee YY, Eggleston DS, and McArdle JV, Elucidation of vancomycin's enantioselective binding site using its copper complex, *Chirality*, 8 (**1996**) 590-595.

87. Nishi H, Nakamura K, Nakai H, and Sato T, Enantiomer separation by capillary electrophoresis using deae-dextran and aminoglycosidic antibiotics, *Chromatographia*, 43 (**1996**) 426-430.

88. Armstrong DW, Rundlett KL, and Chen J-R, Evaluation of the macrocyclic antibiotic vancomycin as a chiral selector for capillary electrophoresis, *Chirality*, 6 (**1994**) 496-509.

89. Armstrong DW, Gasper MP, and Rundlett KL, Highly enantio-selective capillary electrophoretic separations with dilute solutions of the macrocyclic antibiotic Ristocetin A, *J. Chromatogr.*, 689 (**1995**) 285-304.

90. Ward TJ, Dann C, and Blaylock A, Enantiomeric resolution using the macrocyclic antibiotics rifamycin B and rifamycin SV as chiral selectors in capillary electrophoresis, *J. Chromatog. A*, 715 (**1995**) 337-344.

91. Novotony M, Soini H, and Stefansson M, Chiral separations through capillary electromigration methods, *Anal. .Chem.*, 66 (**1994**) 646A-655A.

92. Terabe S, Otsuka K, and Nishi H, Review. Separation of enantiomers by capillary electrophoretic techniques, *J. Chromatogr.*, 666 (**1994**) 295-319.

93. Bereuter TL, Enantioseparation by capillary electrophoresis, *LC-GC Int.*, 7 (**1994**) 78-93.

94. Wren SAC, The theory of chiral separation in capilalry electrophoresis, *J. Chromatogr.*, 636 (**1993**) 57-6.

95. Nishi H and Terabe S, Review. Optical resolution of drugs by capillary electrophoresis techniques, *J. Chromatogr.*, 694 (**1995**) 245-276.

96. Rogan MM, Goodall DM, and Altria KD, Enantioselective separations using capillary electrophoresis, *Chirality*, 6 (**1994**) 25-40.

97. Kuhn R and Hofstetter-Kuhn S, Chiral separation by capillary electrophoresis, *Chromatographia*, 34 (**1992**) 505-512.

98. Otsuka K and Terabe S, Enantiomeric separation by micellar electrokinetic chromatography, *TrAC*, 12 (**1993**) 125-130.

99. Altria KD and Bryant S, Survey into the status of capillary electrophoresis, *LC-GC Int.* 10 Jan. (**1997**) 26- 30.

100. Altria KD and Kersey M, Capillary electrophoresis and pharmaceutical analysis : a survey of the industrial application and their status of in the united states and united kingdom, *LC-GC*, Jan. (**1995**) 40-46.

101. Fanali S and Camera E, Use of methylamino-beta-cyclodextrin in capillary electrophoresis – resolution of acidic and basic enantiomers, *Chromatographia*, 43 (**1996**) 247-253.

102. Wang F and Khaledi MG, Chiral separations by nonaqueous capillary electrophoresis, *Anal. Chem.*, 68 (**1996**) 3460-3467.

103. Valko IE, Siren H, and Riekkola ML, Chiral separation of dansyl amino acids by capillary electrophoresis – comparison of formamide and n-methylformamide as background electrolytes, *Chromatographia*, 43 (**1996**) 242-246.

104. Mori Y, Ueno K, and Umeda T, Enantiomeric separations of primary amino compounds by non - aqueous capillary zone electrophoresis with a chiral crown ether, *J. Chromatogr. A*, 757 (**1997**) 328-332.

105. D'Hulst A and Verbeke N, Chiral separation by capillary electrophoresis with carbohydrates, *J.Chromatgr.*, 608 (**1992**) 275-287.

106. Nishi H, Nakamura K, Nakai H, and Sato T, Enantiomer separation by capillary electrophoresis using deae-dextran and aminoglycosidic antibiotics, *Chromatographia*, 43 (**1996**) 426-430.

107. Sun P, Wu N, Barker G, and Hartwick RA, Chiral separations using dextran and bovine serum albumin as run buffer additives in affinity capillary electrophoresis, *J. Chromatogr.*, 648 (**1993**) 475-480.

108. Peterson AG, Ahuja ES, and Foley JP, Enantiomeric separations of basic pharmaceutical drugs by micellar electrokinetic chromatography using a chiral surfactant n-dodecoxycarbonylvaline, *J. Chromatogr. B*, 683 (**1996**) 15-28.

109. Lelievre F, Gareil P, Bahaddi Y, and Galons H, Intrinsic selectivity in capillary electrophoresis for chiral separations with dual cyclodextrin systems, *Anal. Chem.*, 69 (**1997**) 393-401.

110. Wren SAC and Rowe RC, Thereotical aspects of chiral separations in capillary electrophoresis: I. Initial evaluation of a model, *J. Chromatogr.*, 603 (**1992**) 234-241.

111 Altria KD, Harkin P, and Hindson M, Validation of a CE method for the quantitative determination of tryptophan enantiomers, *J. Chromatogr. B*, 686 (**1996**) 103-110.

112. Liu L, Osborne LM, and Nussbaum MA, Development and validation of a combined potency assay and enantiomeric purity method for a chiral pharmaceutical compund using capillary electrophoresis, *J. Chromatogr. A*, 745 (**1996**) 45-52.

113. Prunonosa J, Obach R, Diez-Cascon A, and Guoesclou L, Determination of cicletanine enantiomers in plasma by high performance capillary electrophoresis, *J. Chromatogr.*, 574 (**1992**) 127-133.

114. Boonkerd S, Detaevernier MR, Heyden YV, Vindevogel J, and Michotte Y, Determination of the enantiomeric purity of dexfenfluramine by capillary electrophoresis – use of a Plackett-Burman design for the optimization of the separation, *J. Chromatogr. A*, 736 (**1996**) 281-289.

115. Wespalec R and Bocek P, Chiral separations by capillary zone electrophoresis - present state of the art, *Electrophoresis*, 18 (**1997**) 843-852.

116. Bojarski J and Aboulenein HY, Application of capillary electrophoresis for the analysis of chiral drugs in biological fluids, *Electrophoresis*, 18 (**1997**) 965-969.

117. Sanger van degriend CE, Wahlstrom H, Groningsson K, and Widahlnasman M, A chiral capillary electrophoresis method for ropivacaine hydrochloride in pharmaceutical formulations - validation and comparison with chiral liquid chromatography, *J.Pharm. and Biomed. Anal.*, 15 (**1997**) 1051-1061.

118. Fillet M, Hubert P, and Crommen J, Enantioseparation of nonsteroidal anti-inflammatory drugs by capillary electrophoresis using mixtures of anionic and uncharged beta-cyclodextrins as chiral additives, *Electrophoresis*, 18 (**1997**) 1013-1018.

119. Huang WX, Xu H, Fazio SD, and Vivilecchia RV, Chiral separation of primary amino compounds using a non-chiral crown ether with beta-cyclodextrin by capillary electrophoresis, *J. Chromatogr. B*, 695 (**1997**) 157-162.

120. Schulte G, Chankvetadze B, and Blaschke G, Enantioseparation in capillary electrophoresis using 2-hydroxypropyltrimethylammonium salt of beta-cyclodextrin as a chiral selector, *J. Chromatogr. A*, 771 (**1997**) 259-266.

121. Ma S and Horvath C, Capillary zone electrophoresis at subzero temperatures .2. Chiral separation of biogenic amines, *Electrophoresis*, 18 (**1997**) 873-883.

5 Determinations of Drug Counter-Ions and Ionic Impurities by CE

5.1 Introduction

The separation and detection of simple small inorganic and organic ions is an important activity in many industries. Ion-exchange chromatography and flame AAS are frequently used as analytical testing techniques to determine levels of these ions. Increasingly CE is being viewed (1–25) as a viable alternative to these established techniques for many applications as CE can offer benefits in terms of simplicity and reduced analysis time and costs. Table 5.1 shows the range of determinations that CE has been applied to in the area. The uptake of CE to determine small ions has been considerable, a recent survey (26) indicated that it constituted 11 % of the current workload of CE in pharmaceutical analysis laboratories. A further survey (27) showed that the determination of small ions was amongst the most frequent uses of CE across all industries.

The most common salts of basic drugs are chloride, hydrochloride and sulphate or simple organic acids such as succinate, maleate or acetate. Aromatic acid salts such as benzoate, tosylate or hydroxynaphthoate are less commonly used. Acidic drugs generally have counter-ions such as metal ions including Na^+, K^+, Mg^{2+}, Ca^{2+} and simple low molecular amines.

During development of a new drug a range of different salts may be synthesised to compare pharmaceutical properties such as solubility, stability and crystallinity of the different salts. The ratio of drug to counter-ion is known as the drug stoichiometry and this needs to be characterised analytically. The typical stoichiometry is a 1:1 drug : counter-ion mixture, however, 2:1 and 1:2 compositions are manufactured dependent upon the ionic nature of the drug and/or counter-ion. There is a clear analytical need to quantify drug counter-ion levels to demonstrate that the correct salt version has been manufactured and that the required stoichiometry can be reliably achieved batch-to-batch when the final drug salt has been selected. Typically counter-ions may constitute 2–20 % w/w of a drug substance and therefore, unless precisely controlled, variable drug potency values for the batches will occur. The analytical technique employed should be capable of giving precise and accurate assay values to confirm the correct input salt and stoichiometric balance.

Typically these counter-ions possess little or no chromophore which generally necessitates use of indirect UV detection. (Table 5.1). However, some larger anionic counter-ions such as benzoates and simple organic acids can possess sufficient UV activity to allow direct UV detection (Table 5.1). Metal ions can be detected by direct UV measurement by on-capillary complexation to form metal chelates which are UV active. This chapter will discuss the various approaches to analyses of cationic or anionic counter-ions. The relative merits and disadvantages of CE will be compared to alternative techniques such as AAS and IEC.

Table 5.1 Range of small ion analysis applications

Analytes	Electrolyte	Ref.	Comments
Anions indirect detection			
Acetate content in acetate drug salt	Phthalate, OFM	1	Full method validation
Acetate residues in drug substance	Phthalate, OFM	2	Trace level detection of acetate contaminant
Drug inorganic counter-ion determination	Chromate, TTAB	3	Quantitative analysis , cross-correlation with microanalysis
Drug organic acid counter-ion determination	Phthalate, MES, TTAB	4	Quantitative analysis of maleate and succinate counter-ions
Inorganic anion contaminant in drug substances	Chromate, OFM	5	Single figure ppm detection limits in drug substance
Quantitative analysis of range of anions	Chromate, TTAB and barbital	6	Good precision through use of an internal standard
Sulphate	Chromate, OFM, pH 8	7	Quantitation in 26 detergent powders
Sulphate, chloride, nitrate	Pyromellitate, pH 7.5	8	Detergent analysis, good correlation with IEC
Various anions	NDC, TTAB, pH 11	9	Analysis of extract filters at ppb levels for organic acids
Various anions	Chromate, OFM, pH 8	10	Low ppb detection using electro-kinetic injection
Anions direct detection			
Drug organic acid counter-ion determination	Borate, pH 9.5	11	Quantitation of drug counter-ions such as benzoate, hydroxy-naphthoate
Maleic and fumaric acids	pH 8.5, boric	12	Direct detection at 200 nm
Organic acids	Borate, pH 10	13	Direct detection at 185 nm
Cations indirect detection			
Calcium, lithium, potassium and sodium counter-ions of glycosaminoglycans	4-aminopyridine buffer, pH 9	14	Range of metal counter-ions quantified in drug substance materials
Calcium in calcium acamprosate drug substance	Imidazole-sulphuric acid electrolyte	15	Method validation including comparison with EDTA titration data
Cation counter-ions content in drug substance	Imidazole, formic acid	16	Experimental design on method robustness
Cations counter-ions content in drug substance	Imidazole, low pH	3	Method validation
Metal ions and amines	Imidazole, HIBA, pH 4	17	Optimisation of electrolyte composition
Metal ions and amines	Imidazole, pH 4.5	18	Quantitation in mineral water

Table 5.1 continued

Analytes	Electrolyte	Ref.	Comments
Metal ions	HIBA, pH 4, CAT1	19	Quantitation of electrolyte solutions for parental use
Potassium in drug substance	Imidazole, formic acid	20	Method transfer exercise between drug companies
Potassium in drug substance	Imidazole, formic acid	21	Full method validation study
Quaternary amine residue in drug substance	Quinine, THF	22	Trace level contaminant determination of tetrabutylammonium ions
Various cations and anions	CAT1/HIBA and chromate/OFM	23	Prenatal vitamin analysis - cross validation with AAS
Cations direct detection			
Calcium, magnesium and colbalt	EDTA, borax, EG	24	Quantitation in flour, direct UV detection
Metals including lead, cadmium and silver	pH 7 phosphate with NaCN	25	Mid ppb detection, in-capillary cyano complexation

where:

TTAB = tetradecyltrimethylammonium bromide

HIBA = hydroxyisobutric acid

CAT 1 = proprietary Waters chemical

OFM = proprietary Waters chemical

MES = 4-morpholineethanesulfonic acid

NDC = napthalenedicarboxylic acid

EG = ethylene glycol

EDTA = ethylenediaminetetracetic acid

IEC = ion exchange chromatography

ICP = inductively coupled plasma spectroscopy

AAS = atomic absorption spectroscopy

5.2 Separation of inorganic anions

Under normal operating conditions the migration direction of anions is against the direction of the EOF which can result in extended analysis times for small inorganic and organic anions. Therefore, a variety of electrolyte compositions have been developed in which cationic surfactant additives are incorporated at low mM concentrations into the electrolyte. These surfactants form bi-layers on the capillary surface (Figure 5.1). The bi-layer effectively generates a positive charge on the capillary wall which causes the reversal in EOF direction. A range of cationic surfactant have been employed at low mM concentrations, including hexamethonium (1,6-bis-(trimethylammonium)hexane) (8, 28) Waters OFM Anion-BT (10) and tetradecyltrimethylammonium bromide (TTAB) (3, 9). When a negative polarity voltage is applied both the migration direction of the anions and the EOF are towards the detector which results in highly efficient and rapid separations.

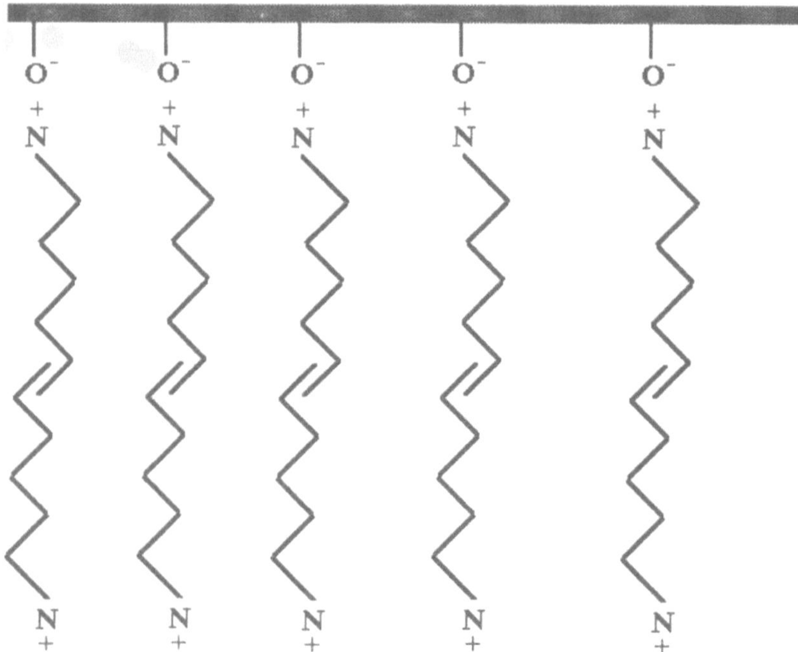

Figure 5.1 Formation of cationic surfactant bi-layer on capillary surface

5.2.1 Indirect detection of anions

A range of anionic UV absorbing species have been employed to provide the background UV signal for indirect UV detection (Table 5.1) including chromate (3,5), 1,2,4,5 benzene tetracarboxylic acid (pyromellitate) (8), and napthalenedicarboxylic acid (9). Chromate is the most widely used UV absorber (Table 5.1) as it possesses good UV activity at 254 nm and its mobility is a good match for small inorganics such as Cl^-, SO_4^{2-} and NO_3^-. Figure 5.2 shows a typical separation (29) of a range of inorganic anions using various UV absorbers. The mobilities of larger anionic solutes such as organic acids including citrate, maleate and acetate are more closely matched to hydroxybenzoate (or alternatively phthalate) and better peak shape and sensitivity is obtained using these additives rather than chromate. Very large non-UV absorbing anions, such as alkylsulphonates, are best detected using electrolytes containing sorbic acid (or barbital).

A wide range of application areas have been reported as these methods offer simple and reliable methods of analysis. Applications include determination of organic acids in wine and coffee (30), determination of vinyl sulphonate in water soluble polymers (31) and profiling of inorganic anion constituents of urine (32).

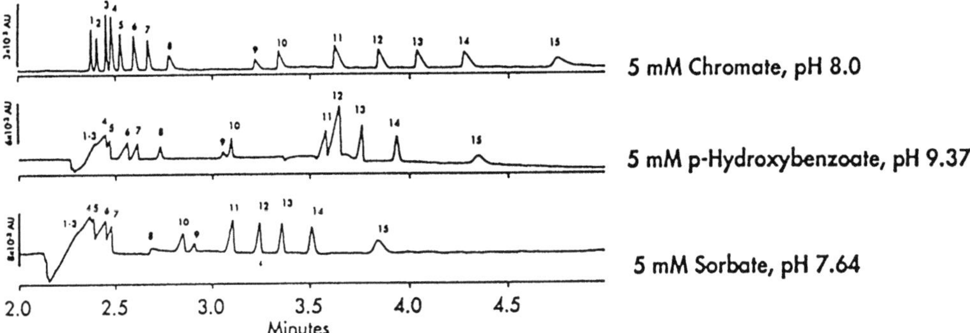

Figure 5.2 Effect of varying the UV co-ion on the separation of a range of anions. Reprinted with permission from reference 29. Peak identity : 1= bromide, 2 = chloride, 3 = sulfate, 4 = nitrite, 5 = nitrate, 6 = molydate, 7 = tungstate, 8 = citrate, 9 = phtalate, 10 = carbonate, 11 = ethanesulphonate, 12 = propanesulphonate, 13 = butanesulfonate, 14 = pentanesulfonate, 15 = hexanesulfonate. Separation conditions: Capillary dimensions are 60 cm (52 cm to the detector) × 75 μm I.D. fused silica, indirect detection at 254 nm (detector signal reversed), with a hydrostatic injection (10 cm for 30 s). 5 mM UV absorber and 0.5 mM OFM

Typically most applications are performed at low-mid ppm concentrations, however, some applications require significantly higher sensitivity. Electrokinetic injection can improve detection limits approximately 10 fold compared to hydrodynamic injection for these small highly mobile ions (10). Good linearities and acceptable precisions were shown (33) using electrokinetic injection. Typical sample loadings are achieved employing 30 second injections at –5 kV. Electrokinetic injection in-conjunction with octanesulphonate addition (75 μm) to the sample solution has been employed (29) to monitor ppb levels of anions in nuclear power plant feed water.

CE, gravimetric analysis and IC have been used (8) to determine levels of inorganic anions in detergent samples. Sulphate results in the 2–40 ppm region were statistically compared and were found to be equivalent (correlation coefficients 0.996–0.998). Injection precision was reported as 1.5 % RSD.

5.2.2 Direct detection of anions

In some instances drugs may have aromatic acid counter-anions such as benzoate or naphthalate. These anions have sufficient UV for direct detection and can be quantified using simple standard high pH buffers such as phosphate or borate. Similar conditions have been used for the detection and quantitation of a range of small organic acids including citric, lactic and acetic with direct UV detection at 185 nm (13).

A borate buffer has been employed (11) to quantify levels of aromatic acid counter-ions for both drug substance materials and excipients with detection at 200 nm. Validation of the method was performed and acceptable linearity and injection precision were obtained using internal standards. Figure 5.3 shows the separation (11) of a denatonium benzoate sample

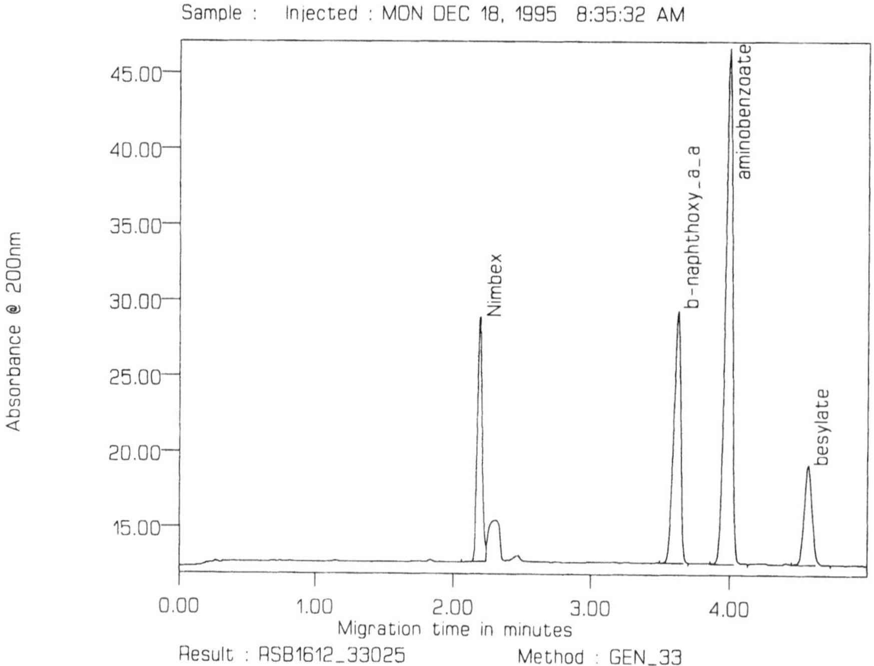

Figure 5.3 Separation of the besylate ion of Nimbex besylate containing two internal standards. Separation conditions: Fused silica capillary 27 cm × 75 μm. Buffer 15mM Borate, voltage 6.5 kV, direct detection wavelength 200 nm, at 30°C.

solution containing aminobenzoate as the internal standard. Injection precision for peak area ratios was 0.23 % RSD. The theoretical benzoate content was 27.3 % w/w and the CE result gave 27.1 % w/w.

5.3 Metal Ion Determinations

5.3.1 Determination of metal ion content by indirect UV detection

Metal ions can be relatively easily separated by simple mobility differences however they cannot be readily detected by UV absorbance measurements. Therefore a range of UV active additives (Table 5.1) are used to enable indirect UV detection. N-N-dimethylbenzylamine (34) and imidazole (3, 15, 16) have been widely employed as these are good mobility matches with metal ions commonly determined.

Co-migration of selected pairs such as ammonium and potassium ions can be a problem as they possess very similar mobilities. On-column chelation of metal ions is often employed to improve selectivity and a range of organic acids such as hydroxyisobutyric acid (HIBA) (17, 19, 23), citric acid (35) and formic acid (20) have been employed at low mM concentrations. The selective chelation of the metal ions with the particular acid reduces the net mobility of the cation resulting in an improved separation. Crown ether can also be added to achieve resolution of NH_4^+ and K^+ as it can selectively bind with the various cations and appropriately alter selectivity. Figure 5.4 shows separation of a range of metal ions using an electrolyte containing imidazole.

These test methods have been applied to monitoring cations in a range of samples including mineral water (18), and drug samples (3). Results obtained (Table 5.2) from the testing of cation contents in vitamin tablets (23) were compared to ICP results.

Levels of metal ions in a range of mineral waters has also been accurately determined by CE (18). Table 5.3 shows the good agreement between CE, HPLC and the label claim on the mineral water bottles.

Levels of K^+, Na^+, Ca^+ and Mg^{2+} have been determined in parenteral solutions (19). Low ppm levels were determined using an electrolyte system containing HIBA. Precisions of < 2 % were obtained for injections of testing containing greater than 30 ppm of the 4 cations. Good linearities (0.997 to 0.9997) and recoveries in the region of 90–100 % were obtained. Cationic nutrients were determined (36) in foods by CE following microwave digestion of the samples. Results obtained were in-line with these achieved by AAS and Ion Chromatography.

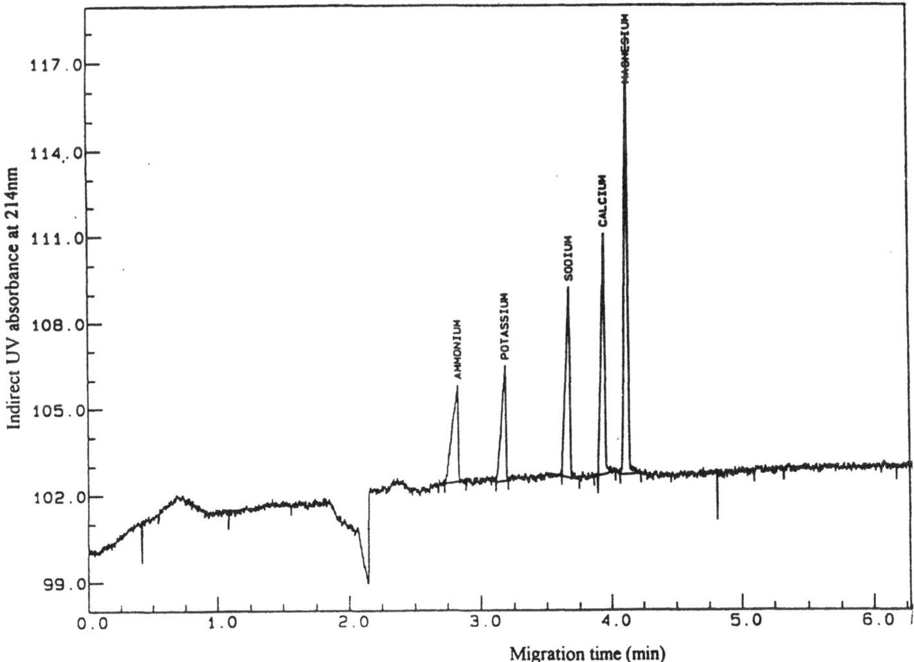

Figure 5.4 Separation of a range of metal ions using a CE method with indirect detection. Separation conditions: Fused silica capillary 57 cm × 75 μm, , voltage +20 kV, indirect UV detection wavelength 214 nm (detector signal reversed), at 30^0C, 4 mM formic acid: 4 mM $CuSO_4$: 3 mM 18-crown-6, +20 kV. Reproduced with kind permission from reference 38.

Table 5.2 Cation content in vitamin tablets (Results reported as mg/tablet)

Species	Tablet 1		Tablet 2		Tablet 3	
	CE	ICP	CE	ICP	CE	ICP
Ca^{2+}	7.2	7.8	7.8	8.6	8.0	9.0
Fe^{3+}	2.3	2.1	2.6	2.4	2.6	2.5
Zn^{2+}	1.0	1.1	1.0	1.2	1.0	1.1

Data reprinted with permission from reference 23

Table 5.3 Metal ion content in water samples. Results in ppm (mg/L)

Sample no.	Method	Potassium	Sodium	Calcium	Magnesium
1	HPLC	1.2	3.2	4.7	0.8
	CE	1.6	3.2	4.1	1.1
	Label	2.0	3.6	3.8	0.6
2	HPLC	5.5	46	64	39
	CE	3.4	49	79	41
	Label	4.6	69	100	44
3	HPLC	5.0	233	138	66
	CE	6.3	236	126	50
4	HPLC	0.5	5.3	15.9	3.3
	CE	0.6	3.1	14.5	2.9

Data reprinted with permission from reference 18

5.3.2 Direct UV detection of metal ion complexes

Metal ions are usually UV transparent and are therefore detected by indirect UV absorbance. However, electrolyte additives such as ethylenediaminetetraacetic acid (EDTA) can be used to chelate (on-capillary) with metal ions to form UV active complexes (Table 5.1). The chelating agents, such as EDTA, are generally polyanionic and instantaneously form UV active negatively charged complexes with appropriate metal ions. Typically, the ligand is added into the CE buffer at low mM concentrations. For example EDTA has been employed for the determination of the divalent cations, Ca^{2+}, Mg^{2+}, Co^{2+} present as impurities in wheat flour (24). The electrolyte contained 20 mM borax, 5 % ethylene glycol and 2 mM EDTA. The anionic complexes were detected at 200 nm. Results obtained by this method were in good agreement with those obtained by AAS. A further example of the use of EDTA has been shown in the determination of alkaline earth ions in water and serum samples (37). Precision values of 1–2 % RSD were obtained with good agreement between the CE results and those achieved using EDTA titration.

5.4 Quantitative Procedures

The general procedure is similar to that involved in main peak assay analysis. Results are calculated using response factors from calibration solutions. These calibration solutions are generally prepared from AnalaR grade ionic salts such as NaCl etc. dissolved in water. Samples and calibrations are prepared at concentrations of 50–100 mg/L (ppm) to generate relatively large peaks with improved precision.

Suitable ions have been employed as internal standards to improve precision. In these cases samples are prepared in distilled water containing 50–100 ppm of the internal standard cation. For example an internal standard solution containing 50 ppm of K^+ could be used in the assay of the Na content in the Na salt of drug.

The sample weight of drug taken should be calculated to give an expected final concentration to match that of the standard (i.e. if a drug substance contains 10% w/w Na^+ then a 50 ppm solution would be obtained by diluting 5 mg of drug substance into 10ml of internal standard solution). It is important, as with analysis of this type, to use purified water containing low levels of residual cations, and to use glassware and vials free from contamination (many vials are cleaned with detergent solutions containing sodium and sulphate salts).

5.5 Determination of Inorganic Anion Counter-Ion Levels in Basic Drugs

Levels of chloride and sulphate have been determined (3) using electrolyte containing chromate and TTAB. Sample solutions were prepared to give 100 ppm of chloride or sulphate as appropriate. AnalaR grade salts such as NaCl were used as reference standards. Peak areas were used to calculate % w/w in samples of three different drug substances. Table 5.4 shows the good agreement between the average CE results, micro-analysis data and the theoretical content. A detection limit of 1 ppm was reported for standard anions. Acceptable linearity and precision for peak areas, response factors and migration times were obtained.

Improved data for injection has been obtained (38) using an internal standard (Table 5.5). For example nitrate is a useful internal standard for determination of chloride levels. Figure 5.5 shows the separation of a 50 ppm test mixture of chloride and the internal standard nitrate.

5.6 Determination of Organic Acid Anion Counter-Ion Levels in Basic Drugs

A CE method for the analysis of simple organic acid drug counter-ions has been validated (4). The method used an electrolyte containing TTAB and a zwitterionic buffer, MES. A pH of 5.2 was selected to maximise differences in the pKa values of the acids separated.

Table 5.4 Anion analysis results. Results as %w/w

Sample	Theoretical content	Micro-analysis	CE results
Chloride			
GRD 1 batch A	8.0	–	8.0, 7.9
GRD 2 batch A	9.6	9.5	9.3, 9.3
GRD 2 batch B	9.6	9.6	9.4, 9.4
GRD 2 batch C	9.6	9.5	9.2, 9.7
GRD 2 batch D	9.6	9.4	9.9, 9.6
GRD 2 batch E	9.6	9.5	9.3, 9.5
Sulphate			
GRD 3 batch A	16.6	–	16.7
GRD 3 batch A	16.6	–	16.8

Reproduced with permission from reference 3.

Table 5.5 Performance data obtained for chloride assay using nitrate as an internal standard

Factor	Result
10 replicate injections of 50 ppm test mixture	
Migration time (mins.)	0.60 % RSD
Relative migration time	0.11 % RSD
Peak area (chloride)	4.19 % RSD
Peak area ratio (using nitrate as internal standard)	0.52 % RSD
Linearity (25–75 ppm Cl^-) calculated using area ratios	
Correlation coefficient	0.9998
Calibration preparation (10 separate preparations)	
Response factor precision ($n = 10$)	1.31 %
Accuracy (assay of KCl, theoretical Cl^- content = 47.7 % w/w)	
Calculated %w/w in sample	47.9 %
Precision of measurement ($n = 4$)	1.10 %

The method was shown (4) to be selective (Figure 5.6) for a range of the commonly analysed low molecular weight acids such as maleic, succinic, acetic and tartaric. The detector response was shown to be linear over a range covering 50–150 % of the nominal organic acid concentration of 50 mg/L. A limit of detection (s/n of 3) of 1 mg/L and a limit of quantitation (RSD of 4.1 % for 10 injections) of 2 mg/L was obtained.

Ten replicate succinate standard solutions were prepared and analysed (4) in duplicate to assess the repeatability of calibration solution preparation. The data in Table 5.6 shows this to be acceptable with an RSD of 0.13 % for the response factors obtained for 10 succinate

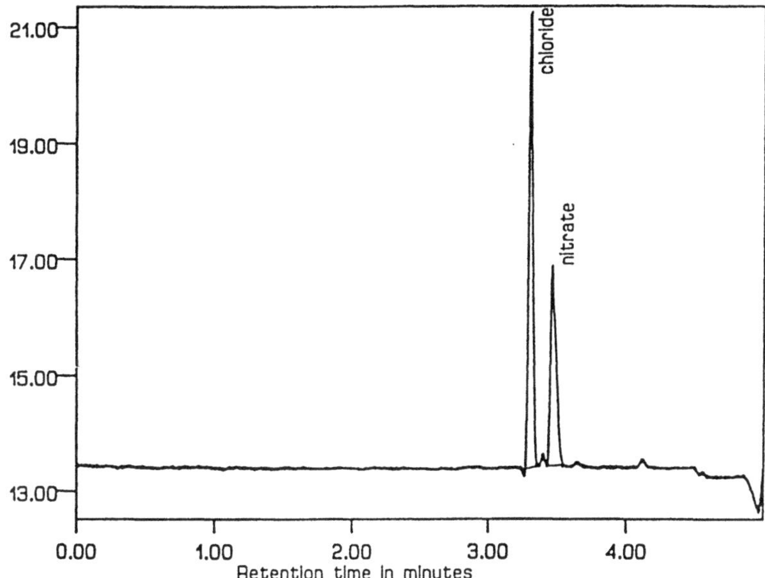

Figure 5.5 Separation of a 50 ppm test mixture of chloride and the internal standard nitrate. Reproduced with kind permission from reference 38. Separation conditions: 1 mM borate : 5 mM chromate : 0.5 mM tetradecyltrimethylammonium bromide, +10 kV, 37 cm (total length) × 75 µm capillary, indirect UV detection at 254 nm, 30°C.

calibration preparations, each solution was injected in duplicate. Ten samples of sumatriptan succinate drug substance were analysed for succinate content. The results in Table 5.6 confirm that the sample solutions could be reproducibly prepared. Ten samples of sumatriptan succinate drug substance were analysed (4) for succinate content. The method was also applied to the determination of maleate content in chlorpheniramine maleate drug substance. The results in Table 5.6 confirm that the sample solutions could be reproducibly prepared.

Table 5.6 Organic acid counter-ion determination results. Precision data for succinate calibration preparation

Calibration number	Response factor	Calibration number	Response factor
1	0.8530	7	0.8541
2	0.8539	8	0.8546
3	0.8532	9	0.8551
4	0.8517	10	0.8541
5	0.8525	Mean	0.853
6	0.8520	RSD ($n = 20$)	0.13%

Table 5.6 continued

Accuracy data for succinate and maleate content

Sumatriptan succinate sample number	Succinate content (%w/w)
1	28.60
2	28.62
3	28.81
4	28.41
5	28.40
6	28.92
7	28.66
8	28.68
9	28.67
10	28.71
RSD	0.42 %
Mean	28.64 %
Theoretical succinate content	28.57 %

Chlorpheniramine maleate sample number	Maleate content (%w/w)
1	29.65
2	29.66
Average	29.66
Theoretical maleate content	29.68 %

Reproduced with permission from reference 4

A method has been reported (1) for the quantitative determination of acetate counter-ion in a novel antifungal lipopeptide with detection using indirect photometric detection at 220 nm. The carrier electrolyte contained 4.0 mM 4-hydroxybenzoic acid (HOC_6H_4COOH) as the UV absorber and 20 mM TTAB. 20mM LiOH was used to adjust the electrolyte to pH 6.0. Linearity for detector response with acetate concentration was shown (1) over the range 0.9 mu g/ml to 46 mu g/ml ($r^2 = 0.9997$). R.S.D. values of about 1.0 % for the injection-to-injection precision were obtained without using internal standards. Comparable data for the acetate content in a batch of drug substance was obtained (1) by an ion-exchange HPLC method and the CE method. The results were 8.75 % w/w and 8.70 % w/w. The CE method gave a 99.7 % recovery for spiked acetate in an aqueous solution of the lipopeptide. The use of the CE method and HPLC method were critically compared – see Table 5.7.

In summary the CE method was considered to be faster, more inexpensive, simpler and more accurate than the HPLC method. The method has been submitted to, and accepted, by the regulatory agencies and is now in routine use.

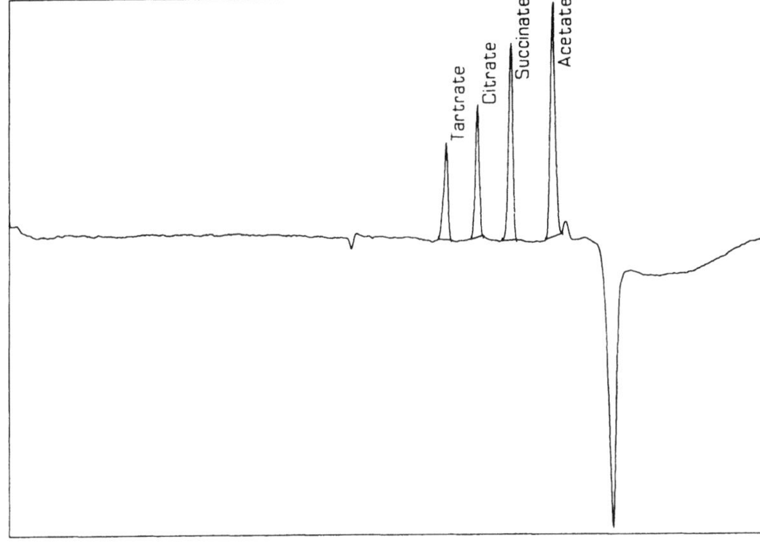

Figure 5.6 Separation of a range of organic acids with indirect UV. Reproduced with permission from reference 4. Separation conditions : 5 mM potassium hydrogen phthalate 0.5 mM tetradecyltrimethylammonium bromide, 50 mM MES, pH 5.2, –3 kV, 27 cm (total length) × 75 µm capillary, indirect UV detection at 254 nm, 30°C, 10 mg L^{-1} in water

Table 5.7 Comparison of CE and HPLC for acetate determinations in drug substance

	CE	HPLC
Capillary/column cost	10 US$	500 US$
Equilibrium time	20 minutes	few hours
Injection precision (%RSD)	1.0 %	2.0 %
Sensitivity (LOD) µg/ml	< 0.1	1–2

Reproduced with permission from reference 1

5.7 Quantitation of Metal Counter-Ion Levels in Acidic Drugs

These analyses have been largely performed using indirect UV detection employing electrolytes containing imidazole. An imidazole – H_2SO_4 electrolyte was used to quantitatively determine levels of Na^+ and K^+ in various drug salts. It was shown that CE assay results were in-line with expected metal ion contents (Table 5.8). Linearity data for the Na^+ concentration range 10–250 ppm gave a correlation coefficient of 0.9999 with an intercept value of 0.8% of the target concentration response. A limit of detection of 1 ppm was reported for both Na^+ and K^+. This report also showed separation of Ca^{2+} and Mg^{2+} under the same operating conditions. (3).

Table 5.8 Determination of metal counter-ions. Results as %w/w

Sample	Theoretical content	HPLC	CE results
Sodium			
GRD 4 batch A	5.2	–	5.2, 5.1
GRD 4 batch B	5.2	–	5.0, 4.8
GRD 4 batch C	5.2	–	4.9, 4.7
GRD 5 batch A	3.6	–	3.3, 3.4
Potassium			
GRD 6 batch A	6.0	6.1	6.1
GRD 6 batch B	6.0	–	5.9
GRD 6 batch C	6.0	–	6.0

Linearity for sodium

Range	correlation coeff.	intercept
$10 - 250$ µg/ml	0.9999	0.8 % of 100 µg/ml peak area

Peak area precision (* denotes calculated using internal standard)

Sample	number of injections	%RSD
250 µg/ml sodium	5	2.2
100 µg/ml sodium	5	2.8
100 µg/ml potassium day 1	45	2.8
100 µg/ml potassium day 1 (*)	45	0.2
100 µg/ml potassium day 2	10	1.3
100 µg/ml potassium day 2 (*)	10	0.3

Response factor precision (* denotes calculated using internal standard)

Standard	number of injections	%RSD
100 µg/ml potassium	10	1.3
100 µg/ml potassium (*)	10	0.5

Migration time precision

Sample	number of injections	%RSD
100 µg/ml sodium	10	0.7
100 µg/ml potassium	10	0.8
100 µg/ml potassium RMT (*)	10	0.1

RMT = relative migration time of potassium compared to sodium
Reproduced with permission from reference 3

Table 5.9 Precision data obtained for different injection times

Sample	Injection time (sec)	No. of injections	RSD (%) K+ area	RSD (%) PAR
Potassium (100 mg/L)	1	9	9.8	–
Potassium and sodium both at 100 mg/L	1	9	3.6	1.1
Potassium and sodium both at 100 mg/L	2	45	2.8	0.24

Note : PAR denotes peak area ratios

Reproduced with permission from reference 21

A CE method has been fully validated (21) for determination of K^+ content in a potassium salt of an acidic drug. The analysis employed a formic acid-imidazole electrolyte. Sodium was employed as an internal standard. A 100 µm bore capillary was used to give large peak areas to minimise integration errors. The use of sodium as the internal standard was shown (21) to improve the injection precision for 45 injections from 2.8 % down to 0.2 % RSD as it effectively eliminated injection related errors (Table 5.9). Other validation criteria successfully evaluated, included repeatability of calibration and sample solutions (21), repeatability of separation by different analysts on different days using alternative instruments, demonstration of appropriate detector linearity, robustness to electrolyte composition variations (which allowed limits to be set for electrolyte preparation) and cross-correlation with theoretical results and ion-exchange chromatography data.

Levels of Na in sodium cephalothin drug substance have been determined (20) by 6 independent pharmaceutical companies in an inter-company exercise. Separation was achieved by all companies using a formic acid - imidazole electrolyte. Table 5.10 shows that acceptable precision for migration times and response factors was obtained by all companies. The theoretical Na content of sodium cephalothin is 5.50 % w/w which was confirmed (20) by the average result of the analyses (5.45 % w/w). This data showed that the method was transferable and was capable of producing accurate and precise data. The average RSD on relative migration times of sodium (0.26 %) was better than for the average actual migration time (0.58 %). Average precision data for calibration response factors was 1.0–1.1 % RSD.

Table 5.10 Assay results for sodium content in sodium salt obtained by independent companies

Company	% w/w Na	% RSD
1	5.50	0.5
2	5.26	0.6
3	5.63	2.4
4	5.49	2.6
5	5.37	0.8
Average	5.45	1.4

Note: $n = 10$ for each company

Average % w/w Na result = 99.0% of theoretical Na content

Reproduced with permission from reference 20

As with all CE methods it is important to conduct robustness testing to establish the effect of small changes in operating parameters upon the system performance. Typically the ranges assessed are 10–20 % above and below the settings specified in the method. For example if the method employs a separation voltage of 20 kV it would be appropriate to check the performance of the method if run using both 18 and 22 kV. Similar ranges would be assessed for all variables within an extended injection sequence. An experimental design could be used to randomise the variations run within the sequence and to allow subsequent statistical evaluation of the data. Robustness studies (16) on a CE method using a formic acid – imidazole electrolyte to separate K^+ and Na^+ ions was conducted using this experimental design approach. A number of factors and ranges were investigated (16) in an initial fractional factorial screening design. Responses such as peak area, migration times and resolutions were recorded for each separation. These responses can then be graphically displayed as a Pareto plot to show the effect that each parameter has on a measured response. The results from this study (16) allowed experimentally determined limits to be set on all operating parameters.

Levels of calcium in the drug substance calcium acamprosate were determined (15) using an imidazole-sulphuric acid electrolyte (pH 4.5). Magnesium was used as an internal standard to give injection precision of less than 1 % RSD. Sample and calibration solutions were both 10 mg/L in calcium and magnesium which gave (15) acceptable resolution and injection precision. Linearity over the range 5–15 mg/L gave a correlation coefficient of 0.9999. A sensitivity of 0.7 mg/L was obtained for Ca^{2+}. The method gave comparable data to that obtained by an EDTA titration method (Table 5.11).

The counter-ion of glycosaminoglycans such as heparin and dermatan were analyzed (14) by CE with indirect UV-detection at 254 nm using a 40 mM 4-aminopyridine buffer. Typically heparin contains between 9.5–12 % sodium or calcium as the counter-ion, other counter-ions are barium and lithium. Calcium, lithium, potassium and sodium were resolved using the CE method. Samples of heparins containing either or both sodium and calcium were successfully analysed (14). Linearity data for calcium and sodium was obtained.

Table 5.11 Calcium content in calcium acamprosate drug substance.

Batch	Theoretical Content	CE data	EDTA Titration
C110	10.01 %	10.02 % (10.00, 10.04)	9.98 %
OTA 37	10.01 %	9.98 % (9.96, 10.01)	10.08 %

Reproduced with permission from reference 15

5.8 Determination of Ionic Contaminants in Drug Substance

It is also possible to monitor levels of ionic contaminants in samples using these simple methods. For example (5) inorganic anion contaminants were determined in drug substance material. Single figure ppm levels of a range of anions (Figure 5.7) were determined in 10 mg/ml drug substance solutions. Levels of various metal ions have been determined in water samples used in chemical processing of pharmaceuticals.

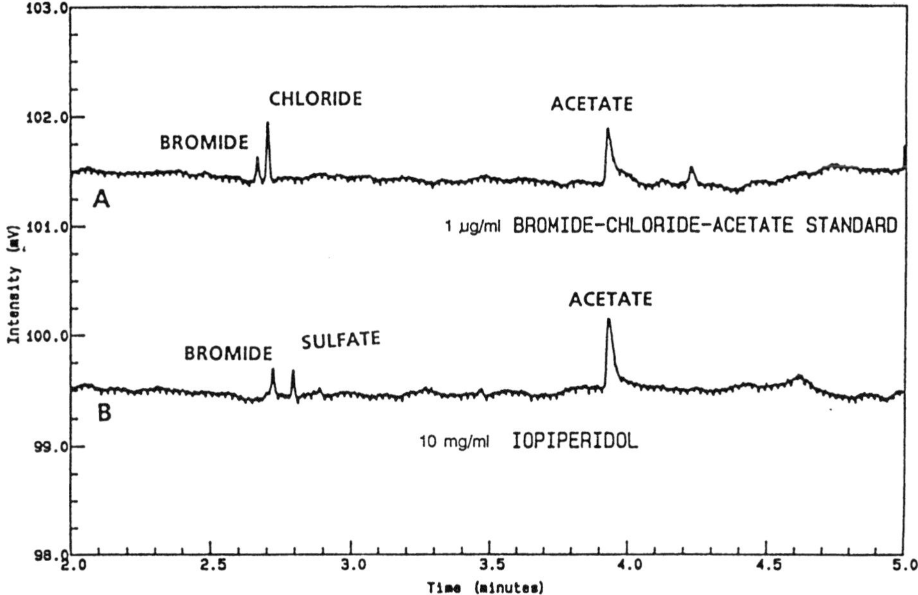

Figure 5.7 Determination of trace level anion impurities. Reproduced with permission from reference 5. Separation conditions: 5 mM chromate : 0.5 mM OFM, indirect UV detection at 254 nm, 30°C.

Trace level residual sodium acetate in antisense oligonucleotides has been determined (2) using a CE method. The sodium acetate is used in the purification of a 20-mer phosphorothioate oligonucleotide drug substance using reverse phase chromatography. A CE method using OFM modifier and phthalate was used with indirect UV detection at 254, chloride was included as an internal standard (RSD values ~2 % RSD). The linear dynamic range for the method was 0.06–240 mu g/mL. A limit of detection of 0.0012 % w/w acetate in drug substance was obtained (2). Interference due to oligonucleotide matrix was eliminated by filtering the olignucleotide solution through a Centricon SR-3 column.

A capillary electrophoretic method was used (22) for the determination of a residual alkyl quaternary amine, tetra-n-butylammonium ion (TBA$^+$), in bulk Indinavir HCl drug substance. The method was also applied to determining TBA$^+$ in process streams. The electrolyte used contained 25 % v/v THF, 3 mM sodium acetate and 12 mM quinine, pH 4.5. The TBA$^+$ was detected by indirect UV at 335 nm. Linearity of 0.9998 was obtained for detector response with TBA$^+$ concentration over the range 1.08 µg/ml to 1.08 mg/ml. An RSD of 1.6 % was obtained at the 1 mg/ml. Spiking experiments were performed (22) at the 140 µg/ml level and recoveries ranged from 96 % to 102 %.

5.9 Benefits and Disadvantages of CE Methods in Stoichiometric Analysis

5.9.1 Disadvantages

Sensitivity – CE is less sensitive when used in standard injection mode than IEC or AAS. However, sensitivity is not of great importance in the testing of drug counter-ion content as concentrations such as 50-100ppm can be selected. If trace level ionic impurity levels are determined then low ppm (mg/l) detection can be obtained using pressure injection. Use of optimised electrokinetic injection procedures can allow the possibility to determine small ions at single figure ppb levels.

Precision – injection precision is poorer than in IEC and use of an internal standard is recommended. Selection of a similar ion is usually sufficient.

5.9.2 Advantages

Simplicity – the inexpensive capillary can be dedicated to analysis of specific anion or cation determinations and can be stored ready for immediate use. Stock solutions of reagents and calibrations and internal standards can be prepared at stored for up to 3 months. The methods and analytical sequences can be programmed into the required CE instruments to reduce the set-up time further in routine analysis. The set-up time prior to analysis is considerably reduced compared to IEC and the costs per analysis are also considerably reduced.

Cost – the expense of capillaries and electrolytes is considered minimal compared to IEC columns and mobile phase, or running costs associated with AAS or ICP operation such as purchase of gases.

Analysis time – sample analysis can be commenced within a few minutes using pre-prepared capillaries and reagents. This compares favourably with preparation of IEC mobile phase, regeneration of IEC columns and equilibration. Overall analysis time is also reduced compared to AAS/ICP where instrument stabilisation times and calibration graph preparation times are to be considered. Analysis is conducted on standard CE equipment and, therefore, requires no additional training or purchase of dedicated equipment. Commercial CE instruments incorporate autosamplers which permits a large number of analyses to be conducted in unattended injection sequences.

Data quality – like IEC, CE produces retrievable data with high precision and accuracy. This compares favourably with micro-analysis techniques such as titration.

References

1. Zhou L and Dovletoglou A, Practical capillary electrophoresis method for the quantitation of the acetate counter-ion , *J. Chromatogr.* A, 763 (**1997**) 279-284.

2. Chen DH, Klopchin P, Parsons J and Srivatsa GS, Determination of sodium acetate in antisense oligonucleotides by capillary zone electrophoresis, *J. Liq. Chromatogr.*, 20 (**1997**) 1185-1195.

3. Altria KD, Goodall DM and Rogan MM, Quantitative determination of drug counter-ion stoichiometry by capillary electrophoresis, *Chromatographia*, 38 (**1994**) 637-642.

4. Altria KD, Assi K, Bryant S and Clarke BJ, Quantitative analysis of organic acid drug counter-ions by capillary electrophoresis, *Chromatographia*, 44 (**1997**) 367-371.

5. Nair JB and Izzo CG, Anion screening for drugs and intermediates by capillary ion electrophoresis, *J.Chromatogr.*, 640 (**1993**) 445-461.

6. Dabek-Zlotorynska E and Dlouhy JF, Capillary zone electrophoresis with indirect UV detection of organic anions using 2,6-napthalene dicarboxylic acid, *J. Chromatogr.*, 685 (**1994**) 145-153.

7. Jordan JM, Moese RL, Johnson-Watts R and Burton DE, Determination of inorganic sulphate in detergent products by capillary electrophoresis, *J. Chromatogr.*, 640 (**1994**) 445-451.

8. Pretswell E L, Morrisson A R and Park J S, Comparison of capillary zone electrophoresis with standard gravimetric analysis and ion chromatography for the determination of inorganic anions in detergent matrices, *Analyst*, 118 (**1993**) 1265-1267.

9. Dabek-Zlotorynska E and Dlouhy JF, Capillary zone electrophoresis with indirect UV detection of organic anions using 2,6-napthalene dicarboxylic acid, *J. Chromatogr.*, 685 (**1994**) 145-153.

10. Jackson P E and Haddad P R, Optimisation of injection technique in capillary electrophoresis for the determination of trace levels of anions in environmental samples, *J. Chromatogr.*, 640 (**1993**) 481-487.

11. Assi KH, Altria KD and Clark BJ, Determination of organic acid counter-ions of basic drugs by capillary electrophoresis, *Pharm. Sci.,* in press.

12. Chadwick RR and Hsieh JC, Separation of cis and trans double bond isomers using capillary zone electrophoresis, *Anal. Chem.*, 63 (**1991**) 2377-2380.

13. Shirao M, Furuta R, Suzuki S, Nakazawa H, Fujita S and Maruyama T, Determination of organic acids in urine by capillary zone electrophoresis, *J. Chromatogr. A*, 680 (**1994**) 247-251.

14. Malsch R and Harenberg J, Purity of glycosaminoglycan-related compounds using capillary electrophoresis, *Electrophoresis*, 17 (**1996**) 401-405.

15. Fabre H, Blanchin MD, Julien E, Segonds C, Mandrou B and Bosc N, Validation of a capillary electrophoresis procedure for the determination of calcium in calcium acamporate, *J. Chromatogr. A*, 772 (**1997**) 265-269.

16. Filbey SD and Altria KD, Robustness testing of a capillary electrophoresis method for the determination of potassium content in the potassium salt of an acidic drug, *J. Cap. Elec.*, 1 (**1994**) 190-195.

17. Quang C and Khaledi MG, Prediction and optimisation of the separation of metal cations by capillary electrophoresis with indirect UV detection, *J. Chromatogr.*, 659 (**1994**) 459-466.

18. Beck W and Engelhardt H, Capillary electrophoresis of organic and inorganic cations with indirect UV detection, *Chromatographia.*, 33 (**1992**) 313-316.

19. Koberda M, Konkowski M, Youngberg P, Jones W R and Weston A, Capillary electrophoretic determination of alkali and alkaline-earth cations in various multiple electrolyte solutions for parenteral use, *J. Chromatogr.*, 602 (**1992**) 235-240.

20. Altria KD, Clayton NG, Harden RC, Hart M, Hevizi J, Makwana J and Portsmouth MJ, Inter-company cross-validation exercise on capillary electrophoresis. Quantitative determination of drug counter-ion level, *Chromatographia*, 40 (**1995**) 47-50.

21. Altria KD, Wood T, Kitscha R and Roberts-McIntosh A, Validation of a capillary electrophoresis method for the determination of potassium counter-ion levels in an acidic drug salt, *J. Pharm. Biomed. Analysis*, 13 (**1995**) 33-38.

22. Johnson BD, Grinberg N, Bicker G and Ellison D, The quantitation of a residual quaternary amine in bulk drug and amine in bulk drug and process streams using capillary electrophoresis, *J. Liq. Chromatogr.*, 20 (**1997**) 257-272.

23. Swartz M E, Capillary electrophoretic determination of inorganic ions in prenatal vitamin formulation, *J. Chromatogr.*, 640 (**1993**) 441-444.

24. Kajiwara H, Sato A and Kaneko S, Analysis of calcium and magnesium ions in wheat flour by capillary zone electrophoresis, *Biosci. Biotech. Biochem.*, 57 (**1993**) 1010-1011.

25. Buchberger W, Semenova O P, Timerbaev A R, Metal ion capillary zone electrophoresis with direct UV detection : separation of metal cyanide complexes, *J. High Resolut. Chromatogr.*, 16 (**1993**) 153-156.

26. Altria KD and Kersey M, Capillary electrophoresis and pharmaceutical analysis : a survey of the industrial application and their status of in the united states and united kingdom, *LC-GC*, Jan. (**1995**) 40-46.

27. Altria KD and Bryant S, Survey into the status of capillary electrophoresis, *LC-GC Int.* 10 Jan. (**1997**) 26- 30.

28. Harrold M P, Wojtusik M J, Riviello J and Henson P, Parameters influencing separation and detection of anions by capillary electrophoresis, *J. Chromatogr.*, 640 (**1993**) 463-471.

29. Jandik P and Jones W R, Optimisation of detection sensitivity in the capillary electrophoresis of inorganic anions, *J. Chromatogr.*, 546 (**1991**) 431-443.

30. Tindall G W, Wilder D R and Perry R L, Optimising dynamic range for the analysis of small ions by capillary zone electrophoresis, *J. Chromatogr.*, 641 (**1993**) 163-167.

31. Ryder DS, Determination of sodium vinyl sulphonate in water-soluble polymers using capillary zone electrophoresis, *J. Chromatogr.*, 605 (**1992**) 143-147.

32. Wildman B J, Jackson P E, Jones W R and Alden PG, Analysis of anion constituents of urine by inorganic capillary electrophoresis, *J. Chromatogr.*, 546 (**1991**) 459-466.

33. Romano J P and Krol J, Capillary ion electrophoresis, an environmental method for the determination of anions in water, *J. Chromatogr.*, 640 (**1993**) 403-412.

34. Chen M and Cassidy R M, Separation of metal ions by capillary electrophoresis, *J. Chromatogr.*, 640 (**1993**) 425-431.

35. Weston A, Brown P R, Jandik P, Jones W R and Heckenberg A L, Factors affecting the separation of inorganic metal cations by capillary electrophoresis, *J. Chromatogr.*, 593 (**1992**) 289-295.

36. Morawski J, Alden P and Sims A, Analysis of cationic nutrients from foods by ion chromatography, *J. Chromatogr.*, 640 (**1993**) 359-364.

37. Motomizu S, Oshima M, Matsuda S-Y, Obata Y and Tanaka H, Separation and determination of alkaline-earth metal ions as UV absorbing chelates with EDTA by capillary electrophoresis. Determination of calcium and magnesium in water and serum samples, *Anal. Sci.*, 8 (**1992**) 619-624.

38. Altria KD, Elgey J, Lockwood P and Moore D, An overview of the applications of capillary electrophoresis to the analysis of pharmaceutical raw materials and excipients, *Chromatographia*, 42 (**1996**) 332-342.

6 Trace Analysis and Residues Determination

6.1 Introduction

Capillary electrophoresis is now beginning to become established as an appropriate means to determine components at trace levels. In the area of pharmaceutical analysis, this may involve determination of residues of drug or detergent following formulation manufacture. Additionally, CE can be used to monitor operator exposure levels or environmental levels. The sensitivity disadvantage of CE, compared to HPLC, can be overcome in various ways. The features of CE testing in these areas such as simplicity, the ability to simultaneously monitor levels of several compounds, reduced analysis time and costs can make analysis very attractive.

The use of capillary electrophoresis to monitor trace level of contaminants is becoming increasingly established in many application areas (1–10). The sensitivity requirements for these applications is often at the low ppm (mg/L) or ppt (µg/L) levels. Several approaches to obtaining these detection levels have been adopted including the use of solid phase extraction and combinations of low wavelength UV detection and wide bore capillaries.

The number of samples involved in trace level determinations is often high and therefore a short analysis time is a necessity. Short capillary lengths and/or high voltages have been widely used to reduce analysis times in CE. Table 6.1 gives details of a range of CE applications (1–10) in trace level determinations. Non pharmaceutical related trace level determinations have included a low pH CE method for aromatic amines in water samples (2), or high pH electrolytes for determination of both acidic herbicides (6) and surfactants (9). Micellar electrokinetic capillary chromatography (MECC) methods have been used to separate complex mixtures of neutral and charged components such as explosive residues (5).

6.2 Drug Residue Analysis

Following use of pharmaceutical manufacturing equipment, such as tablet presses, there is a regulatory requirement to demonstrate that the equipment has been appropriately cleaned. Extensive cleaning is necessary to avoid any possibility of cross-contamination of products. Dependent upon the solubility of the drug, the manufacturing equipment is cleaned by use of water, detergent solutions and/or organic solvents as appropriate. The cleanliness of the equipment is then tested using an appropriately sensitive method to determine residual levels for the drug. For example, in the case of a water soluble drug, pre-weighed portions (swabs) of cotton wool, or other suitable material, are moistened with a few drops of water. Specified areas of the equipment are wiped with the damp swabs to quantitatively remove any remaining drug. The drug present on the swab is then extracted into solution. Typically, the swab

Table 6.1   Trace analysis applications of CE

Application	Comments	Ref.
Acidic drugs and detergent residues on pharmaceutical manufacturing equipment	Low ppb detection levels for drugs and sub ppm LOD for metal ions	1.
Aromatic amines	Solid phase extraction for determination in water samples	2
Basic drug residues on pharmaceutical manufacturing equipment	Routine method applicable to wide range of basic drugs	3
EDTA residues on pharmaceutical manufacturing equipment	Trace level detection possible	4
Explosives residues	Determination in soil samples	5
Herbicides	Determination in water samples	6
Mirtazarpine	Determination of drug residues in cleaning qualification studies	7
Sulphonamide residues in pork meat extracts	2–9 ppm LODs for 16 sulphonamides in extracted samples	8
Surfactant residues (SDBS) on pharmaceutical manufacturing equipment	Method in routine use using detection at 200 nm, run-time 1.5 minutes	9
Occupational hygiene monitoring	Low pH method to quantify range of basic drugs	1
Oxytetracycline in fishmeat	0.1 ppm LOD with 90 % recoveries	10

is added to 5.0 ml of water and sonicated for 5 minutes. The resultant solution is then analysed for drug content. The drug concentration, if any, can then be related back to the amount of residual drug remaining on the equipment.

The sensitivity requirements in this type of application are quite demanding as cross-contamination is to be avoided. Typically, it is necessary to demonstrate that levels of 1µg or less of the drug remain on a specific area that has been swabbed. Typically, the extraction volume is 5 ml which equates to a minimum detection level of 1 µg/5 ml (200 ppb) in the test solution. These detection levels, and lower, are possible using commercial CE instruments by adopting a possible combination of wide bore capillaries (i.e. 100 µm) and UV detection around 200 nm or similar. Generally many conventional pharmaceutical possess highest UV activities in the region 200–220 nm at wavelengths which are generally inaccessible in HPLC due to UV absorbance of the solvents employed. The use of wider bore capillaries increases sensitivity considerably for two reasons. The on-capillary detection used means that use of a wider bore capillary effectively increases the detection cell volume resulting in larger peaks. In addition, the volume of sample solution injected per second using wider bore capillaries is considerably increased again resulting in larger signals. An increase in current is the main disadvantage when increasing capillary bore. This increase, if too excessive, can be compensated for by reducing the electrolyte concentration. Alternatively, a combination of lower voltage and shorter capillary will also serve to reduce current.

Recent advantages in capillary design have lead to the development of modified capillaries with enhanced detection area. The most frequently used modified capillary has a bubble formed in the area used for detection (11). The bubble area may have a pathlength of 150 micron compared to the capillary internal diameter of 50 micron. The wider pathlength gives a significantly better sensitivity.

The use of general CE methods for analysis of basic or acidic drugs make this activity attractive. A single set of operating conditions can often be applied to determine a wide range of drugs with no modification.

6.2.1 Basic drugs residues

Figure 6.1 shows use of a general method for basic drugs to achieve separation (1) of a test mixture containing 4 basic drugs and an internal standard (imidazole). A pH 2.3 phosphate buffer was used for separation with detection at 200 nm. Detection limits as low as 25 ppb have been reported (1, 3) for drug residues. HPLC is widely used for this type of determination and Table 6.2 shows (3) a cross-comparison of CE and HPLC data for analysis of residues of a basic drug.

Early work (3) on a low pH general method was extended (1) to improve injection precision and sensitivity for basic drug residue analysis. Aminobenzoate and imidazole were used as internal standards to improve injection precision. These internal standards were added at 100 µg/L levels into the solvent used to prepare both the calibration solutions and to extract

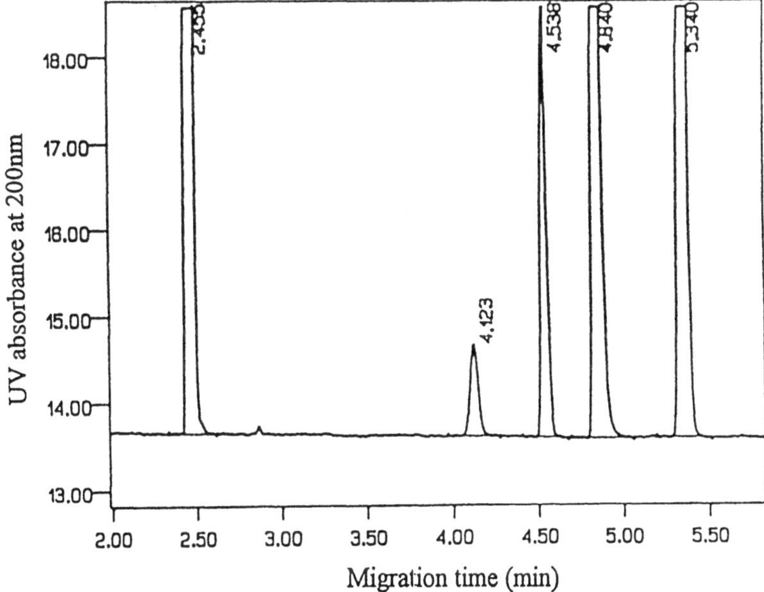

Figure 6.1 Separation of the multi-drug mixtures used in the extraction procedure optimisation experiments. Separation conditions: 27 cm × 100 µm capillary (20cm to detector), 200 nm, 7 kV, 25 mM NaH₂PO₄ pH 2.3, 30°C. Peak identities : 2.455 min imidazole, 4.123 min. ranitidine, 4.538 min. lamiduvine, 4.840 salbutamol, 5.340 alosetron. Reproduced with permission from reference 1.

Table 6.2   Comparison of HPLC and CE data for the residue content of a drug present on manufacturing equipment

	CE*	HPLC*
Swab 1	100	123
Swab 2	ND	ND
Swab 3	80	86
Swab 4	57	50
Swab 5	ND	ND

* expressed as µg/swab

ND denotes not detected (< 1 µg/swab)

Reproduced with permission from reference 3.

the drug from the swab sample. Precision values considerably below the required 5 % RSD were routinely obtained for applications when using the internal standards The sensitivity of the method was indicated (1) by the basic drug alosetron where limits of detection were 0.02 µg/ml detected at 200 nm, this is equivalent to 0.2 µg/swab as an extraction volume of 10 ml was used. This sensitivity was obtained using a low UV detection wavelength and a 100 µm capillary. The limit of quantitation was determined as 0.07 µg/ml. Imidazole was used as the internal standard for alosetron. Table 6.3 shows precision data obtained (1) for calibration solutions of a number of basic drugs at two concentration levels all containing imidazole as an internal standard. Acceptable injection precision was obtained for all components and was improved at higher drug concentrations. Linearity was demonstrated (1) by analysing, in duplicate, 5 standard solutions covering the range 0.1–10 µg/ml. A correlation coefficient of 0.99997 was obtained with an intercept value of –0.8 % of the 10 µg/ml standard.

The method has been successfully employed (1) in a number of laboratories using a range of instrument types and locally sourced reagents and capillaries. Water insoluble drugs were extracted and prepared using either acidified water, diluted electrolyte (typically 1 to 10 dilution with water) or aqueous-organic solvents (listed in terms of preference as the choice of sample diluent has a pronounced effect on the quality of separation obtained).

The ability to simultaneously separate and quantify a range of basic drugs is helpful when optimising extraction approaches. For example a volume (e.g. 100 µl) of a solution contai-

Table 6.3   Injection precision for a range of basic drugs at two different concentrations ($n = 10$)

50 µg			
Lamivudine	Ondansetron	Salbutamol	GG167
1.15 %	1.76 %	0.97 %	1.34 %
250 µg			
Sumatriptan	Salbutamol	Ranitidine	Lamivudine
0.70 %	0.49 %	1.57 %	0.11 %

Reproduced with permission from reference 1.

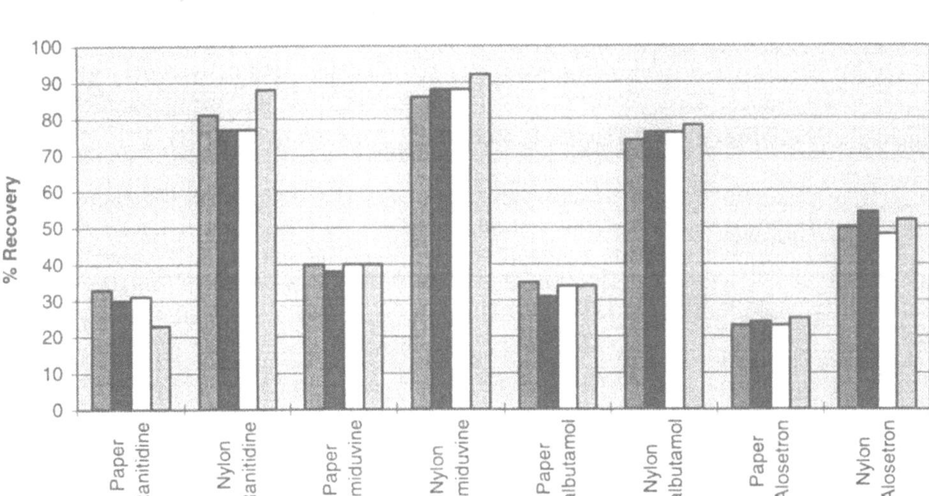

Figure 6.2 Comparison of the drug recoveries obtained with different swab materials for a range of basic drugs.

ning a mixture of drugs was spiked (1) onto pre-cleaned surfaces of a known volume and material (for example stainless steel plates 20 cm × 20 cm). The plates were left to dry and then the drug residues removed by various extraction procedures in order to determine the optimal extraction procedure. The ability to simultaneously quantify a range of drugs with different solubility's allowed an improved assessment of the procedures rather than use of a single drug. Imidazole was used as an internal standard. The drug was removed by wiping the surface of the stainless steel plate using a dampened portion of either nylon filter, paper filter or cotton wool.

The impact of the type of material used for the swab was assessed by spiking a range of basic drugs onto plates at either 50 or 250 µg levels. The residues were removed by wiping the plates with either nylon filters, paper filters or cotton wool swabs moistened with methanol. The amounts removed was highly dependent upon the material used (Figure 6.2). Nylon was the most effective and cotton wool and paper gave similar performance. The recovery levels showed good repeatability (Figure 6.2) as 4 extraction's were performed using each material to assess precision. It was suggested that the nylon filter is more flexible and has higher coefficient of drag which generated better mechanical removal of the drug compared to the cotton wool or paper. In addition the level of drug binding onto the nylon filter may be lower than the alternatives, which would again improved recovery from the swab. Ondansetron gave the lowest recoveries as it is the least water soluble of the test solutes.

An additional experiment involved a double procedure in which cotton wool, heavily wetted with methanol, was used (1) to wipe the surface and a further dry cotton wool swab was used to wipe the wet surface of the plate. The two cotton wool swabs were then independently analysed. The results (Table 6.4) indicated (1) the total drug residue extracted was higher using this double-swab technique. The % recovery of total drug was equivalent at the two concentrations investigated. To reduce analysis time it would be possible to extract both swabs together to produce a single solution for analysis.

A low pH electrolyte (80 mM phosphate) in water:methanol (adjusted to pH 2) has also been used to determine trace levels of the basic drug, mirtazarpine, following manufacture of granulate (7). A wavelength of 210 nm was used to give maximum sensitivity.

Table 6.4   Recovery data for basic drugs at two concentration levels using a 'double-swab' extraction procedure

	µg recovered on swab			
50 µg	Ranitidine	Lamivudine	Salbutamol	Sumatriptan
First swab	24	24	24	24
Second swab	14	17	16	14
Total	38	41	40	38
% Recovery	76 %	83 %	80 %	76 %
250 µg	Ranitidine	Lamivudine	Salbutamol	Sumatriptan
First swab	120	116	117	119
Second swab	70	81	80	69
Total	190	197	197	188
% Recovery	76 %	78 %	79 %	75 %

Reproduced with permission from reference 1.

6.2.2 Acidic drug residues

A 10 mM borate buffer has been found (1) to be suitable for a wide range of water soluble and insoluble acidic drugs. A low voltage of 6.5 kV in combination with a short (27 cm) 100 µm capillary enabled a short analysis time (3–5 minutes) with an acceptable level of operating current. Aminobenzoate and β-napthoxy acetic acid were used as internal standards.

Figure 6.3 shows a separation for a calibration solution for the determination of an acidic drug (Troglitazone). Use of β-napthoxy acetic acid as the internal standard enabled sub 5% RSD values to be obtained for calibration solution injection repeatability throughout routine analytical sequences. Linearity for Troglitazone was demonstrated (1) by analysing in duplicate 5 standard solutions covering the range 0.3–30 µg/ml. A correlation coefficient of 0.99959 was obtained with an intercept value of –0.6 % of the 30 µg/ml method value. A limit of detection of 0.02 µg/ml was obtained which equates to 0.2 µg/swab as 10 ml of acetonitrile:water (1:1) was used for extraction of the swabs. A limit of quantitation of 0.07 µg/ml (0.7 µg/swab) was obtained (1).

Oxytetracycline (OTC) residues exceeding 0.1 ppm tolerance level were detected (10) by CE in catfish fillets 18 h after oral feeding with 37.5, 75.0, and 150.0 mg OTC/kg in medica-

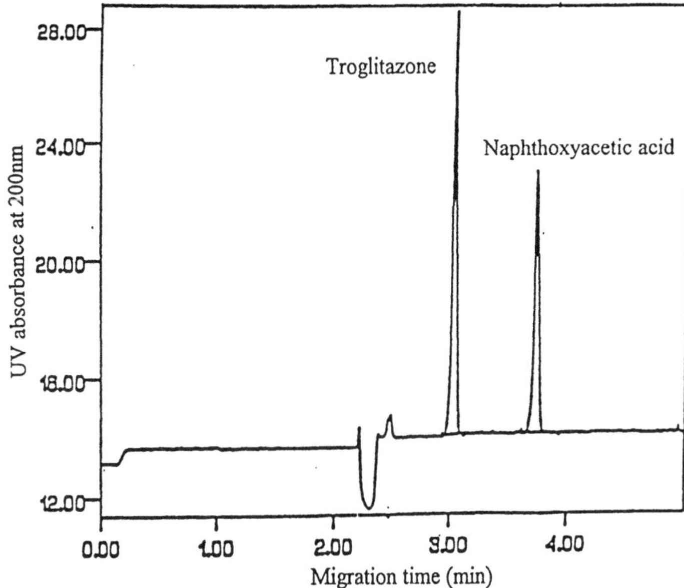

Figure 6.3 Separation for a Troglitazone solution with the internal standard β-naphthoxy acetic acid. Separation conditions: 27 cm × 100 μm capillary (20 cm to detector), 200 nm, 6.5 kV, 15 mM borate, 30°C, sample 0.01 mg/ml Troglitazone in 0.02 mg/ml standard β-naphthoxy acetic acid (50:50 %v/v Acetonitrile:water). Reproduced with permission from reference 1.

ted feed for 10 days. The mean OTC recovery rates in spiked catfish were 92.9 % over the concentrations of 0.1–25 ppm, Cooking procedures (frying, baking, and smoking at 190 degrees C) reduced OTC levels in catfish fillets but did not completely eliminate OTC residues.

6.3 Detergent Solution Residue Analysis

6.3.1 Surfactant residues

Manufacturing equipment is washed with water to remove residual traces of water soluble drugs. Detergent solutions are however employed to remove traces of water insoluble drugs. Regulations are in place which stipulate that it is necessary to determine levels of surfactant residues remaining on the manufacturing equipment following cleaning with detergent solution. Similar to drug residue analysis, areas of the manufacturing equipment are swabbed and the swab extract solutions are analysed for surfactant content. Conventionally, this analysis involves titrimetry, however, CE has been employed (9) for this purpose. Figure 6.4 shows separation of a 10 μg/ml dodecylbenzenesulphonate (SDBS) solution. The separation is achieved using a pH 9.5 phosphate-borate electrolyte with detection at 200 nm.

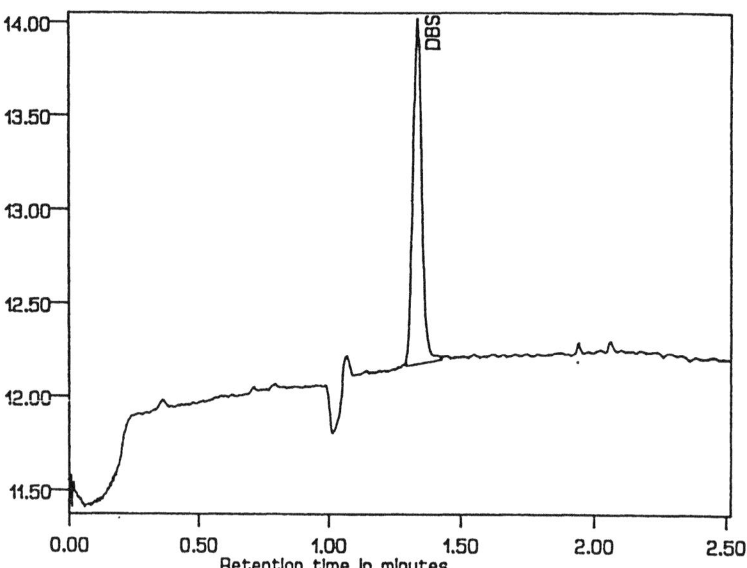

Figure 6.4 Separation of 10 μg/ml DBS solution. Separation conditions: 2.5 mM borax with 5 mM Na$_2$HPO$_4$ +25kV, 37 cm × 75 μm, 200 nm. Reproduced with permission from reference 9.

Validation of the method for SDBS residues include acceptable measurements of precision (1–2 % RSD), linearity over the range 0–20 μg DBS/ml (corr. coef. 0.9992), reproducibility of standard and sample preparation, robustness, recoveries (50–63 %), detection limit of 4 μg detergent residue/swab and limit of quantification of approximately 8 μg detergent residue/swab (0.6 μg DBS/ml). Ten preparations of standard solutions (10 μg DBS/ml) were analysed in duplicate and gave an acceptable RSD value of 3.7 % for the 20 response factors. Buffer, swab and DBS standard stability were established. An experimental design was used in the robustness studies.

Alternative detergent solutions may not contain SDBS but may contain sodium dodecyl sulphate (SDS) which has no chromophore and cannot be detected by UV absorbance. Indirect UV absorbance has been used for determination of SDS using barbital or dinitrobenzene containing electrolytes. Sensitivities of low bbp levels for SDS levels in river water have been reported (12).

6.3.2 Metal ion residues from detergent solutions

Residues of a caustic detergent solution containing KOH have been monitored (1) by determining the levels of potassium by an indirect UV detection CE method. The electrolyte contains formic acid and crown ether to adjust the selectivity and CuSO$_4$ to provide the background UV signal. Magnesium was the preferred internal standard. Figure 6.5 shows separation of a potassium standard with the magnesium internal standard.

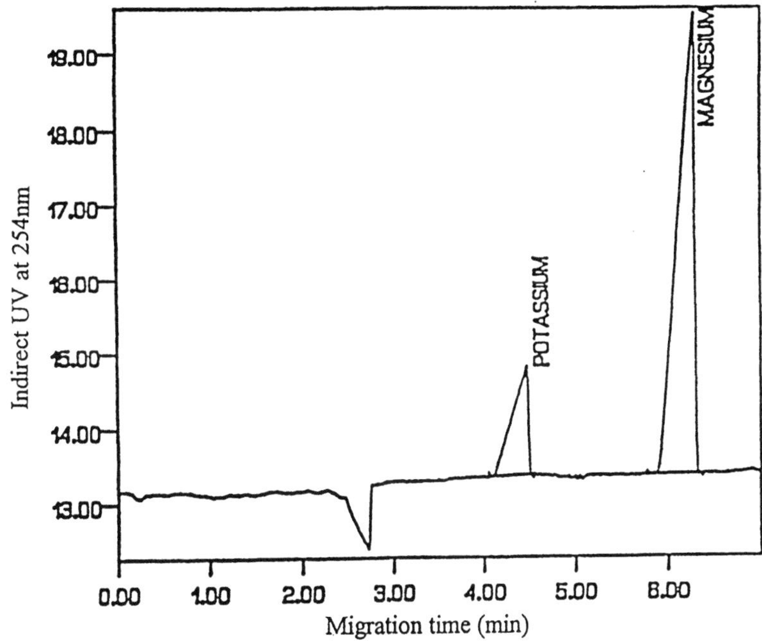

Figure 6.5 Separation of a potassium with magnesium as an internal standard. Separation conditions: 57 cm × 75 µm capillary (50 cm to detector), indirect UV at 254 nm, 20 kV, 5 mM $CuSO_4$/ 4 mM formic acid / 3 mM 18-crown-6, 30°C, sample 50 mg/l potassium in 50 mg/L magnesium. Reproduced with permission from reference 1

A sensitivity of 0.5 µg/ml was obtained for potassium which was considered (1) an acceptable level. The sensitivity corresponds to 5 µg detergent residue/swab. The residue sensitivity was calculated using the % w/w of solid residues formula of the detergent solution and the extraction volume of 5 ml of water-based internal standard solution. The sensitivity obtained was considered adequate as it was a considerable improvement over the titration method previously employed. Linearity for potassium was assessed (1) over the range 1.25–125 µg/ml and correlation coefficient of 0.99970 was obtained with an intercept value of 0.4 % of the 125 µg/ml value.

6.3.3 EDTA residues

A number of detergent solutions containing levels of EDTA (ethylene diamine tetracarboxylic acid) are used in pharmaceutical equipment cleaning. EDTA is non-chromophoric and therefore detection is difficult. However, EDTA forms negatively charged highly UV active complexes with divalent metal ions. EDTA residue levels have been determined (4).

6.4 Drug Doping Levels

CE has also been used to determine the levels of sulphonamides added to pork meat. Separation of 16 sulphonamides by CZE at pH 7 was optimised (8). The sulphonamides were extracted with acetonitrile and the effect of extraction technique on the background was investigated. Homogenisation of meat with acetonitrile proved best for extracting the veterinary residues. Detection limits were 2–9 ppm for a 10 s injection (18 nl) at 254 nm. Low ppm detection levels were achieved with no sample preconcentration. Linearities of > 0.999 were obtained over the ranges of 0.1–0001 mg/ml for 5 selected sulphonamides.

6.5 Environmental Analysis

To-date, there have been no reports of the use of CE in this area for the determination of pharmaceuticals. However several reports have appeared covering the determination of ppb and ppt levels of herbicides (6), explosive residues (5), aromatic amines in water and soil samples (2). Generally, solid-phase extraction is used to preconcentrate species prior to CE analysis. The most comprehensive study has focused (2) on determination of aromatic amines in water samples. 18 amines were separated using a pH 2.35 phosphate buffer containing 7 mM 1,3-diaminopropane with detection at 280 nm. Recoveries were in the region of 60–82 % with detection limits ranging from 0.07–1.6 ppm. Detection levels of 0.1 ppb were obtained (6) for metsulfuron and chlorsulfuron in tap water with solid-phase extraction preconcentration.

6.6 Advantages and Disadvantages of the Use of CE in Residues Analysis

6.6.1 Disadvantages

HPLC is the method generally employed to determine drug residues. The performance levels for CE precision and sensitivity are generally poorer than in HPLC. In CE it is general practise to use internal standards to achieve the required injection precision whilst this is not the case in HPLC. Equivalent sensitivity of CE and HPLC can only be achieved through application of both low UV wavelength detection and wider bore (or modified) capillaries in CE.

6.6.2 Advantages

There a number of compelling advantages for CE in the determination of residues compared to other analytical alternatives. These include versatility, costs, freedom from interference and data quality improvements.

Versatility

Specific separation conditions can be used (1, 3) to quantify a wide range of basic or acidic drugs. This situation means that CE instruments can be dedicated to this type of testing and there is a minimum of changes required when switching between the drugs being monitored. Conversely in HPLC analyte specific mobile phases and columns need to be purchased, prepared and maintained. The ability to simultaneously monitor a wide range of drugs is useful when monitoring residues of combination products or cross-contamination between drug types. The use of indirect UV detection on standard CE instruments also allows monitoring of surfactant and inorganic contaminant residues from detergent solutions.

Costs

There are generally a high number of samples generated in residues analysis testing and therefore cost is an important issue. The use of inexpensive aqueous buffers and cheap capillaries compares very favourably with HPLC costs. Additional cost and time savings are accrued through the use general CE methods as there are no requirements for changing when switching between drug types. Additionally there is no need to prepare solute specific mobile phases and maintain a range of HPLC columns.

Freedom of interference

There are a number of sources of interference's in residue analysis – these include species from the materials used for swabbing and compounds extracted from the equipment being tested. For example cotton wool contains a number of chemicals which can be extracted and produce peaks on an analytical separation. Additionally peaks can occur due to compounds that are extracted from lubricants or excipients present on manufacturing equipment. In the majority of cases these interfering species are uncharged and do not generate interfering peaks in CE separations.

Data quality improvements

CE offers generally an improvement in both data quality and level of automation when compared to non-separative techniques such as titrimetry or colourmetric analysis of non-chromophoric species such as surfactants. Retrievable peak areas can be obtained in CE to allow automated calculation of residue levels.

References

1. Altria KD, Creasey E, and Howells JS, Routine trace level determinations of pharmaceutical and detergent residues *J. Liq. Chromatogr.* in press.

2. Cavallaro A, Piangerelli V, Nerini F, Cavalli S, and Reschiotto C, Selective determination of aromatic amines in water samples by capillary zone electrophoresis and solid phase extraction, *J. Chromatogr. A*, 709 (**1995**) 361-366.

3. Altria KD and Hadgett T, An evaluation of the use of capillary electrophoresis to monitor trace drug residue levels following the manufacture of pharmaceuticals, *Chromatographia*, 40 (**1995**) 23-27.

4. Altria KD, Bryant S, and Kooyman O, unpublished data.

5. Mussenbrock E and Kleibohmer W, Separation strategies for the determination of residues of explosives in soils using micellar electrokinetic capillary chromatography, *J. Microcol. Sepn.*, 7 (**1995**) 107-116.

6. Dinelli G, Bonetti A, Catizone P, and Galletti G C, Separation and detection of herbicides in water by micellar electrokinetic capillary chromatography, *J. Chromatogr. B*, 656 (**1994**) 275-280.

7. Wynia GS, Windhorst G, Post PC, and Mars FA, Development and validation of a capillary electrophoresis method within a pharmaceutical quality control environment and comparison with high-performance liquid chromatography, *J. Chromatogr. A*, 773 (**1997**) 339-350.

8. Ackermans MT, Beckers JL, Everaerts FM, Hoogland H, and Tomassen MJH, Determination of sulphonamides in pork meat extracts by capillary zone electrophoresis, *J. Chromatogr.* , 596 (**1992**) 101-109.

9. Altria KD, Gill I, Howells J, Luscombe CN, and Williams RZ, Trace analysis of detergent residues by Capillary Electrophoresis, *Chromatographia*, 40 (**1995**) 527-531.

10. Huang TS, Du WX. Marshall MR, and Wei CI, Determination of oxytetracycline in raw and cooked channel catfish by capillary electrophoresis, *J. Agric. Food Chem.*, 45 (**1997**) 2602-2605.

11. Leone A M, Francis P L, Rhodes P, and Moncada S, A rapid and simple method for the measurement of nitrite and nitrate in plasma by high performance capillary electrophoresis, *Biochem. Biophys. Res. Comm.*, 200 (**1994**) 951-957.

12. Gibbons J M and Hoke S H, Capillary zone electrophoresis with indirect UV detection: determination of sodium dodecyl sulfate in simulated stream water, *JHRCC*, 17 (**1994**) 665-667.

7 Pharmaceutical Raw Materials and Excipients Analysis

7.1 Introduction

Raw materials are input substances used in both chemical synthesis and processing and include buffers, cleaning agents, common solvents and commonly used synthetic starting materials such as amino acids. Excipients are substances which are used as ingredients in pharmaceutical formulations such as tablets and capsules. Typical excipients include simple salts such as NaCl, solvents, carbohydrates, detergents, flavouring agents and dyes. When these materials are delivered they need to be tested to confirm that they are indeed what they claim to be and have not been inappropriate labelled. It may also be necessary to use a quantitative separative method to demonstrate the quality and content of selected materials. Failure to check on the identity and quality of these materials could result in very costly wastage's.

Excipients are also analysed during development of formulation manufacturing process. For example solid formulations such as tablets may contain a number of constituents such as lubricants, carbohydrates and fillers. Granulate is generally prepared by mixing quantities of drug and excipients. Tablets are then prepared by compressing granulate into the required shape. Prior to production the manufacturing process needs to be optimised and of particular importance is that the excipients and drug is distributed sufficiently well through a batch of granulate. These work is called a homogeneity study and the quality of the batch can be assessed by measuring the amount of drug and/or excipients in various subsamples taken from the batch. These subsamples may be taken at timepoints during manufacture and/or following processing. Acceptable analytical methods are therefore necessary to support homogeneity testing as the numbers of subsamples tested can be relatively high.

Table 7.1 shows that a wide range of CE application have been performed in this area (1–53). The table provides details of the nature and use of the substance, an example material and some appropriate CE references. In many instances reference is made to use of CE in other industries as the number of reports is still somewhat limited within the pharmaceutical industry.

Classical methods such as titrimetry, colour reactions and gravimetric analysis are conventionally used to analyse excipients and raw materials. Within the pharmaceutical industry (54) there is a tendency to replace, or supplement, this type of testing with more modern instrumental, and increasingly automated analytical techniques such as HPLC, GC and Near Infra-Red spectroscopy (NIR). CE, like HPLC and NIR, has considerable advantages over classical analytical techniques in terms of increased automation and speed of analysis. Particular benefits of concern to the pharmaceutical industry, that are common to the three techniques, are the improved quality of the data produced compared to classical testing methods and that retrievable raw data is generated.

Table 7.1 Range of applications of CE to the analysis of raw materials and excipients

Material type	Function	Examples	Electrolyte	Ref(s)
Alcohols	Preservative Solvent	Ethanol 2-Propanol	MECC with SDS and barbitone	1
Carbohydrates	Sweeteners	Glucose	Various	2–8
Cyclodextrins	Solubiliser	β-cyclodextrin	High pH, benzoate	9, 10
Dyes	Colouring agents	Various	Various	11–17
Fatty acids	Lubricant	Stearate	MECC with solvents	18–21
Flavouring	Sweetener Taste masker	Aspartame Bitrex	MECC High pH	22
Inorganic anions	Tablet excipient	Phosphate buffer	Various	23–25
Lecithin	Solubiliser	USNF Lecithin	MECC with IPA	26
Metal ions	Tablet excipients	Carmellose Na NaCl	Various	27, 28
Organic acids	Preservative Buffer or drug counter-ion	Propionate, benzoate, acetate, citrate	High pH, TTAB with chromate or phthalate	29, 30
Polycarboxylic acids	Chelating agents	EDTA	Electrolyte or sample containing metal ions	31–33
Preservatives	Preservative	BKC, PHB's	Various	34–37
Starting materials	Synthesis	Imidazole Amino acids	MECC Various	38 39–42
Surfactants	Emulsifiers and preservatives	SDBS, SDS	Various	43–51
Water	Diluent	Purity determination	Various	52, 53

BKC   =   benzylalkonium chloride
PHB   =   parahydroxybenzoate
TTAB =   tetradecyltrimethylammonium bromide
USNF =   United States National Formulary

Many excipients and starting materials such as inorganic buffers and surfactants have only limited UV activity and cannot be directly analysed by HPLC. However the use of either low UV wavelengths, such as 190–210 nm, where many substances have increased absorbances, or the use of indirect UV detection is widely established in CE (see Chapter 5) which allows effective analysis of analytes with poor UV absorbances. Various modes of CE, coupled with use of organic solvents, have been developed which allow analysis of a wide range of compounds having differing solubilities and polarities.

This chapter provides an overview of the potential applications of CE to the analysis of pharmaceutical excipients and raw materials. The performance levels of CE in many of these applications is described with illustrative examples. The potential advantages and disadvantages of adopting CE in preference to classical techniques, or other modern analytical techniques, is also discussed. Each particular application area is discussed with reference to the methodology employed and to the performance levels achieved wherever appropriate. Many examples employ indirect UV detection, (Chapter 5 – Determinations of Drug Counter-ions, gives full details on this method of detection).

7.2 Alcohols

Simple alcohols and organic solvents such as acetonitrile, methanol, ethanol and acetone are widely used in chemical processing. Identity confirmation of the solvent is usually performed by GC prior to their use. This function can be achieved with an MECC method using indirect detection (1) to confirm the solvent identity. The method separated a range of organic solvents using a high pH barbitone buffer with 180 mM SDS concentration. The method has been used to quantify (1) levels of ethanol in a syrup formulation using isopropanol as an internal standard (Figure 7.1). The method was simple, rapid, inexpensive to operate and is capable of good precision (RSD for migration times of 1 % and peak area ratios of less than 1%) and accuracy.

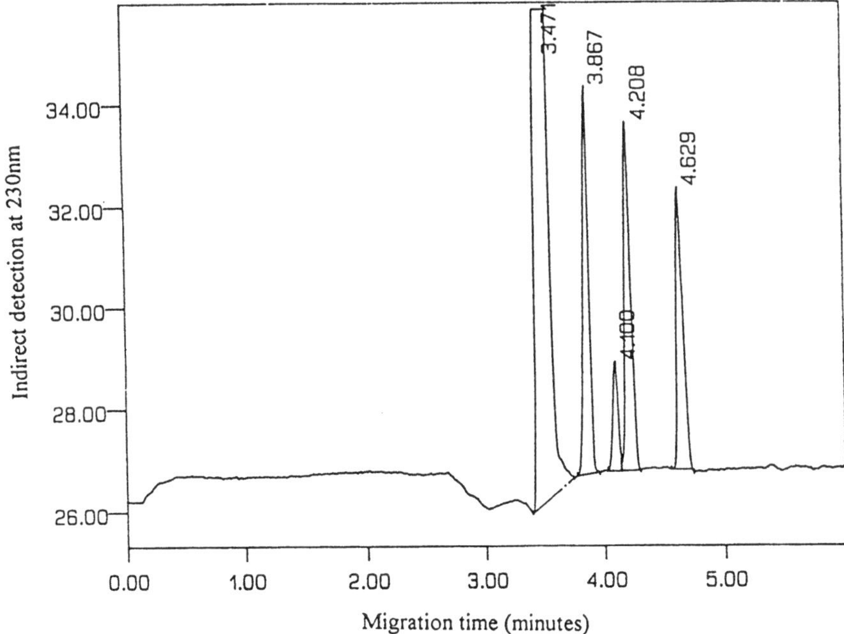

Figure 7.1 Separation of ethanol in a syrup formulation. Separation conditions : indirect UV at 230 nm, +5 kV, 200 mM SDS with 20 mM barbitone, 30°C. Peak identification: 3.471 min. = water, 3.867 and 4.100 min. = excipient peaks, 4.208 min. = ethanol, 4.629 min = 2-propanol (internal standard). Reproduced with permission from (1)

7.3 Carbohydrates

A variety of carbohydrates such as glucose and sucrose are employed as both sweeteners and viscosity modifiers. CE has been widely used for the analysis of carbohydrates and comprehensive reviews (2, 3) have summarised the performance and approaches possible in this area. The separation approaches have included use of a high pH electrolyte containing tryptophan for indirect detection and the use of CE to resolve carbohydrates following the simple one-stage derivatisation with reagents such as aminobenzoate (4). Many carbohydrates can complex in solution with borate ions which permit their separation as anions. The complexation is favoured at higher temperature and higher borate concentrations (5) for example underivatised carbohydrates were resolved using 60 mM borate at 60°C. Figure 7.2 shows use of a modified form (24) of this method for the separation of lactose which is an important pharmaceutical excipient.

Operation at pH values of approximately 12 ensures the ionisation of the weakly acidic hydroxyl groups on underivatised saccharides (6). Addition of 6 mM sorbic acid to the pH 12 electrolyte provided the background UV signal for indirect detection. An alternative approach is to employ a pH 12 electrolyte containing tryptophan as the UV absorber with indirect detection at 280 nm (7).

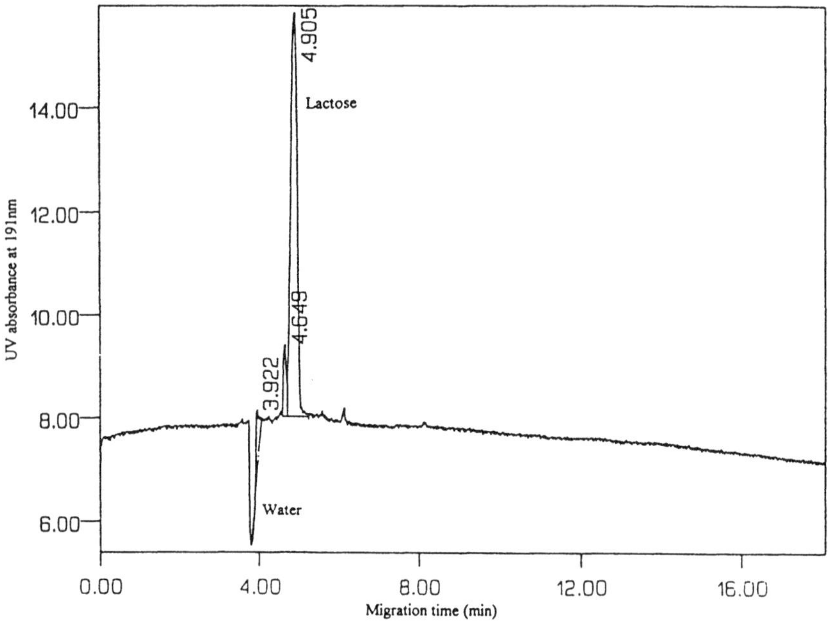

Figure 7.2 Separation of lactose using a borate electrolyte. Separation conditions : 50 mM borate, +20 kV, 56 cm (total length) × 50 μm capillary, 191 nm, 30°C, 20 mg ml^{-1} lactose in water. Reproduced with permission from (24).

7.4 Cyclodextrins

These macrocyclic carbohydrates are added to liquid pharmaceutical formulations to improve solubility of poorly soluble drugs. Cyclodextrins have poor UV activity and identity confirmation is usually performed using optical rotation measurements which are time-consuming and also involve use of relatively large amounts of sample. Separation of the different cyclodextrin forms has been reported (9) by CE with an electrolyte containing benzoate ions which also provides the background signal for indirect UV detection. The cyclodextrin interact with the benzoate ions and the method resolved α, β and γ cyclodextrins. Alpha-, beta-, and gamma-CD were separated (10) using 2 mM 1-naphthylacetic acid (NAA) or 5 mM sorbate (pH 12.2) with indirect detection. The detection limit for the NAA system was 0.1 mM.

Chemically derivatised cyclodextrins such as hydroxypropyl- or dimethyl-β-cyclodextrin are widely used as these have improved water solubility when compared to underivatised cyclodextrins. The chemical derivatisation process is not total and synthetic products can often contain a range of modified and unmodified forms of the cyclodextrins. A CE method employing (10) indirect absorbance method detection has been reported which allows profiling of the derivatised forms. Mixtures of alpha- and beta-CDs, and dimethyl- and trimethyl-derivatives of beta-CD could also be analyzed by CZE, using 50 mM salicylic acid or benzylamine solution (pH 6.0) as BGE with indirect absorbance detection at 230 and 210 nm, respectively.

7.5 Dyes

There have been a number of reports of the use of CE in the analysis of dyes (11–16). Quantitative and qualitative has been reported for both water soluble and insoluble dyes with UV absorbance detection. For example low pH electrolytes are suitable for cationic dyes and MECC (with possible addition of organic solvents) is appropriate for water insoluble dyes. Acidic dyes have been separated using a high pH phosphate or borate buffer. Comparative data between CE and HPLC have been reported (14) in this area.

Non-aqueous solvents have been used (16) for the separation of acidic dyes (Figure 7.3). This is especially useful for water insoluble dyes. Chapter 13 covers the use of non-aqueous CE electrolytes in great detail).

A (pH 6.8) carrier electrolyte containing beta-cyclodextrin was used (17) to determine dyes with limits of detection for the dyes of 11–300 ppb at 254 nm. RSD's of 0.4–3.0 % were typical for the determinations of the dyes present in samples at 16 ppm concentrations. Dyes were determined in soft drink concentrate and liquers and the stability of aqueous solutions of indigo carmine was monitored.

7.6 Fatty Acids

Fatty acids such as magnesium stereate are widely incorporated into pharmaceutical formulations to act as lubricants. Analysis of fatty acids is generally performed by GC following

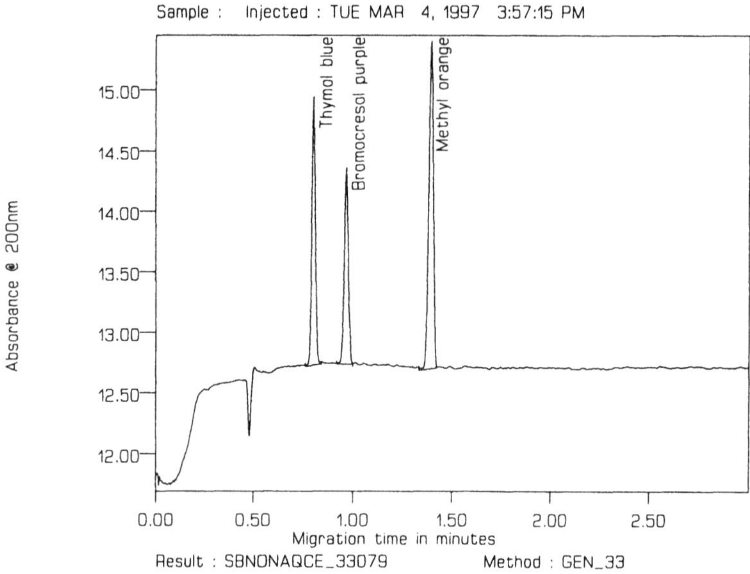

Figure 7.3 Separation of acidic dyes using 2 mM Na Acetate in 50/50 ACN/MeOH. Separation conditions: 27 cm × 50 µm capillary (20 cm to detector), 200 nm, 30 kV, 1 second injection, using 50/50% v/v ACN/MeOH containing 2 mM sodium acetate pH* 9.3 (unadjusted). Reproduced with permission from (16).

their derivatisation to more volatile esters. The longer chain fatty acids are particularly water insoluble and possess no appreciable UV activity. The free fatty acids have been determined by CE (18–21) using indirect UV detection and surfactant solutions containing high (30–60 % v/v) levels of organic solvents such as acetone or acetonitrile.

For example it has been shown (18) that it is possible to resolve fatty acids with chain lengths of C4 to C18 within 16 minutes using an electrolyte containing 40 % acetonitrile and with indirect UV detection at 254 nm using trinitrobenzenesulfonic acid as the UV background agent. Detection limits of 6 µM were recorded (21) for fatty acids and the relative standard deviation of the method was in the range of 4 to 10 %. The method was successfully applied (21) to the analysis of free fatty acids in human stratum corneum after solid phase extraction of the lipid fraction.

7.7 Flavouring Agents

A number of basic and acidic compounds are employed as flavouring agents in pharmaceutical formulations which can be separated using standard CE conditions. For instance a pH 9.5 borate buffer can be used to determine levels of organic acids such as saccharin, and aspar-

Table 7.2 Analysis of various components in cola drink (content as mg/L)

Cola sample	Aspartame		Benzoic acid		Caffeine	
	MECC	HPLC	MECC	HPLC	MECC	HPLC
1	510	530	170	165	140	130
2	440	405	170	160	100	100
3	450	430	175	175	90	95
4	470	450	170	165	85	85
5	335	335	150	150	60	60
6	N/D	N/D	320	320	80	80

N/D – denotes none detected – this sample contained saccharin (MECC 75 mg/L, HPLC 80 mg/L)
Reproduced with permission from (22).

tame. Whilst a pH 2 phosphate buffer can be used to determine levels of cationic flavouring agents such as the denatonium benzoate (Bitrex) and aspartame. Table 7.2 shows the quantitative results from the analysis for aspartame, benzoic acid and caffeine content in various cola drink samples (22) by both HPLC and an MECC method.

7.8 Inorganic Anions

Many raw materials and starting materials are ionic salts and it is necessary to confirm the correct salt is being used prior to pharmaceutical manufacture. In the case of anionic counter-ions such as chloride, sulphate or phosphate, titration or ion-exchange chromatography is widely used for assay purposes. However, CE methods are available to successfully perform analysis in this areas and the CE methods can have benefits in terms of ease of use, reduced costs and quicker analysis times. The application of these methods are described in greater detail in Chapter 5 where CE has been used to quantitatively determine the levels of anions such as sulphate and chloride, or metal ions in various drug substance salts. A similar approach has been used to quantitatively determine levels of sulphate in detergent products with good agreement obtained (23) between a gravimetric assay method and CE data.

Trace levels of various anions such as chloride and sulphate can have significant impact on the crystallinity of excipients such as sucrose. Therefore it is important to have a suitable analytical method to establish levels of these contaminants. CE can offer the require sensitivity and selectivity. For example low ppm levels of various contaminant anions such as chloride, and acetate were determined (25) in drug substance sample by CE. Concentrated samples of drug substance (~10 mg/ml) were directly injected and trace levels of inorganic anionic contaminants such as chloride, sulphate and various organic acids were determined.

7.9 Lecithins

These water insoluble naturally occurring phospholipid products are employed as solubilisers and are widely used in inhaled products. There are problems with the analysis of lecithins as they possess limited UV chromophores and are mixtures of different homologues. Therefore the typical analysis is performed by TLC. Analysis of phospholipids is possible (24, 26) by CE using micellar solutions containing relatively high levels of organic solvents such as isopropanol with detection at 200–210 nm. Constituent Pharmaceutical USNF grade lecithin components has been separated (24) by CE. Recently the use of capillary electrochromatography (Chapter 12) has been shown (55) for the analysis of lecithins and lipids in steroid formulations. The method used a column filled with ODS packing material, a non-aqueous electrolyte and detection at 200 nm.

7.10 Metal Ions

CE can be used to quantify levels of metal ions in a variety of pharmaceutical raw materials and excipients using indirect UV detection. Typical excipients include sodium saccharin, sodium and potassium buffers, sodium glycolate, sodium carbonate and sodium chloride. In addition the majority of acidic drugs are manufactured as metal salts with sodium and potassium being the most widely used drug counter-ions (Chapter 5). Other cations of interest include ammonium, calcium and magnesium. Identity confirmation of the test substance cation content can be achieved by calculation of % w/w cation level and by agreement of peak migration time with a standard. Typically standards are prepared from authentic AnalaR grade materials such as Na or KCl, dissolved in water. An appropriate internal standard is often used to improve precision to 1–2 % RSD for peak area ratios.

A range of typical cations have been quantified (24) using a copper sulphate based electrolyte containing levels of crown ether. The $CuSO_4$ is used to provide the UV background at 214 nm whilst the crown ether assists in the separation of ammonium and potassium through selective complexation. Table 7.3 shows that suitable quantitative performance could be obtained when applying methods of this type to various excipients and raw materials. The examples analysed cover selected examples of commonly encountered ionic excipients and raw materials. To discriminate between the two phosphate buffers using classical testing it may be necessary to perform both a sodium fusion test to confirm presence of sodium and then to perform a titration to determine the levels of phosphate. In this work magnesium was employed as an internal standard to give the required precision. For example the precision for injection ($n = 10$) for peak area ratios for NH_4^+ and Na^+ was 1.22 and 1.04 % RSD respectively. Table 7.3 also confirms that suitable detector linearity was obtained over a range of 50–150 % of the target sample concentration (50 µg/ml). Therefore analysis of the cation content of raw materials can be considerably more efficient than performing classical testing as CE provides a quicker, more automated approach with a similar degree of precision.

Pharmacopeail testing often requires use of methods to monitor trace levels of various cations that may be present as contaminants in samples present at low ppm levels. This type

Table 7.3 Quantitative analysis results from testing of ionic excipients

Linearity results (25–75 µg/ml)	Correlation coeff.	Slope	Intercept
K^+	0.998	0.022	0.004
Na^+	0.999	0.020	0.009
Ca^{2+}	0.999	0.018	0.004

Sample	Theoretical %w/w	Calculated %w/w
Na_2CO_3	43.4 %w/w Na	43.9 %w/w Na
Na_2HPO_4	32.4 %w/w Na	32.0 %w/w Na
$NaH_2PO_4,2H_2O$	14.7 %w/w Na	14.9 %w/w Na
KI	24.5 %w/w K	24.5 %w/w K

Reproduced with permission from reference 24.

of application has also been performed by CE. For example trace metal ion levels have been determined in water (27) and trace levels of metal ions including Ca^{2+} and Mg^{2+} have been determined in various foods (28) following micro-wave digestion of the samples.

7.11 Organic Acids

A variety of CE applications are possible in this area as organic acid salts are widely used in chemical synthesis as drug counter-ions, preservatives or as buffering agents in liquid or solid formulations. Relatively small organic acids such as acetic and citric possess very limited chromophores and therefore detection is generally performed using indirect UV detection. Larger organic acids such as benzoic acid can be directly detected using direct UV absorbance in CE and HPLC.

The most frequent determination is of simple acids such as acetic, formic and citric present in ionic input materials such as sodium acetate or citrate buffer salts. In addition, simple organic acids are common counter-ions for basic drugs with acetate, maleate and succinate being frequently used (Chapter 5). Phthalate is often added at low millimolar quantities into the electrolyte to provide the high UV background signal at 254 nm. TTAB is also added to coat the capillary surface to reverse the EOF direction to coincide with the migration direction of the acids. Negative signal are obtained for the acids but for ease of integration the polarity of the detector output is reversed to give apparently positive peaks. The method can be used to confirm the correct identity of the organic acid constituent of the test material by comparing the results obtained with those generated by analysing an appropriate standard material such as AnalaR grade Na citrate. The migration times of the organic acid peak in both the standard and sample should be concordant. Further confirmation may be performed by calculating the % w/w of the acid in the test material using response factors obtained from analysis of standards of the appropriate acid. The calculated % w/w of organic acid in the sample can then be compared to the theoretical % w/w content.

Organic acids which have a reasonable chromophore can be directly quantified by normal UV absorbance measurements. The compounds are usually analysed using a high pH electrolyte where the acids are fully ionised and are separated as anions. A wide range of organic acids have been separated using pH 10 borate electrolyte with a flow reversal surfactant in the electrolyte (29) with direct UV detection was possible at 195 nm. The levels of organic acids in foods have been determined (30) by CE.

7.12 Polycarboxylic Acids

Polycarboxylic acids such as ethylendiamine tetracarboxylic acid (EDTA) are widely used chelating agents and are often added to liquid formulations. There have been a number of reports of the use of CE for their analysis (31–33). Approaches have included formation of metal complexes by the addition of metal ions into the sample solution and/or the separation electrolyte. Figure 7.4 shows separation (56) of a range of polycarboxylic acids separated as metal chelates formed by the addition of copper ions to the sample solution. Benzoic acid was added as an internal standard.

7.13 Preservatives

CE has been used (34–37) to analyse a wide range of preservatives. For example the cationic preservative, benzylalkonium chloride (BKC), has been determined (35) using a pH 2.5 phosphate buffer. BKC is a mixture of homologues and Figure 7.5 shows the separation of

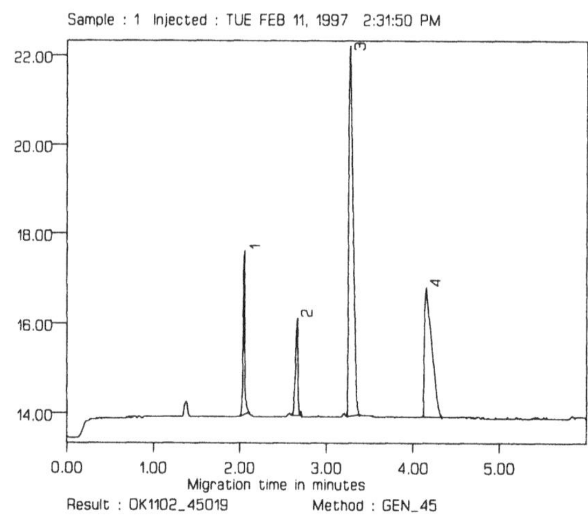

Figure 7.4 Separation of a range of polycarboxylic acids and an internal standard (benzoic acid). 20 mM borate; Detection: direct UV at 254 nm; Voltage: +10 kV. Capillary: 27 cm × 75 µm. 5 mM $CuSO_4$ added to sample solutions. Peak identities 1 = HEDTA; 2 = benzoic acid; 3 = EDTA; 4 = DTPA

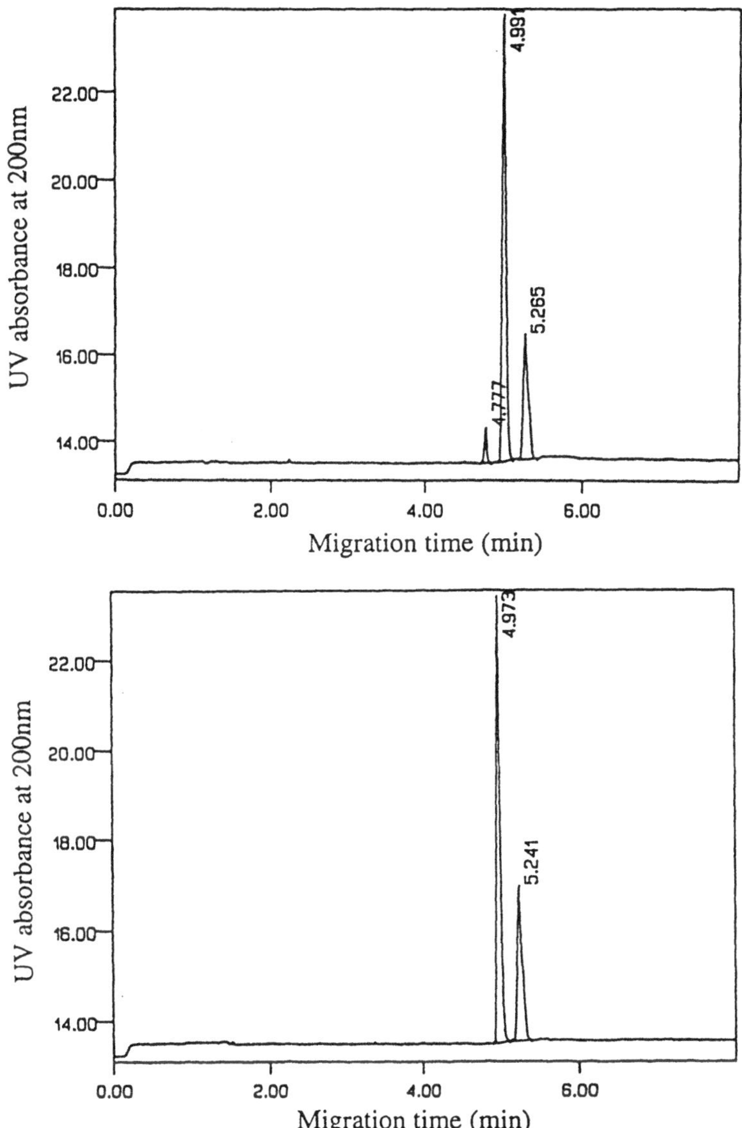

Figure 7.5 Separation of BKC homologue constituents in two separate input batches. Separation conditions : 25 mM NaH_2PO_4 (pH 2.5), +20 kV, 37 cm (total length) × 50 µm capillary, 200 nm, 30°C, 0.1 mg ml^{-1} in water. Reproduced with permission from reference 35.

BKC homologue constituents. CE has been used to quantify levels of BKC in pharmaceutical solutions (35). A CE assay figure of 0.11 % w/v was in good agreement with the manufacture claim of 0.10 % w/v. The method was sensitive with detection limits of 0.05 µg/ml when monitoring at 200 nm. The CE method was also used to simultaneously determine both the level of BKC and the basic drug present in the sample solutions (Table 7.4)

Table 7.4 Results from the simultaneous testing of drug and preservative content

Test	Factor		Result
Precision	Normalised area ($n = 10$)		1.60 %
Accuracy (two samples	BKC	0.010 %w/v	0.011 %w/v
at each concentration	HAP	0.12 mg/ml	0.12 mg/ml
injected in duplicate)		0.25 mg/ml	0.25 mg/ml
		0.50 mg/ml	0.49 mg/ml
		1.00 mg/ml	0.98 mg/ml
		2.00 mg/ml	1.99 mg/ml
		4.00 mg/ml	4.03 mg/ml
		8.00 mg/ml	8.07 mg/ml
		16.00 mg/ml	15.42 mg/ml
Sensitivity (LOQ deter-	HAP	LOD	0.5 µg/ml
mined from 10 injections)	HAP	LOQ	1.5 µg/ml
	BKC	LOD	0.05 µg/ml
	BKC	LOQ	0.15 µg/ml
Response factor	BKC		2.2 % RSD ($n = 12$)
precision	HAP		0.5 % RSD ($n = 8$)

HAP = histamine acid phosphate
Reproduced with permission from reference 35.

CE is also applicable to the analysis of a range of acidic preservatives such as benzoic acid, sorbic acid and parahydroxybenzoates (PHB's). A specific example involves the analysis of PHB's in cosmetic samples. Methyl-, ethyl-, propyl- and butyl-PHB were separated using a MECC method. The butyl-derivative was employed as an internal standard. Assay figures obtained were in-line (36) with HPLC data and the label claim for PHB content. Non-aqueous CE buffers have been used (16) to give an enhanced separation of PHB's (Figure 7.6).

Acidic preservatives such as benzoic acid and sorbic acid can be determined (37) using a borate electrolyte. Benzoic acid levels in foods were determined (37) with migration time precisions of 1 % and peak area precision of 2.5 % RSD.

7.14 Starting Materials

Generally organic synthesis involves reaction of a series of smaller compounds to form the desired product. Many of these starting materials are common to a number of reactions. Material is generally purchased from approved suppliers and only a limited amount of testing is performed to ascertain quality and identity. Typical starting materials may be imidazoles, or amino acids such as tryptophan. These small organic materials are generally suitable for analysis by CE. For example (38) imidazole and a number of its closely related derivatives have been resolved using an MECC method involving SDS micelles and an ion-pair reagent with detection at 214 nm. HPLC analysis of imidazoles is problematic as they have poor UV responses above 220 nm and are very basic compounds which can lead to peak tailing problems.

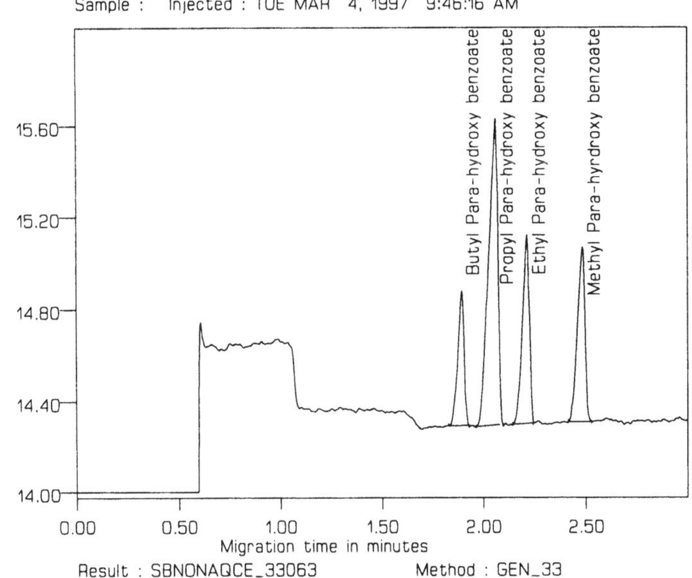

Figure 7.6 Separation of parahydroxybenzoates using a non-aqueous CE buffer Reproduced with permission from reference 16.

Many intermediates are chiral compounds and require use of a chiral specific method. CE is especially useful for the separation of chiral species (Chapter 4). For example a number of chiral imidazole derivatives have been chirally resolved (39) using a range of cyclodextrins.

Amino acids are also very widely used starting materials and synthetic reagents. Amino acids can be successfully analysed by CE (40, 41) with direct or indirect UV detection. The limited chromophore of the majority of amino acids requires that they are derivatised with a UV active species prior to analysis by HPLC. Amino acids are chiral species and have been separated chirally by CE. For example racemic tryptophan has been resolved using a low pH electrolyte containing α-cyclodextrin (41). This method also allowed quantitation of the tryptophan content as aspartame was employed (Figure 7.7) as an internal standard to give an improved precision.

7.15 Surfactants

This type of compound covers a wide range of chemicals which are used in the pharmaceutical industry as both emulsifiers and preservatives. A range of cationic (43, 44), neutral (45–47) and anionic (48–51) surfactants have been separated using CE. Detection has been by direct or indirect UV depending upon the chromophores present.

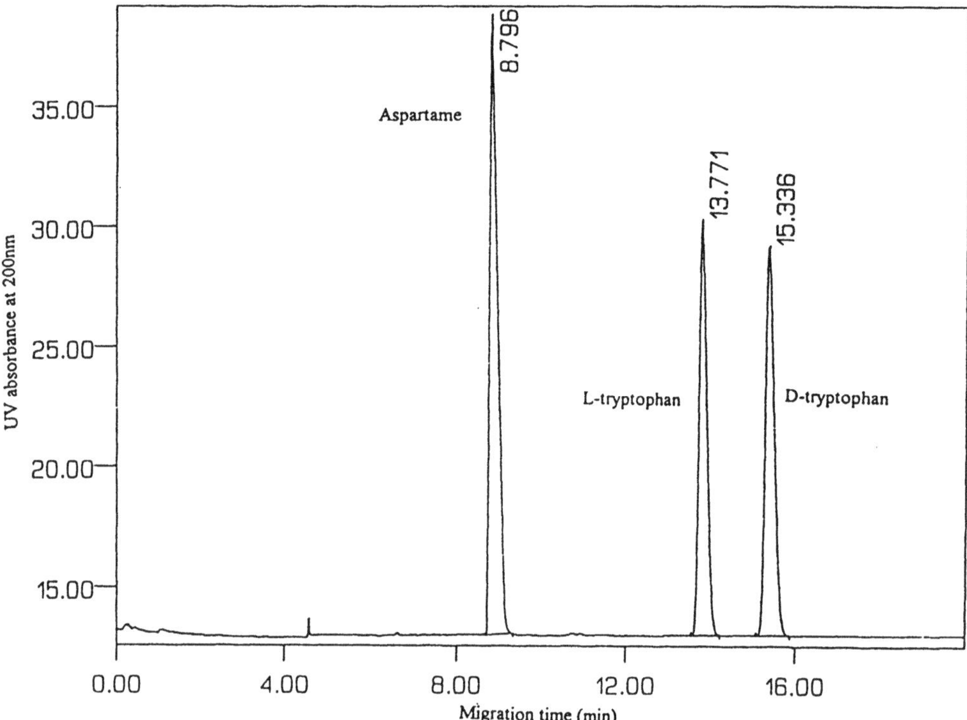

Figure 7.7 Separation of a racemic tryptophan sample with aspartame internal standard. Separation conditions : 75 mM α-cyclodextrin dissolved in 25 mM triethanolamine-phosphoric (pH 2.5), +20 kV, 37 cm (total length) × 50 μm capillary, 200 nm, 30°C, 0.1 mg ml^{-1} in water. Reproduced with permission from reference 41.

Cationic N-benzyl-N-alkyl-N,N-dimethylammonium chloride compounds with different alkyl chain lengths (C_{12}-C_{18}) have been separated by CE and monitored (44) by direct UV detection. Tetrahydrofuran was added to the electrolyte to prevent the surfactants from forming micelles and producing problems with the assay.

Ionic and nonionic ethoxylated surfactants and poly(ethylene glycol) (PEG) oligomers have been resolved by CE (45) using polyacrylamide gel filled CE columns. The non-ionic surfactants were derivatized with phthalic anhydride in order to provide a charge and UV activity. Nonionic octyl- and nonylphenol polyethoxylates have been separated as their ethoxylate homologues by CE (46).

Ionic surfactants such as sodium dodecyl benzene sulphonate (SDBS) have good UV response and their analysis is readily achieved by CE (48, 49). Commercially available SDBS is composed of a mixture of homologues long chain derivatives. If a more complete resolution is required the methods could be further optimised by increasing electrolyte strength and lowering the separation voltage or by the addition of specific divalent metal ions such as magnesium.

Anionic surfactants such as sodium dodecyl sulphate (SDS) possess no appreciable UV activity and are usually analysed by titrimetry. However, analysis of SDS and similar anionic surfactants by CE can be readily (50) performed by employing indirect detection. Figure 7.8

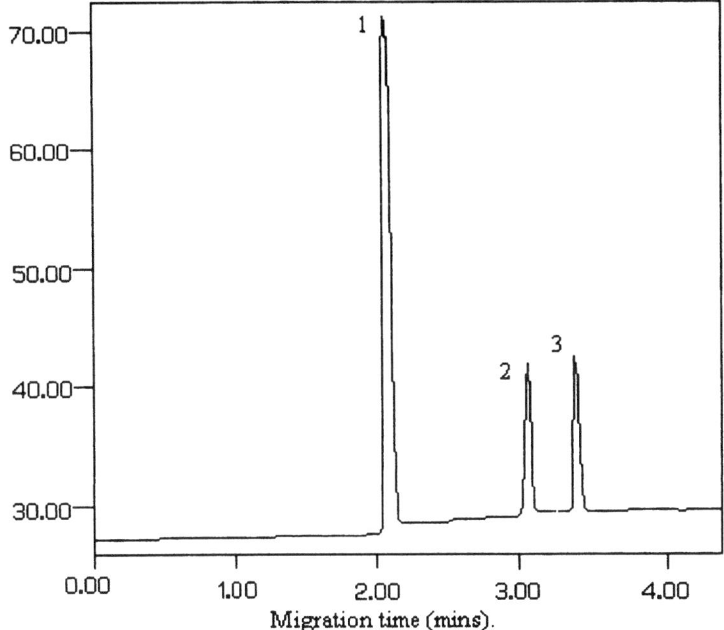

Figure 7.8 Separation of SDS and an internal standard. Peaks 1) Water, 2) Sodium dodecyl sulphate and 3) Hexane sulphonate (I.S). Separation conditions: 8 mM Barbitone buffer (unadj.), 37 cm × 75 µm capillary length, indirect detection at 214 nm, voltage 10 kV and injection 1 second. Reproduced with permission from (51).

shows separation of SDS using a barbitone buffer (51) which gives a natural pH of around 9.5, gives stable separations, and provides sufficient background UV signal for indirect UV detection. An internal standard (heptane sulphonate) is included in Figure 7.8 which allows improved precision (1–2 % RSD) to be achieved. The method is also capable of resolving homologues of SDS from each other and from the SDS peak. The levels of SDS in various tablet samples were measured by this method and found to be in-line with the expected content (Table 7.5).

Table 7.5 Quantitative CE results for the SDS content in various tablets

Strength (As Cefuroxime)	SDS content per tablet (mg)	
	Actual	Obtained
125 mg	2.25	2.18
250 mg	4.5	4.35
500 mg	9.0	8.87

Reproduced with permission from reference 51.

7.16 Water Purity

The levels of inorganic anions, metal ions and organic acid levels can be readily determined in CE at single figure ppm (mg/ml) and better sensitivity levels. Single figure ppb (ng/ml) sensitivity can be achieved (52) using optimised electrokinetic injection procedures. The quality of nuclear power station feed water (53) and boiler feed water (24) has been monitored by CE with single figure ppb detection levels for various anions.

Trace levels of inorganic contaminates in input water can have deleterious effects on synthetic processes as these can poison catalysts. For example Figure 7.9 shows analysis of water from two water purification units. Difficulties had been experienced when attempting to perform a particular synthesis in the laboratory having water which produced the separation given in Figure 7.9a. However the process had been successfully achieved in the laboratory whose water produced the separation given in Figure 7.9b. Adjustment of the water purification unit produced (24) acceptable water and the process was then successfully achieved in both laboratories.

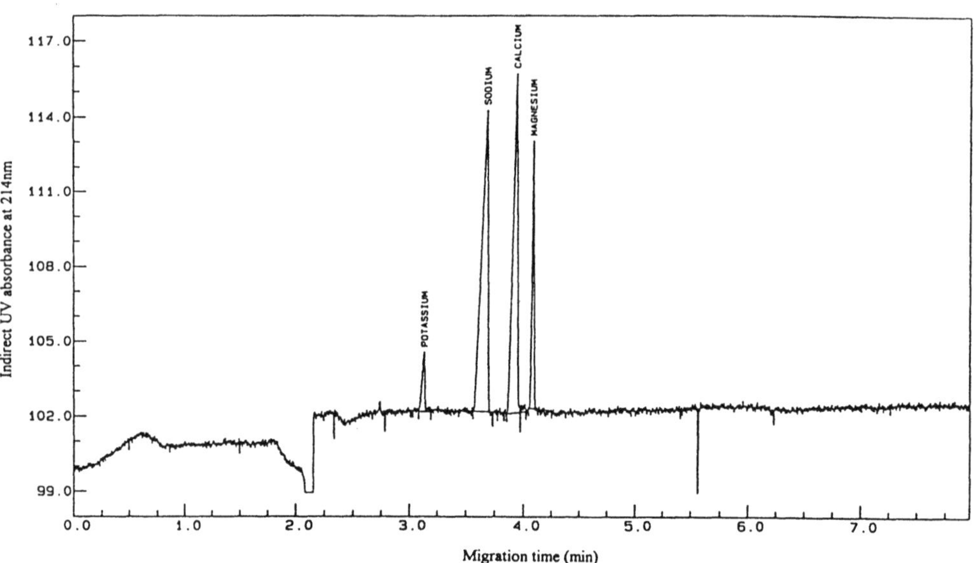

Figure 7.9a Metal ion content in laboratory water

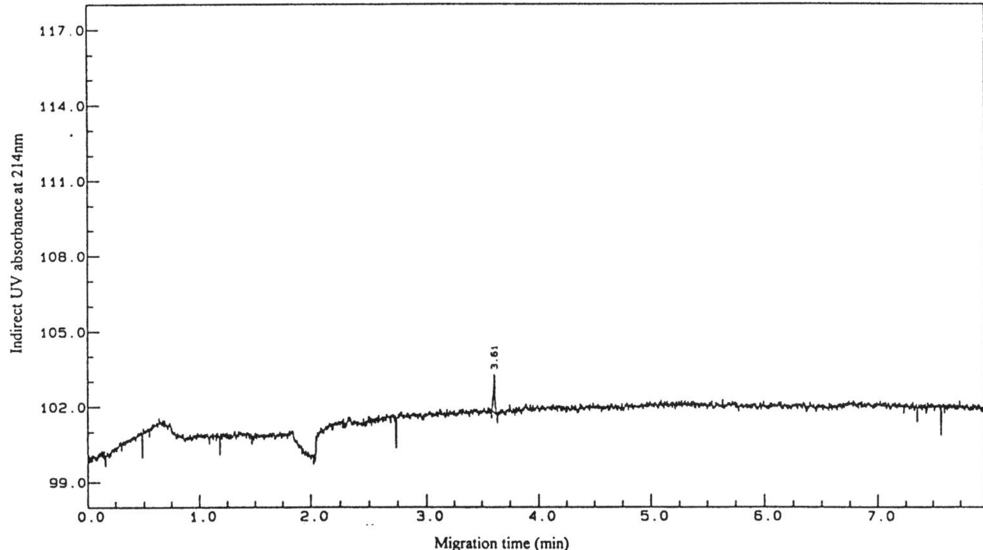

Figure 7.9b Metal ion content in laboratory water

Figure 7.9 Analysis of water from two water purification units. Separation conditions: 1mM borate : 5 mM chromate : 0.5 mM tetradecyltrimethylammonium bromide, +10 kV, 37 cm (total length) × 75 μm capillary, indirect UV detection at 254 nm, 30°C. Reproduced with permission from reference 24.

References

1. Altria KD and Howells JS, Quantitative organic solvent determination by capillary electrophoresis using indirect UV detection, *J. Chromatogr.*, 696 (**1995**) 341-348.

2. El Rassi Z and Mechref Y, Recent advances in capillary electrophoresis of carbohydrates, *Electrophoresis*, 17 (**1996**) 275-301.

3. Oefner P, Chiesa C, Bonn G, and Horvath C, Development in capillary electrophoresis of carbohydrates, *J. Cap. Elec.*, 1(**1994**) 5-26.

4. Klockow A, Paulus A, Figueiredo V, Amado R, and Widmer HM, Determination of carbohydrates in fruit juices by capillary electrophoresis and high-performance liquid chromatography, *J. Chromatogr.*, 680 (**1994**) 187-200.

5. Hoffstetter-Kuhn S, Paulus A, Gassman E, and Widmer H M, Influence of borate complexation on the electrophoretic behaviour of carbohydrates in capillary electrophoresis, *Anal. Chem.*, 63 (**1992**) 1541-1547.

6. Vorndan A G, Oefner P J, Scherz H, and Bonn K, Indirect UV detection of carbohydrates in capillary zone electrophoresis, *Chromatographia*, 33 (**1992**) 163-168.

7. Lu B and Westerlund D, Indirect UV detection of carbohydrates in capillary zone electrophoresis by using tryptophan as a marker, *Electrophoresis*, 17 (**1996**) 325-332.

8. Lee YH and Lin TI, Determination of carbohydrates by high-performance capillary electrophoresis with indirect absorbance detection, *J. Chromatogr.* , 681 (**1996**) 87-97.

9. Nardi A, Fanali S, and Foret F, Capillary zone electrophoretic separation of cyclodextrins with indirect UV photometric detection, *Electrophoresis*, 11 (**1990**) 774-776.

10. Lee YH and Lin TI, Capillary electrophoretic analysis of cyclodextrins and determination of formation constants for inclusion complexes, *Electrophoresis*, 17 (**1996**) 333-340.

11. Burkinshaw SM, Hinks D, and Lewis DM, Capillary zone electrophoresis in the analysis of dyes and other compounds in the dye industry and dye-using industries, *J. Chromatogr.*, 640 (**1993**) 413-417.

12. Hinks D and Lewis DM, Capillary electrophoresis of dyes, *Chromatogr. and Analy.*, Aug/Sept., (**1993**) 9-11.

13. Suzuki S, Shirao M, Aizawa M, Nakazawa H, Sasa K, and Sasagawa H, Determination of synthetic food dyes by capillary electrophoresis, *J. Chromatogr.* , 680 (**1994**) 541-547.

14. Evans KP and Beamont GL, Role of capillary electrophoresis in speciality chemical research, *J. Chromatogr.* , 636 (**1993**) 153-169.

15. Tapley KN, Capillary electrophoretic analysis of the reactions of bifunctional reactive dyes under various conditions including a study of the analysis of the traditionally difficult to analyse phthalocyanine dyes, *J. Chromatogr.* , 706 (**1995**) 555-562.

16. Altria KD and Bryant SM, Highly selective and efficient separations of a wide range of acidic species in capillary electrophoresis employing non-aqueous media, *Chromatographia*, 46 (**1997**) 122-130.

17. Masar M, Kaniansky D, and Madajova V, Separation of synthetic food colourants by capillary zone electrophoresis in a hydrodynamically closed separation compartment. *J. Chromatogr. A.* , 724 (**1996**) 327-336.

18. Erim FB, Xu X, and Kraak JC, Application of micellar electrokinetic chromatography and indirect UV detection for the analysis of fatty acids, *J. Chromatogr. A*, 694 (**1995**) 471-479.

19. Collet J and Gareil P, Capillary zone electrophoretic separation of C14-C18 linear saturated and unsaturated free fatty acids with indirect UV detection, *J. Cap. Elec.*, 3 (**1996**) 77- 82.

20. Buchberger W and Winna K, Determination of free fatty acids by capillary zone electrophoresis, *Mikrochim. Acta*, 122 (**1996**) 45-52.

21. Neubert R, Raith K, and Schiewe J, Capillary zone electrophoresis in skin fatty acid analysis, *Pharmazie*, 52 (**1997**) 212-215.

22. Thompson CO, Trennery VG, and Kemmery B, Micellar electrokinetic capillary chromatographic determination of artificial sweeteners in low-joule soft drinks and other foods, *J. Chromatogr.*, 694 (**1995**) 507-514.

23. Jordan JM, Moese RL, Johnson-Watts R, and Burton DE, Determination of inorganic sulphate in detergent products by capillary electrophoresis, *J. Chromatogr.* , 640 (**1994**) 445-451.

24. Altria KD, Elgey J, Lockwood P, and Moore D, An overview of the applications of capillary electrophoresis to the analysis of pharmaceutical raw materials and excipients, *Chromatographia*, 42 (**1996**) 332-342.

25. Nair JB and Izzo CG, Anion screening for drugs and intermediates by capillary ion electrophoresis, *J. Chromatogr.* , 640 (**1993**) 445-461.

26. Ingvardsen E, Michaelsen S, and Sorensen H, Analysis of individual phospholipids by high performance capillary electrophoresis, *JAOCS*, 71 (**1994**) 183-188.

27. Kajiwara H, Sato A, and Kaneko S, Analysis of calcium and magnesium ions in wheat flour by capillary zone electrophoresis, *Biosci. Biotech. Biochem.*, 57 (**1993**) 1010-1011.

28. Morawski J, Alden P, and Sims A, Analysis of cationic nutrients from foods by ion chromatography, *J. Chromatogr.*, 640 (**1993**) 359-364.

29. Shirao M, Furuta R, Suzuki S, Nakazawa H, Fujita S, and Maruyama T, Determination of organic acids in urine by capillary zone electrophoresis, *J. Chromatogr. A*, 680 (**1994**) 247-251.

30. Lalljie SPD, Vindevogel J, and Sandra P, Quantitation of organic acids in sugar refinery juices with capillary zone electrophoresis and indirect UV detection, *J. Chromatogr. A*, 652 (**1993**) 563-569.

31. Buchberger, W and Winna, K, Optimization of the separation of polycarboxylic acids by capillary zone electrophoresis, *J. Chromatogr. A*, 739 (**1996**) 389-397.

32. Wiley JP, Determination of polycarboxylic acids by capillary electrophoresis with copper complexation, *J. Chromatogr. A*, 692 (**1995**) 267-274.

33. Harvey SD, Capillary zone electrophoresis and micellar electrokinetic capillary chromatographic separations of polyaminopolycarboxylic acids as their copper complexes, *J. Chromatogr. A*, 736 (**1996**) 333-340.

34. Waldron KC and Li JJ, Investigation of a pulsed-laser thermo-optical absorbance detector for the determination of food preservatives separated by capillary electrophoresis *J. Chromatogr. B*, 683 (**1996**) 47-54.

35. Altria KD, Elgey J, and Howells JS, Validated capillary electrophoresis method for the simultaneous determination of histamine acid phosphate and benzalkonium chloride *J. Chromatogr. B*, 686 (**1996**) 111-117.

36. Geise RJ and Machnicki NI, A study of parabens as model hydrophobic compounds by capillary electrophoresis and their determinations in cosmetic formulations, *J. Cap. Elec.*, 2 (**1995**) 69-75.

37. Ng CL, Lee HK, and Li SFY, Analysis of food additives by ion-pairing electrokinetic chromatography, *J. Chrom. Sci.*, 30 (**1992**) 167-170.

38. Ong CP Ng CL, Lee HK, and Li SFY, Separation of imidazole and its derivatives by capillary electrophoresis, *J. Chromatogr.*, 686 (**1994**) 319-324.

39. Chankvetadze B, Endresz G, and Blaschke G, Enantiomeric resolution of chiral imidazole derivatives using capillary electrophoresis with cyclodextrin-type buffer modifiers *J. Chromatogr.*, 700 (**1995**) 43-49.

40. Lee Y-H and Lin T-I, Capillary electrophoretic determination of amino acids with indirect absorbance detection, *J. Chromatogr.A*, 680 (**1994**) 287-297.

41 Altria KD, Harkin P, and Hindson M, Validation of a CE method for the quantitative determination of trytophan enantiomers, *J. Chromatogr. B*, 686 (**1996**)103-110.

42. Skocir E, Vindevogel J, and Sandra P, Separation of 23 danyslated amino acids by micellar electrokinetic chromatography at low temperatures, *Chromatographia*, 39 (**1994**) 7-10.

43. Shamsi SA and Danielson ND, Capillary electrophoresis of cationic surfactants with tetrazolium violet and of anionic surfactants with adenosine monophosphate and indirect photometric detection, *J. Chromatogr. A*, 739 (**1996**) 405-412.

44. Piera E, Erra P, and Infante MR, Analysis of cationic surfactants by capillary electrophoresis, *J. Chromatogr. A*, , 757 (**1997**) 275-280.

45. Wallingford RA, Oligomeric separation of ionic and nonionic ethoxylated polymers by capillary gel electrophoresis, *Anal. Chem.*, 68 (**1996**) 2541-2548.

46. Shamsi SA, Weathers RM, and Danielson ND, Capillary electrophoresis of phosphate ester surfactants with adenosine monophosphate and indirect photometric detection, *J. Chromatogr. A*, 737 (**1996**) 315-324.

47. Heinig K, Vogt C, and Werner G, Separation of nonionic surfactants of the polyoxyethylene type by capillary electrophoresis, *Fres. J. Anal. Chem.*, 357 (**1997**) 695-700.

48. Altria KD, Gill I, Howells J, Luscombe CN, and Williams RZ, Trace analysis of detergent residues by Capillary Electrophoresis, *Chromatographia*, 40 (**1995**) 527-531.

49. Desbene PL, Rony C, Desmazieres B and Jacquier JC, Analysis of alkylaromatic sulphonates by high-performance capillary electrophoresis, *J. Chromatogr.*, 608 (**1992**) 375-383.

50. Nielen M W F, Quantitative aspects of indirect UV detection in capillary zone electrophoresis, *J. Chromatogr.*, 588 (**1991**) 321-326.

51. Kelly MA, Altria KD, and Clark BJ, Quantitative analysis of sodium dodecyl sulphate by capillary electrophoresis, *J. Chromatogr. A*, 781 (**1997**) 67-71.

52. Jackson PE and Haddad PR, Optimisation of injection technique in capillary electrophoresis for the determination of trace levels of anions in environmental samples, *J. Chromatogr.*, 640 (**1993**) 481-487.

53. Bondoux G, Jandik P, and Jones WR, A new approach to the analysis of low levels of anions in water, *J. Chromatogr.*, 602 (**1992**) 79-88.

54. York P, Pharm. *Tech. Europe*, June (**1994**) 17-23.

8 Analysis of dissolution test sample solutions

8.1 Introduction

A variety of analytical methods are available to perform analysis of sample solutions generated during tablet dissolution testing. These analytical methods have recently been reviewed (1) which shows that the majority of analyses are being performed by on-line UV absorbance measurements using flow-through UV cells. UV measurements are generally used as these are simple and rapid to perform and can be used to establish a real-time release profile. However, simple UV measurements are often insufficient when analysing dosage forms containing excipients that are strongly UV active or when pharmaceutical products contain 2 or more active ingredients. In these circumstances diode-array UV spectrometers may be useful (1) in-conjunction with chemometric treatment of the UV absorbance data generated. Alternatively, a separation technique is used to allow clear quantitation of the analyte(s) of interest. Predominantly the separation technique used has been HPLC (1, 2) which is capable of the required sensitivity and high degree of automation necessary to process relative high numbers of sample solutions.

CE methods can offer distinct advantages over alternative HPLC methods in terms of reduced costs, improved ease of operation and use of low UV detection wavelengths. The number of papers in this area is relatively low but will expand as awareness of the possibilities increases and interfaces become available between automated dissolution equipment and CE instruments.

Two forms of dissolution testing can be performed. A total amount of drug released within a given time can be measured, or several data points can be generate over a set dissolution time which enables a dissolution profile to be established. CE has been applied (3–6) to both forms of this dissolution analysis (Table 8.1).

Table 8.1 Application of CE in dissolution testing analysis

Analyte	Application	Comment	Ref.
Lamiduvine	Total drug release assay	Low UV detection	3
Ranitidine	Dissolution profile	Comparison with UV data	3
Levothyroxine	Dissolution profile	Comparison with HPLC method	4
Clenbuterol	Dissolution profile	Solid phase extraction sample pre-treatment	4
Salbutamol	Dissolution profile	Monitoring of release of enantiomers	5
Betamethasone	Dissolution profile	Method validation	6

8.2 Total Drug Content Release Testing

A validated (7) CE method has been used to monitor the dissolution of various basic drugs using a low pH phosphate buffer. The CE data generated compares favourably with that generated by on-line UV absorbance measurements. The CE method was capable of resolving a range of basic drugs with detection at 200 nm.

Figure 8.1a shows the HSCE separation (3) of a test mixture containing 5 basic compounds to be completed within 2 minutes, which was the first of 10 replicate injections performed to test injection precision. Figure 8.1b shows the tenth injection and confirms the reproducibility of the separation. Table 8.2 contains the precision data obtained for the 10 injections and confirms that HSCE is capable of generating highly consistent data within short analysis times. The data also confirms the utility of an internal standard to improve injection precision for both area and migration time measurements.

Rapidly dispersing 100mg lamiduvine tablets were tested (3) for total drug released after 15 minutes. On-line UV absorbance was measured and compared (Table 8.3) to data generated by off-line analysis of the sample solutions by CE.

The injection precision for the lamiduvine standard solution was 2.4 % RSD ($n = 8$). This was obtained with no internal standard. It was concluded that the data generated by on-line UV and CE was comparable and that the average % release was in good concordance with expected % release.

Table 8.2 Precision of HSCE analysis (% RSD for $n = 10$)

	Peak 1	Peak 2	Peak 3	Peak 4	Peak 5
Time	0.57	0.50	0.37	0.53	0.49
RMT	0.43	0.36	0.27	–	0.05
Area	1.32	1.10	1.19	1.19	1.01
PAR	0.39	0.27	0.70	–	0.37

RMT =   migration time relative to peak 4
PAR =   peak area ratio relative to peak 4

Table 8.3 % total release of lamiduvine at 15 minute time point

Tablet number	UV % release	CE % release
1	98.8	97.5
2	105.3	103.1
3	99.1	101.9
4	96.8	106.0
5	96.8	96.9
6	96.2	96.6
Mean	98.9	100.3

Tables 8.2 and 8.3 reproduced with permission from ref. 3

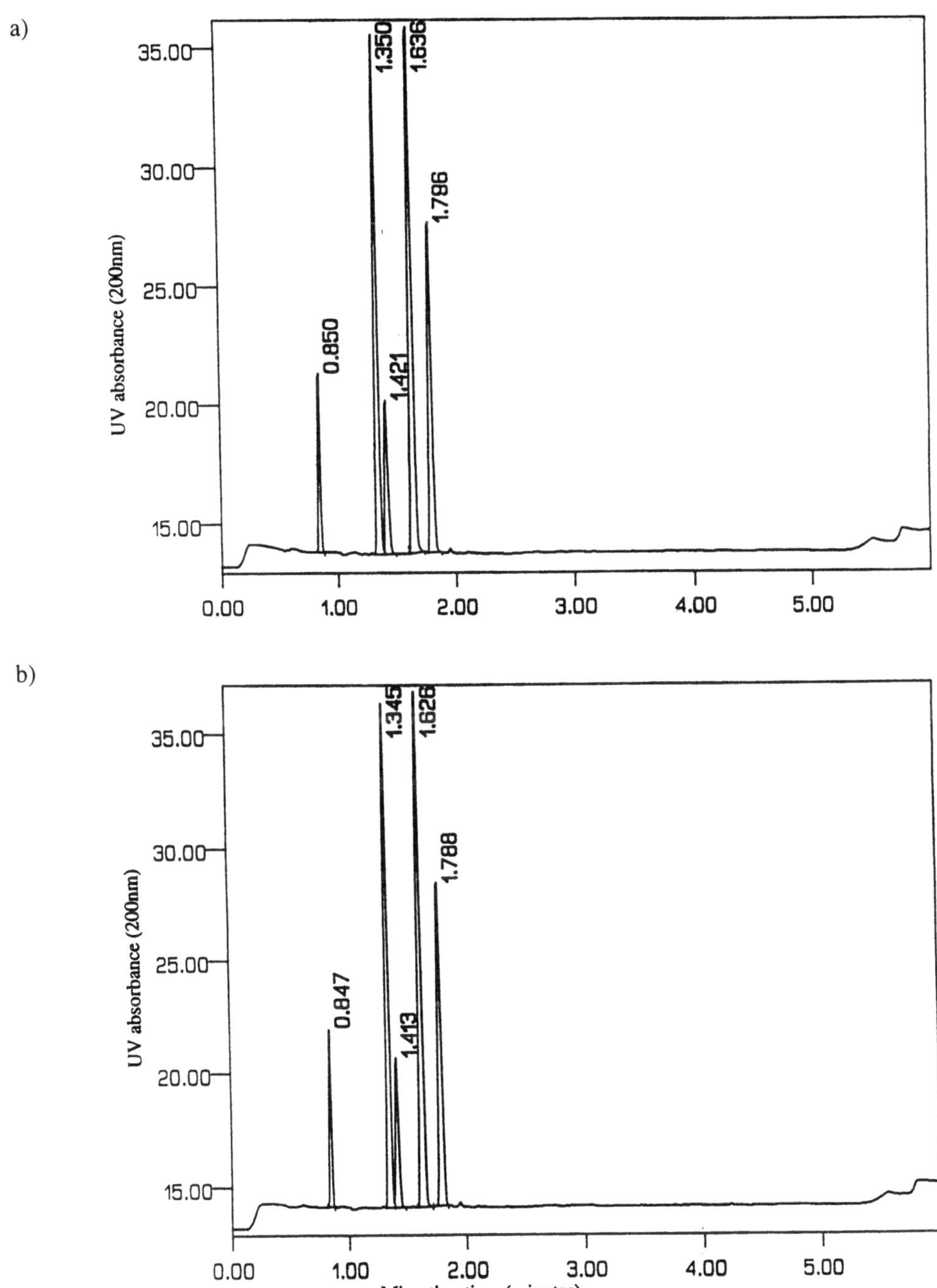

Figure 8.1 HSCE separations of a 5 component test mixture using a 27 cm capillary; a) First injection of precision measurement sequence; b) Tenth injection of precision measurement sequence. Separation conditions: 27 cm lengths (20 cm to detector) and 50 micron, 25 mM NaH_2PO_4 pH 2.5, 200 nm. Reproduced with permission from (3)

8.3 Dissolution Profile Monitoring

The testing involves generation of several result sets at different time-points throughout the dissolution test duration. Six individual 150 mg ranitidine hydrochloride tablets were tested (3) with a total dissolution time of 60 minutes. On-line UV absorbance measurements were performed throughout the 60 minutes. Samples were removed for CE analysis at 15, 30, 45 and 60 minute time-points. The samples for CE were diluted 1:1 with a 0.1 mg/ml imidazole internal standard solution. The precision (calculated from peak area ratios) for calibration response factor was 1.9 % RSD ($n = 10$). Figures 8.2a and 8.2b show the comparable % release dissolution profiles obtained for the 3 tablets using the UV and CE data respectively. The computer package interpolated intermediate dissolution time-points for CE as measurements at only 4 time-points were obtained by CE. Both data sets give a good indication of the dissolution characteristics of the tablets tested.

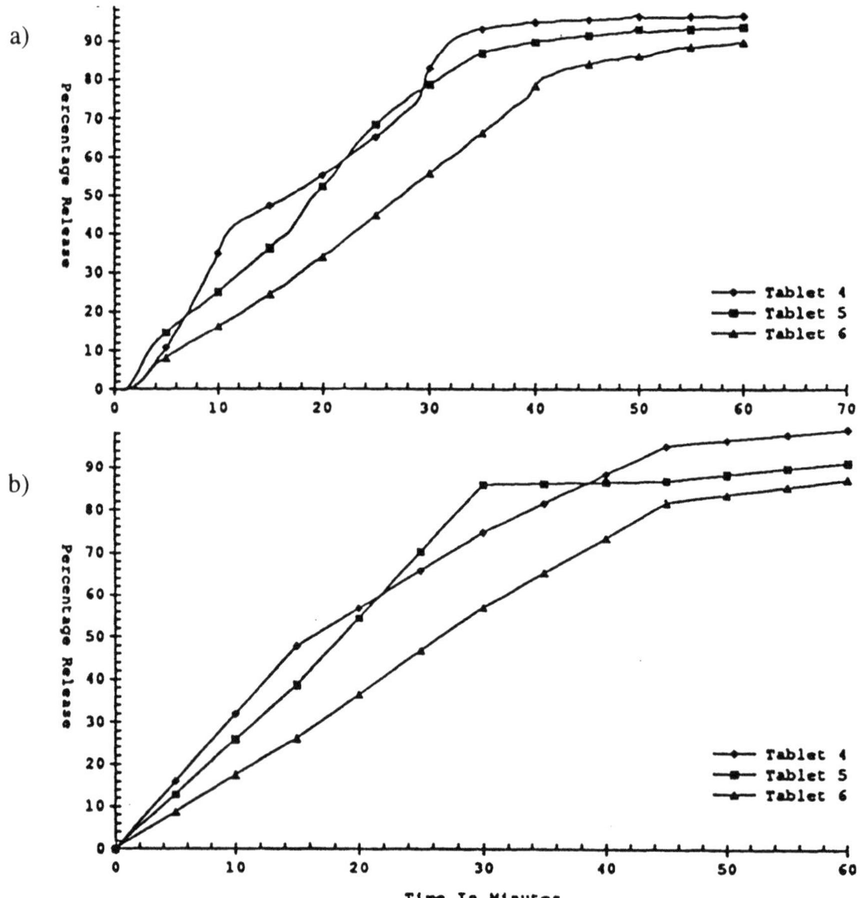

Figure 8.2 Dissolution profiles obtained from CE and UV data; a) Dissolution profiles for 3 ranitidine tablets from UV data; b) Dissolution profiles for 3 ranitidine tablets from CE data. Reproduced with permission from (3)

The dissolution analysis of low dose strength tablets containing clenbuterol (20 µg/dose) or levothyroxine (100 µg/dose) was performed by CE. Sample solutions were pre-concentrated (up to 50 fold increase) using extraction disks. Elution from the disks was performed using methanol as an eluent. Sample solutions at the low ng/ml level were successfully analysed. A range of validation criteria were assessed for both clenbuterol and levothyroxine determinations. Internal standards were not employed in this work. Table 8.4 shows that the analytical methods gave acceptable performance levels.

Solid phase extraction (SPE) was also used (6) in the analysis of betamethasone and ergotamine dissolution sample solutions. The sample solutions were enriched 20 fold using SPE. Simulated dissolution media was prepared by adding excipients into the dissolution liquids. Known amounts of the drug materials were spiked into the simulated media and the recovery data from the resulting solutions was measured by both HPLC and CE. Drug was spiked in a concentrations covering 50–150 % of the likely amounts in real samples. Recovery data for both techniques was similar with average recovery data of 97 % for both HPLC and CE.

Accuracy of the HPLC and CE methods was demonstrated (4) by preparing simulated dissolution samples containing levothyroxine and clenbuterol. Good agreement was obtained between the two data sets (Table 8.5) which confirmed that both methods were accurate.

8.4 Monitoring of the Dissolution of Chiral Drugs

When solid formulations contain a chiral excipient such as cellulose there is the potential (5) that chiral drug enantiomers will be differentially released from the tablet matrix. A simple UV measurement dissolution test would be inappropriate. The use of CE to achieve chiral separations is well established (Chapter 4) as a inexpensive and reliable alternative to HPLC. A chiral CE method has been employed (5) to monitor the dissolution release of salbutamol enantiomers from a tablet formulation. The method achieved chiral separation of the salbutamol enantiomers using a dimethylated-beta-cyclodextrin additive in Tris-phosphate pH 2.5. Figure 8.3 shows the chiral separation achieved.

Validation parameters for the chiral method were assessed (5) for the chiral separation method. Linearity for the enantiomers was demonstrated over the range (0.5–20 mg/l), corre-

Table 8.4 Performance of CE method for dissolution sample testing

	Clenbuterol	Levothyroxine
CE sensitivity (µg/ml)	0.16	0.3
Linearity correlation coeff.	0.9997	0.9990
Linearity range	(0.5–80 µg/ml)	(1–30 µg/ml)
Migration time precision ($n = 6$)	0.7	1.8
Peak area precision ($n = 6$)	1.4	0.8
Sample solution LOD (ng/ml)	22	200

Reproduced with permission from (4)

Table 8.5 Recovery data obtained from the analysis of simulated dissolution samples

		HPLC	CE
Clenbuterol			
75 %	Recovery	98.3	96.3
	RSD	1.3	2.0
100 %	Recovery	95.5	96.6
	RSD	1.6	1.2
125 %	Recovery	98.3	99.0
	RSD	1.2	1.6
Levothyroxine			
75 %	Recovery	95.1	94.4
	RSD	2.0	1.4
100 %	Recovery	98.5	95.8
	RSD	97.6	1.3
125 %	Recovery	97.6	98.5
	RSD	0.9	1.2

Reproduced with permission from (4)

lation coefficients of greater than 0.999 were obtained. Accuracy data was determined by assaying samples of placebo mixture spiked with salbutamol enantiomers – recoveries in the region of 99–103 % were obtained. A limit of quantitation of 1 mg/l was confirmed.

The CE method was applied to monitoring the dissolution rate of individual salbutamol enantiomers and racemic drug from the tablet matrix. Figure 8.4 shows that the individual enantiomers were released at the same rate and at the same rate as the racemic salbutamol.

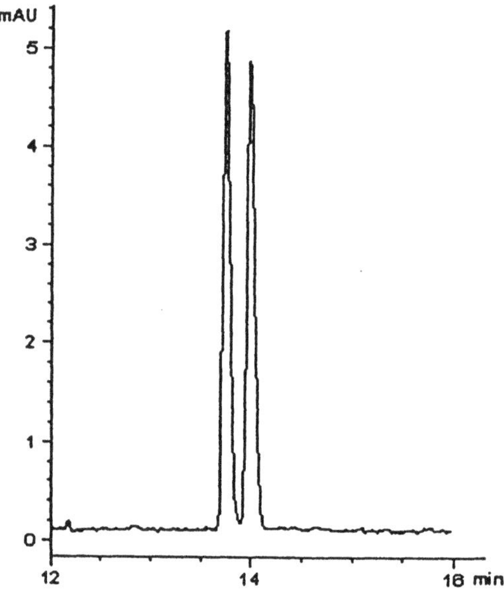

Figure 8.3 Chiral separation of salbutamol enantiomers. Separation conditions: 40 mM Tris pH 2.5 containing 20 mM dimethyl-beta-cyclodextrin, 40 cm (48.5 cm total) × 50 µm capillary, 15 kV, 15 degC. Reproduced with permission from (5).

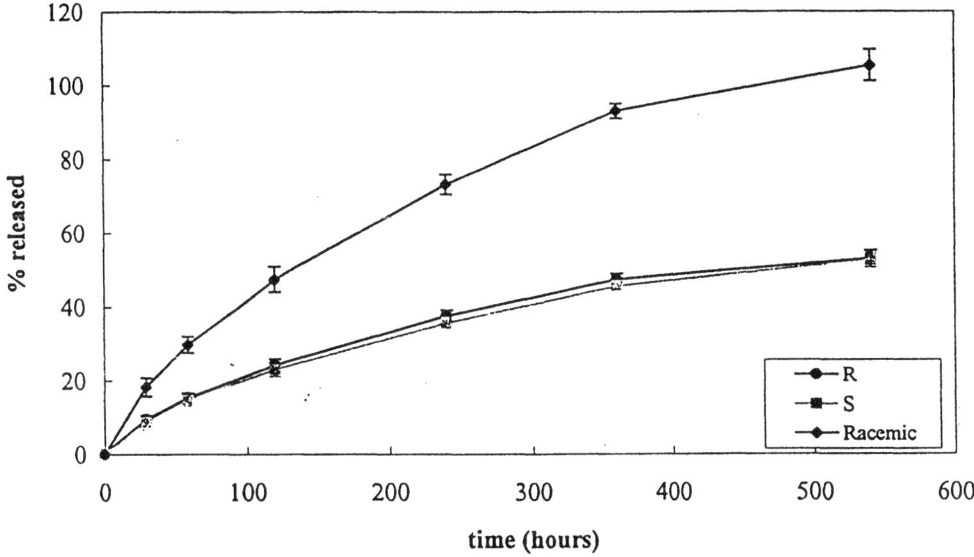

Figure 8.4 Dissolution profile for racemic salbutamol and the individual enantiomers. Reproduced with permission from (5)

8.5 Multi-Component Analysis

Combination formulations which contain a number of active constituents can represent problems for UV based measurements. Although statistical treatment of the UV signal can allow assessment of dissolution rates, generally a separative method is employed to give categorical data. The use of CE methods to allow simultaneously quantitation of a number of components is a key feature of the technique. Chapter 2 covers the use of general analysis type CE methods (low pH phosphate for basic drugs and borate buffer for acidic drugs). Therefore it may be possible to relatively rapidly develop CE methods capable of monitoring a number of components. For example if the formulation contains several basic drugs these could be separated using a simple low pH electrolyte.

The use of diode arrays is widespread in CE and the information can be manipulated in the determination of constituents in formulations.

8.6 Benefits and Disadvantages of Adopting CE for Dissolution Analysis

The main benefit of CE over HPLC is that a simple operating conditions can be used to assay a variety of drugs which makes routine CE assay extremely attractive compared to use of several compound specific HPLC methods. These generic-type CE methods also allow the possibility of simultaneously quantifying components from combination products. Since the separation in CE is based on the charge-to-mass ratios of the solutes the separations can be extremely robust. The electrolytes are invariably aqueous based and buffer requirements may be in the order of 20 ml per day. Shelf-lives for standard electrolytes are typically 3 months. Electrolyte considerations minimise costs in terms of reduced preparation times and elimination of solvent purchase and disposal costs. Capillary costs are also minimal compared to the purchase price of HPLC columns. Further advantages include the relative freedom from excipient peak interferences compared to HPLC where this can be problematic resulting in extended analysis times or additional sample pre-treatments. The use of CE to achieve highly efficient and robust chiral separations is well documented and CE may become the method of choice to monitor dissolution of the individual enantiomers of a racemic drug present in formulations.

The use of short capillaries can allow rapid separations which can be used to increase sample throughput. Generally, analysis times are similar in both HPLC and CE and are in the order of 5–10 minutes. The use of short HPLC columns has been shown to be useful in reducing the analysis time for analysis of captopril dissolution sample solution analysis (8)

The interfacing of automated dissolution equipment with CE instrument has not been reported but should not represent an insurmountable technical problem.

Excipient peak interferences are generally considerably reduced in CE compared to HPLC. For example, (9) a diluted syrup solution can be directly analysed by CE using a low pH electrolyte. Under these conditions the basic drug is positively charged and migrates to the detector whilst the preservatives and colouring agents are uncharged and do not migrate. The unquantified material is rinsed out of the capillary between injections.

The disadvantages of CE compared to HPLC are generally those of reduced precision and sensitivity. Injection precision for CE is generally poorer than in HPLC but this can be largely overcome through use of internal standards and/or high sample concentrations (9) The injection volumes in CE are (10) to the viscosity of the sample solution. Internal standards are especially useful in dissolution testing as the tablet excipients released during dissolution may continuously alter the viscosity of the sample solution. Incorporation of an internal standard alleviates most of the imprecision associated with CE. The use of low UV detection wavelengths (below 220 nm) can often make CE detection limits equivalent to those obtained by HPLC.

References

1. Mehta AC, Review of analytical methods used in the dissolution testing of pharmaceuticals, *Anal. Proc.*, 31, (**1994**) 245-248.

2. Lau-Cam CA, Rahman M, and Roos R, Rapid reversed phase high performance liquid chromatographic assay method for ranitidine hydrochloride in dosage forms, *J. Liq. Chromatogr.*, 17 (**1994**) 1089-1104.

3. Altria KD, Traylen E, and Turner N, Analysis of dissolution test sample solutions using high speed capillary electrophoresis, *Chromatographia*, 41 (**1995**) 393-397.

4. Carducci CN, Lucangioli SE, Rodriquez VG, and Otero GCF, Application of extraction disks in dissolution tests of clenbuterol and levothyroxine tablets by capillary electrophoresis, *J. Chromatogr. A*, 730 (**1996**) 313-319.

5. Esquisable A, Hernadez RM, Gascon AR, Igartua M, Clavo B, and Pedraz JL, Determination of salbutamol enantiomers by high performance capillary electrophoresis and its application to dissolution testing, *J. Pharm. Biomed. Anal.*, 16 (**1997**) 357-366.

6. Lucangioli SE, Rodriquez VG, Otero GCF, and Carducci CN, Development and validation of capillary electrophoresis methods for pharmaceutical dissolution assays, *J. Cap. Elec.*, 4 (**1997**) 27-31.

7. Altria KD, Frake P, Gill I, Hadgett T, Kelly MA, and Rudd D R, Validated capillary electrophoresis method for the assay of a range of basic drugs and excipients, *J. Pharm. Biomed. Anal.*, 13 (**1995**) 951-957.

8. Timmins P, Dissolution testing, *Drug Dev. Ind. Pharm.*, 12 (**1986**) 2031-2040.

9. Altria KD and Rogan MM, Reductions in sample pretreatment requirements by using high performance capillary electrokinetic separation techniques, *J. Pharm. Biomed. Anal.*, 8 (**1990**) 1005-1008.

10. Altria KD and Fabre H, Approaches to optimisation of precision in capillary electrophoresis, *Chromatographia*, 40 (**1995**) 313-320.

9 Determination of Vitamins by Capillary Electrophoresis

9.1 Introduction

The majority of vitamin determinations are currently performed by HPLC with UV detection. The majority of HPLC methods used involve gradient elution and often extensive sample work-up is required prior to analysis to remove sample matrix interferences. The majority of vitamins are water soluble acidic compounds and can be readily determined by free solution CE using high pH electrolytes. Water insoluble and neutral vitamins require use of MECC separation conditions. Table 9.1 lists the range of applications that have been reported to date and show there to be a similar number of FSCE and MECC applications.

The use of low pH electrolytes is appropriate for vitamins containing aminofunctions such as thiamine, nicotanide and vitamin B_{12} and analogues (1). Many of the vitamins are acids or contain groups that can be ionise at high pH and therefore borate or phosphate buffers in the range 7–9 have been used extensively. Use of the higher pH range is recommended to ensure presence of a negative charge (10, 12) as several of the B vitamins have both positive and negatively charged groups around pH 6–7. Effective and accurate analysis of uncharged vitamins requires use of MECC separation conditions (9). Typical MECC conditions are a pH 9 electrolyte containing 30–100 mM SDS (10, 12).

Figures 9.1A and 9.1B shows the CE and MECC separations (9) of a range of vitamins and an internal standard (IS, paracetamol). In Figure 9.1a, thiamine (B_1) is positively charged whilst nicotinamide (PP) is neutral and migrates with the EOF, all other compounds are negatively charged and migrate later. The addition of 100 mM SDS to the electrolyte causes a beneficial alteration to the migration order. In the MECC separation (Figure 9.1b) the cationic B_1 migrates lastly due to its strong ion-pair interaction with the negatively charged SDS micelle. The MECC conditions allow determination of PP as it is resolved from the EOF front at 5 minutes. For comparison (9) Figure 9.1c shows separation of some of these components by a standard HPLC separation. The HPLC method has poorer peak shape and longer analysis times.

The range of vitamin analysis reports mirrors the range of drug analyses and includes assay, related impurities determinations, stability evaluations, clinical determinations and identity confirmation. The number of vitamin applications is limited in comparison to the extensive drug literature but is anticipated to expand as the analysis of vitamins by CE offers the same benefits in terms of reduced costs and improved operational simplicity.

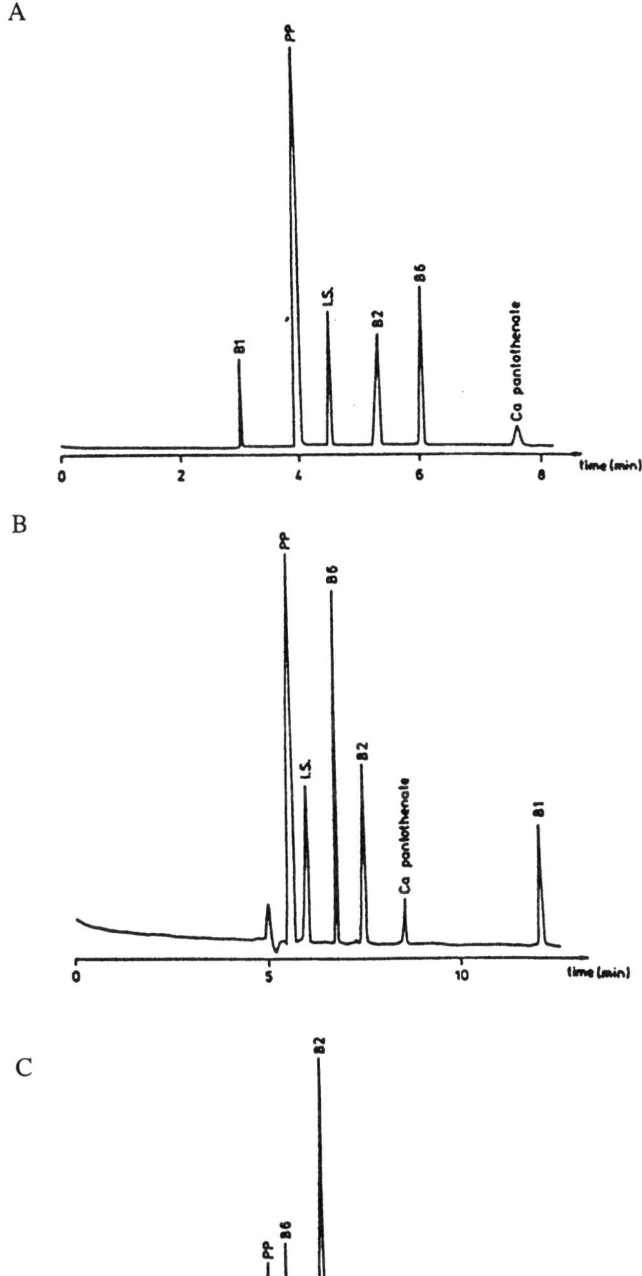

Figure 9.1 Separations of a range of vitamins by FSCE (1A) and MECC (1B) and HPLC (1C). Separation conditions: A): 20 mM borate pH 9 detection at 214 nm; B): 20 mM borate phosphate pH 7 with 100 mM SDS and 13 % acetonitrile detection at 214 nm; C): USP HPLC method (C18) water-methanol-glacial acetic acid (73: 26: 1), detection at 280 nm. Reproduced with permission from reference 9.

Table 9.1 Vitamin analysis CE applications

Analytes	Mode	Application	Ref.
FSCE			
B_{12} forms: hydroxo-cobalamin (OH-cbl), cyano-cobalamin (CN-cbl), 5'-deoxyadenosyl-cobalamin (Ado-cbl) methyl-cobalamin (CH_3-cbl), $CN\alpha$-cobinamide, $CN\beta$-cobinamide	FSCE, pH 2.5	Multi-vitamin formulation	1
Vitamin A	FSCE, pH 7.8	Human blood content – LIF detection	2
Vitamin C	FSCE	Urine and plasma content – UV detection	3
Thiamine (CB), nicotamide (PP), pyridoxine (PN), pantothenate, ascorbic acid (C), folinate (B_8) orotic acid (B_{13}) nicotoric acid (B_3).	FSCE, pH 7	Over-the-counter formulation	4
Riboflavin-5'phosphate	pH 7 phosphate	Purity control	5
Proflavine	FSCE, pH 7	Stability experiment	6
Thiamine, nicotamide, biotine, ascorbic acid, nicotinic acid, riboflavine phosphate and pyridoxine	FSCE, pH 8	Formulations, citrus juices and fruit beverage	7
Biotin	FSCE, pH 8	Capsules	8
B_1, B_6 (pyridoxine), B_2, PP	FSCE, pH 9 and MECC	Tablets, capsules, syrup	9
MECC			
B_1, B_6 (PL, PN and PM), PL phosphate, PM-phosphate, B_2, B_3, B_{12}, niacin B_2-phosphate	MECC, SDS (ion-pair)	Test mixtures	10
7 Water soluble vitamins	MECC, SDS	Vitamin injection	11
B_1, B_3,cyanocobalamin (B_{12}),C, riboflavine phosphate (B_2P), PN	MECC, SDS	Vitamin – enriched drink	12
B_6 vitamers: pyridoxamine (PM), pyridoxine (PN), pyridoxal (PL).	MECC, SDS	Tablet-electrochemical; detection	13
B_6 vitamers	MECC	Urine – LIF detection	14
B_6 vitamers	MECC	Whole blood – LIF detection	15
B_1, B_2, B_3, B_5, B_{12}, PP, H, C, P, Pyroxidine	MECC, SDS	Tablets assay – full method validation	16
Water and fat – soluble vitamins	MECC, SDS	Method development study	17

where:

A	retinol		B_8	folinate
B_1	thiamine		B_{12}	cyanocobalamine (and analogues)
B_2	riboflavin (and B_2 phosphate denoted as B_2P)		B_{13}	orotic acid
B_3	nicotinic acid		C	ascorbic acid
B_5	calcium pantothenate		H	Biotin
B_6	mixture of 3 vitamers (pyridoxine, PN, pyridoxamine, PM, and pyridoxal, PL, also PL-5-phosphate, PM-5-phosphate)		P	Rutin
			PP	Nicotinamide

9.2 Assay

The levels of specific vitamins in various solid and liquid pharmaceutical formulations have been determined by CE. Levels of vitamin C (ascorbic acid) have been determined in fruit juices (7).

The most extensive quantitative paper (9) concerns the comparison of both FSCE and MECC (Figure 9.1) with the USP HPLC method to determine various B group vitamins in capsules, tablets and syrups. Table 9.2 shows the data obtained by the three techniques to be equivalent. The use of an internal standard (paracetamol) improved the CE and MECC precision from 7–10 % RSD to 1%. Analysis of the soft gelatin capsules could not be performed by HPLC due to the presence of oily droplets in the sample solutions. The capsules could be analysed by both CE and MECC. The preferred technique being MECC as the oil droplets are solubilized by the micelle. Accuracy and repeatability of the MECC method was demonstrated (9) by spiking appropriate levels of the vitamins into artificially prepared mixtures of the tablet excipients. Six replicate samples were analysed on each of four separate days, Table 9.3 shows that acceptable data was obtained for average recovery and precision for the pooled assay results. The good migration time precision (<0.5 % RSD) allowed clear identification of the active components. Sample preparation was simpler for the MECC and FSCE methods than for the HPLC analysis. For example the syrup sample was simply diluted with internal standard solution prior to CE analysis.

In a similar study (12) a MECC method (borate pH 9, 30 mM SDS) was used to analyse a range of vitamins in a vitamin enriched drink. Ortho-ethoxybenzamide was used as an in-

Table 9.2 Cross-validation of vitamin assay results by FSCE, MECC and HPLC

| Sample | Analyte | Results as % label claim | | |
		FSCE	MECC	HPLC
Tablet	B_1 (15 mg)	118.7 ± 1.7	123.6 ± 2.6	123.8 ± 3.6
	PP (50 mg)	110.8 ± 3.1	108.0 ± 1.2	108.7 ± 2.1
	B_2 (15 mg)	99.0 ± 2.2	99.4 ± 2.1	103.9 ± 0.7
	B_6 (10 mg)	110.9 ± 3.3	113.7 ± 1.7	112.4 ± 3.2
Syrup (5ml)	B_1 (10 mg)	117.2 ± 4.0	112.4 ± 1.3	111.6 ± 1.6
	PP (20 mg)	111.2 ± 1.4	109.4 ± 0.9	111.5 ± 3.4
	B_2 (1 mg)	115.4 ± 1.5	119.3 ± 2.9	117.2 ± 2.2
	B_6 (5 mg)	109.9 ± 1.4	106.2 ± 3.1	113.2 ± 3.9
Soft capsule	B_1 (10 mg)	122.0 ± 2.2	126.6 ± 1.7	n. a.
	PP (30 mg)	111.3 ± 1.8	108.6 ± 1.7	n. a.
	B_2 (7 mg)	112.1 ± 3.1	114.9 ± 1.6	n. a.
	B_6 (5 mg)	108.6 ± 1.8	108.2 ± 1.6	n. a.

Reprinted with permission from reference 8.

Table 9.3 Accuracy and repeatability data

	Component			
	PP	B_6	B_2	B_1
Recovery %	100.4	99.8	99.7	100.9
Repeatability % RSD	0.8	1.2	1.8	2.8

Reprinted with permission from reference 9

ternal standard to allow the performance data given in Table 9.4 to be obtained. Linearity data was generated over the operating range of 10–1000 mg/L for each vitamin. The sensitivity (detection limit) reported for each component gave a peak with a signal-to-noise ratio of 3 at 220 nm. The vitamin enriched drinks were diluted 1:5 with internal standard solution and directly analysed. Good agreement with label claim was obtained without sample matrix interferences.

Fotsing et al (16) separated 10 water-soluble vitamins (Figure 9.2) using a pH 8.5 boric acid buffer. The addition of 25 mM SDS was required (16) to separate cyanocobalamine and nicotinamide which are uncharged at this high pH. Extensive between-analysis rinses with NaOH were necessary to remove adsorbed excipients such as cellulose from the capillary surface. Poor migration time precision was obtained with only limited rinses. Six of the vitamins were determined in tablets using nicotinic acid as an internal standard to obtain good injection precision. Tablets were ground and extracted at 65 degC with the samples protected from light. Correlation coefficients of greater than 0.998 were obtained for detector linearity for all six vitamins determined over the range 50–150 % of the nominal concentration of sample solutions. Within-run precision data less than 1 % RSD was reported for all 6 vitamins. Combined use of both internal standards and peak area normalisation permitted between-day precisions of less than 3 % RSD. Recovery/accuracy data was obtained by spiking placebo preparations. The recoveries from placebos ranged from 98.2 to 101.8 %. Good agreement between the CE results and % label claim for the multi-vitamin tablets tested was demonstrated.

Table 9.4 Validation data for vitamin determinations in vitamin-enriched drink

	RSD % ($n = 5$)			
Compound	Migration time	Peak area ratio	Linearity Correlation	Detection Limit (μg/L)
B_3	0.5	1.6	0.99998	240
Caffeine	0.7	1.7	0.99998	160
B_6	0.4	1.0	0.99980	190
B_2 phosphate	0.3	1.3	0.99995	650
B_1	0.3	1.8	0.99968	660

Reprinted with permission from reference 12.

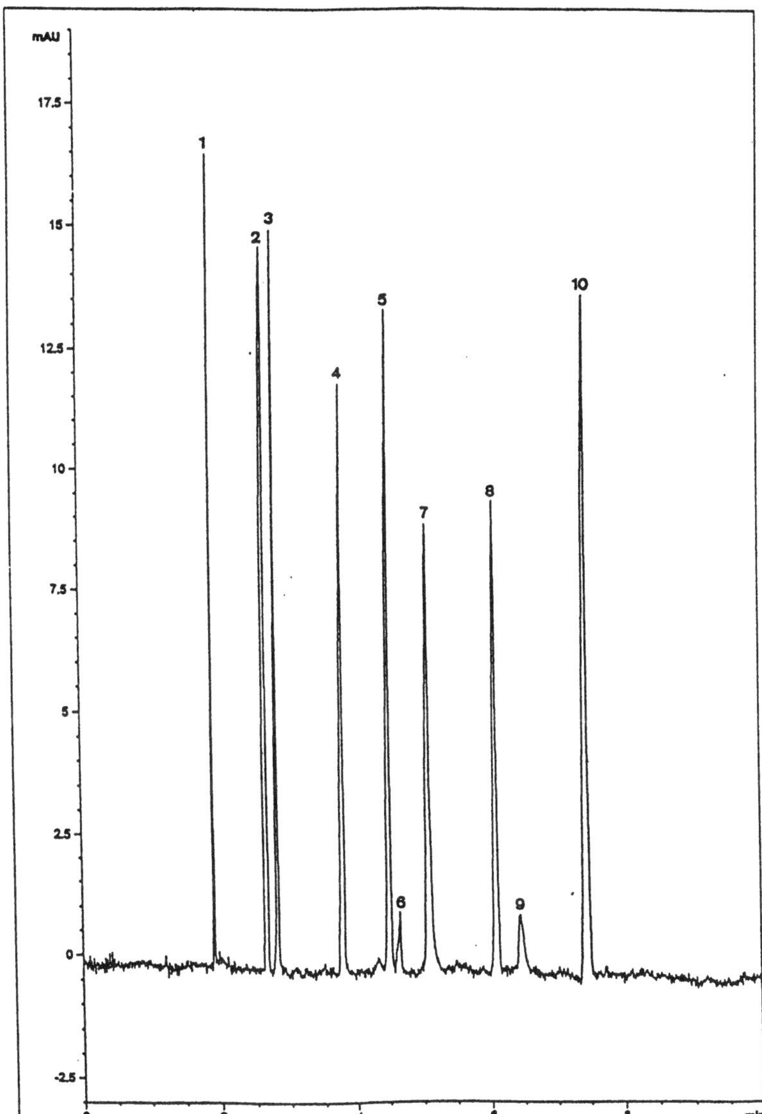

Figure 9.2 Separation of 10 water soluble vitamins by FSCE. Operating conditions: 50 mM boric acid pH 8.5, 48 cm × 50 μm, detection at 225 nm, 25 kV. Peak identities: (1) thiamine (2) nicotinamide (3) adenine (4) riboflavine (5) pyridoxine (6) biotin (7) rutin (8) ascorbic acid (9) panthotenic acid (10) nicotinic acid. Reproduced with permission from reference 16.

Schiewe et al (7) used phosphate pH8 electrolyte to analyse levels of various vitamins in a tablet formulation and both orange and tangerine juices with detection at 200 nm. Assay results of 4.63 and 4.79 mg were obtained for capsules containing 5 mg of biotine with a 6 minute analysis time.

Levels of vitamins B_1,B_3,B_6,B_2,P and C were determined in Injection Solutions using a MECC method (11) with ethylaminobenzoate as an internal standard to give sub 2 % RSD values for precision. Agreement with label claim ranged from 98.8–104 % for the 5 components.

Levels of B_6 vitamers in a tablet formulation were determined (13) using electrochemical detection. The analysis gave a result of 98.3 mg (0.98 % RSD) for a 100 mg tablet. Sample preparation consisted simply of dissolving the tablet in buffer followed by filtration.

Figure 9.3 shows separation of a range of vitamins in a multi-vitamin enriched syrup sample. The separation was optimised (18) by use of a SDS electrolyte containing both acetonitrile and cyclodextrin as selectivity modifiers. The method was successfully applied to a range of pharmaceutical formulations. Table 9.5 shows assay data obtained (18) for multi-vitamin tablet.

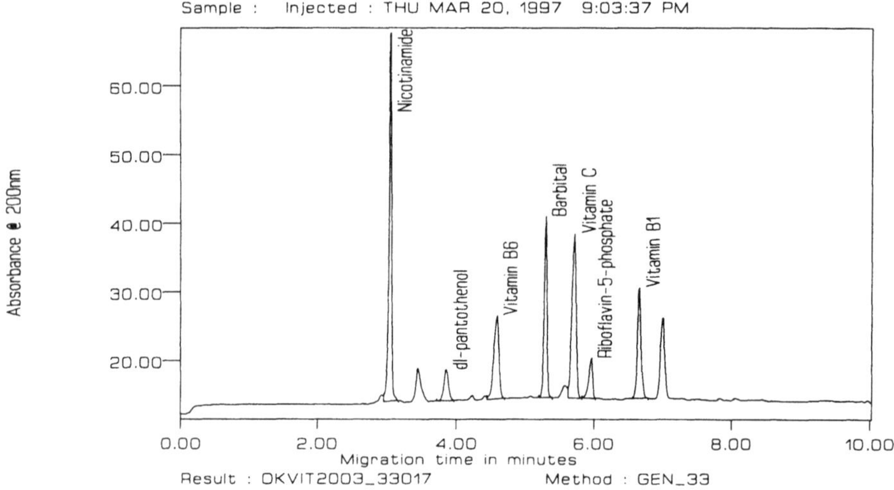

Figure 9.3 Separation of a range of water soluble and insoluble vitamins using an MECC electrolyte containing acetonitrile and cyclodextrin. Separation conditions:75 mM SDS, 20 mM Borate, 3 mM γ-cyclodextrin, 10 % acetonitrile at pH 9.5, 200 nm, 27 cm × 75 μm, 25°C , +7.5 kV. Reproduced with permission from reference 18.

Table 9.5 Quantitation of vitamins in multi-vitamin tablets.

vitamin	relative migration time, RSD (%), $n = 14$	amount label	amount measured
nicotinamide	0.22	10	10.58
riboflavin	0.14	1.08	1.10
thiamine	0.40	1.2	1.15

Reprinted with permission from reference 18

9.3 Related Impurities Determinations

The purity and chemical stability of test samples can be readily assessed using CE. For example various analogues of vitamin B_{12} (corrinoids) were separated using a pH 2.5 phosphate buffer with detection at 214nm (1). Figure 9.4a shows separation of a test mixture of 5 corrinoids and Figure 9.4b shows separation of the same sample solution following its exposure to a high intensity light. The OH-, Ado- and CM_3- cbl have converted to CN-cbl whilst the level of CN-cobinamide remains constant. It would be possible to follow the rate of reaction by CE if required.

Kenndler (5) determined the purity and stability of riboflavin-5-phosphate using free solution CE with UV detection. The purity determination of proflavine had earlier been demonstrated (6) using fluorimetric detection.

9.4 Clinical Determinations

Chapter 10 describes more fully the use of CE in the area of clinical analysis. There have been only a limited number of clinical applications involving vitamin determinations, however these have illustrated the possibilities in this area. The potential benefits of adopting CE for clinical determinations are the reduced analysis times and both sample volumes and pretreatments required, compared to other techniques such as HPLC and immunoassays.

Free solution CE has been used (7) to determine levels of vitamin A (retinol) in blood by laser excited fluorescence with a LOD of 3 μ/L. Linearity over the range employed gave a correlation coefficient of 0.997. Results for CE and HPLC for retinol concentrations in 19 serum samples gave a statistical correlation coefficient of 0.97. The recoveries, measured by CE, for 7 serum samples spiked with retinol, were in the range 83–113 %. The CE sample pretreatment involved a dilution and centrifugation whilst the HPLC sample pretreatment involved a longer and more elaborate liquid-liquid extraction procedure.

Levels of vitamin C (absorbic acid) were determined (3) in both urine and plasma with UV absorbance detection. Sample pretreatment was either avoided or consisted of a simple protein precipitation step.

9.5 Identity Confirmation

Identity confirmation can be performed by concordance of the migration time of peaks from the analysis of the sample with those peaks obtained from analysis of standard solutions. Improved performance is obtained when internal standards are employed to obtain relative migration times. Additional confirmation of the identity of solute peaks can be achieved through use of diode array detectors to obtain UV spectra (12) or to display the same separation at a number of different wavelengths (4).

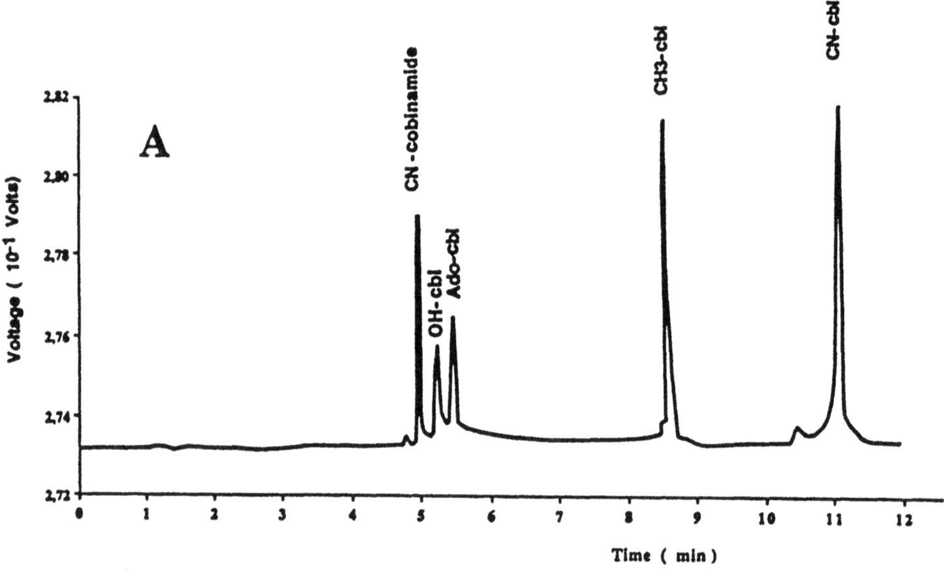

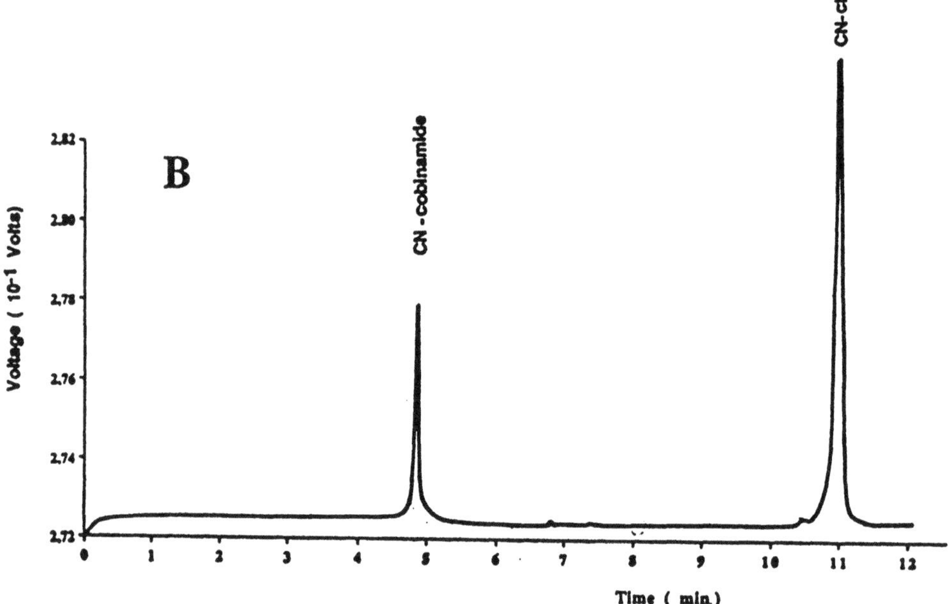

Figure 9.4 Separation of 5 corrinoids before and after exposure to high intensity light. Operating conditions: 25 mM phosphate pH 2.5, detection at 214 nm, 100 cm × 75 μm, 30 kV. Reproduced with permission from reference 1.

9.6 Microemulsion Electrokinetic Capillary Chromatography (MEEKC)

MEEKC is a separation technique similar to MECC in which high pH buffers containing surfactants are used to resolve compounds. Typically in MEEKC a micro-emulsion is formed of tiny drops of a water immiscible solvent in water. High concentrations of SDS (typically 3.3 % w/w) are added. The hydrophobic portion of the SDS penetrates into the core of the oil droplet whilst the hydrophilic ionic groups remain in the aqueous buffer. The presence of the SDS therefore generates a negative charge on the droplet and the droplet is able to migrate against the EOF in a similar fashion to that of a micelle. Solutes can partition with the droplet and those having the strongest partitioning will be detected last. A recent review (19) covers developments in the technique. MEEKC applications are limited thus far but the technique has been successfully employed (20) in the separation of water insoluble vitamins.

9.7 Conclusions

Vitamins are well suited to analysis by CE as many are water soluble and have ionisable functionalities. A range of application areas including assay of vitamin content, determination of related impurities, clinical applications and identity confirmation. The performance of these methods is similar to that obtained for analysis of drugs and it is therefore expected that the scope and number of applications of CE to vitamin analysis will continue to expand.

References

1. Lambert D, Adjalla C, Felden F, Benhayoun S, Nicolas J P and Gueant J L, Identification of vitamin B_{12} and analogues by high-performance capillary electrophoresis and comparison with high-performance liquid chromatography, *J. Chromatogr.*, 608 (**1992**) 311-315.

2. Shi H L, Ma Y F, Humphrey J H and Craft N E, Determination of vitamin A in dried human blood spots by high-performance capillary electrophoresis with laser-excited fluorescence detection, *J. Chromatogr. B*, 665 (**1995**) 89-96.

3. Koh E V, Bisell M G and Ito R K, Measurement of vitamin-C by capillary electrophoresis in biological fluids and fruit beverages using a stereoisomer as an internal standard, *J. Chromatogr.*, 633 (**1993**) 245-250.

4. Jegle U, Separation of water-soluble vitamins via high-performance capillary electrophoresis, *J. Chromatogr. A*, 652 (**1993**) 495-501.

5. Kenndler E, Schwer C and Kaniansky D, Purity control of riboflavin-5-phosphate (vitamin B_2 phosphate) using capillary zone electrophoresis, *J. Chromatogr.*, 508 (**1991**) 203-207.

6. Altria K D and Simpson C F, Analysis of some pharmaceuticals by capillary zone electrophoresis, *J. Pharm. Biomed. Anal.*, 6 (**1988**) 801-805.

7. Huopalahti R and Sunell J, Use of capillary zone electrophoresis in the determination of B-vitamins in pharmaceutical products, *J. Chromatogr.*, 636 (**1993**) 133-135.

8. Schiewe J, Gobel S, Schwarz M and Neubert R, Application of capillary zone electrophoresis for analysing biotin in pharmaceutical formulations – a comparative study, *J. Pharm. Biomed. Anal.*, 14 (**1996**) 435-439.

9. Boonkerd S, Detaevernier M R and Michotis Y, Use of capillary electrophoresis for the determination of vitamins of the B group in pharmaceutical preparations, *J. Chromatogr.*, 670 (**1994**) 209-214.

10. Nishi H, Tsumagari N, Kakimoto T and Terabe S, Separation of water-soluble vitamins by micellar electrokinetic chromatography. *J. Chromatogr.*, 465 (**1989**) 331-343.

11. Fujiwara S, Iwase S, and Honda S, Separation of water-soluble vitamins by micellar electrokinetic chromatography. *J Chromatogr*, 465 (**1989**) 331-343

12. Soya T, "Simultaneous analysis of water-soluble vitamins using capillary electrophoresis", *Hewlett-Packard application note*, Publication number 12-5962-9812E

13. Yik Y F, Lee H K, Li S F Y and Khoo S B, Micellar electrokinetic capillary chromatography of vitamin B_6 with electrochemical detection, *J. Chromatogr.*, 585 (**1991**) 139-144

14. Burton D E, Sepaniak M J and Maskarinec M P, Analysis of B_6 vitamers by micellar electrokinetic capillary chromatography with laser-excited fluorescence detection, *J Chromatogr*, 24 (**1986**) 347-351

15. Swaile D F, Burton D E, Balchunas A T and Sepaniak M J, Pharmaceutical analysis using micellar electrokinetic capillary chromatography, *J. Chromatogr. Sci.*, 26 (**1988**) 406-409,

16. Fotsing L, Fillet M, Bechet I, Hubert Ph and Crommen J, Determination of six water-soluble vitamins in a pharmaceutical formulation by capillary electrophoresis, *J. Pharm. Biomed. Anal.*, 15 (**1997**) 1113-1123

17. Ong C P, Ng C L, Lee H K and Li S F Y, Separation of water and fat-soluble vitamins by micellar electrokinetic chromatography, *J. Chromatogr.*, 547 (**1991**) 419-428

18. Kooymans O and Altria KD, unpublished work

19. Watari H, Microemulsions in separation science, *J. Chromatogr. A*, 780 (**1997**) 93-102

20. Boso R L, Bellini M S, Miksik I and Deyl Z, Microemulsion electrokinetic chromatography with different organic modifiers: Separation of water- and lipid-soluble vitamins, *J. Chromatogr. A*, 709 (**1995**) 11-19

10 Overview of Application of CE to determine drugs in biofluids

10.1 Introduction

The use of CE in the clinical and biomedical areas is now becoming established as an alternative and compliment to existing standard techniques such as HPLC, immunoassays or GC-MS. CE can have particular advantages for certain assays including reduced sample preparation requirements and less expensive analysis and reduced interference in the assay. Also of note is the ability to perform effective enantioselective separations of chiral compounds present in biofluids. Table 10.1 shows that CE has been used (1–40) for a wide range of clinical and biomedical applications. Other advantages of CE compared to other analytical techniques used in bioassays is that the analysis times can be rapid (1–5 minutes). Another advantage of CE is that the separation can be performed until the peak(s) of interest have been detected. After this point the separation can then be stopped and a rinse step can be used to remove unquantified peaks remaining in the capillary. These unquantified peaks would need to be removed from the HPLC column which may cause unnecessarily long analysis times or use of gradient elution. An additional feature of CE is that typical injection volumes are in the order of 10–50 nl which is a fraction of the volumes injected in HPLC (10–100 µl). This small volume ability means that several injections are possible in CE from tiny sample volumes which can lead to significant savings (9).

Recent reviews (41, 42) have covered both pharmaceutical and non-pharmaceutical biomedical/clinical applications of CE.

Non-pharmaceutical applications include aspects such as inorganic ions, diagnosis of metabolic disorders by biofluid analysis, and therapeutic protein determinations (43). For example CE has been used to measure the amounts of nitrate and nitrite present in the gastrointestinal tracts of rabbits following nerve stimulation with nitric oxide (44) and the determination of phenylalanine levels in serum as an indication of phenylketonuria (45).

10.2 Sample Pretreatment Procedures

The extent and complexity of sample preparation required is of critical importance when determining drugs in biofluids as the drugs can be bound to components of the sample matrix such as proteins and give an incorrect assay result. Often the separation requirements are minimal and the emphasis is placed on preparation of the sample into a suitable state for analysis. Constituents of the sample solution matrix can also cause problems with fouling of the capillary and suitable capillary rinsing regimes are established during method develop-

Table 10.1 Range of applications of CE to the analysis of drugs in biofluids

Application	Conditions	Pre-treatment	Ref. No.	Comments
Therapeutic Monitoring				
Antipyrine in biofluids	MECC, SDS	Direct injection	1	4 minute analysis, correlation of 0.99 for CE results with HPLC for 75 samples
Aspoxicillin in plasma	MECC, SDS	Direct injection	2	Linearity 0.999, recoveries 94–104 %
Cefpiramide in plasma	MECC, SDS	Direct injection	3	Antipyrine used as internal standard
Cefuroxime in serum	MECC, SDS	Direct injection	4	Recovery data of 98–101%
Creatinine, uric acid in plasma and urine	MECC, SDS, IPA	Direct injection	5	Good correlation with enzymatic assays
Cicletanine in plasma	MECC, SDS	Solvent extraction	6	20 ppm LOD, excellent correlation with HPLC
Cimetidine in plasma	Phosphate pH 2	SPE	7	LOD of 250 ng/ml, peak identity by UV-DAD
Cimetidine in serum	HTAB, Tris, pH 6	SPE	8	Ranitidine used as an internal standard, 20–50 µl sample aliquots used
Cyclic guanine monophosphate phosphodiesterase inhibitor in rat serum	Phosphate pH 6.0	Deproteinized with ACN	9	Low sample volumes used in CE reduced the number of animals used in drug testing
Cytosine-β-D-arabinosine in plasma	pH 2.5 acetate	SPE	10	Micromolar sensitivity with 4.5 minute assay, validated assay
Drugs in serum	MECC, SDS	None	11	Poor peak shape if good protein binder
Fosfomycin in serum	Borate and phenylphosphonic acid	Deproteinized or direct injection	12	Indirect UV detection with LOD of 10 µg/ml
heparinoid mimetics in human and rat plasma	Citrate-methanol buffer, pH 4.0	Dilution and centrifugation	13	Analysis of more than 350 plasma samples from pharmacokinetic studies
Iohexol in serum	Borate pH 8	Deproteination or direct injection	14	Internal standard – good precision, linearity, recovery
Naproxen in serum	Tricine pH 8	Solvent-solvent extraction	15	LIF detection limit 3 fmol compared to UV LOD of 100 fmol
Pentobarbital in serum	Borate, pH 8.5	Deproteinized with ACN	16	Results correlated with HPLC data, isobutyl-1-methyl-xanthine used as IS, 4 min assay

Piracetam in human plasma	Borate and alpha-CD	Various	17	LOD of 1 mg/ml with detection at 200 nm
Purines in biofluids	MECC, SDS	Direct injection	18	Serum, salvia, urine results compared to immunoassay
Suramin in serum	CAPSO pH 9.7	Deproteinized with ACN	19	Validation including 93 % recovery, high ionic strength buffer used
Theophylline in capillary ultrafiltration probes	MECC, SDS	Direct injection	20	Consistent CE/HPLC results, 15 nl injection volume
Thiopental in serum and plasma	MECC, SDS	Solvent extraction	21	66 serum sample results correlated with HPLC
Tricyclics in plasma	pH 8, TTAB, urea	Solvent extraction	22	Extensive validation exercise, 5 ng/ml LOD
Phenotyping Disease Diagrnostic				
Caffeine metabolites in urine	MECC, SDS	Liquid extraction	23	Acetylator phenotyping via analysis of 4 caffeine metabolites
Dextromethorphen	Borate pH 9.3	None	24	Validation, used for debrisoquin-oxidation metabolic phenotyping
Dihydrocodeine and metabolites in human urine	MECC, SDS	Direct injection or SPE	25	Phenotyping of patients based on metabolites produced
Hypoxanthine and xanthine in urine	Phosphate pH 8	5 fold dilution	26	Diagnosis of xanthinuria by monitoring hypoxanthine and xanthine levels
Methylmanolic acid in serum	Tris-citrate pH 6.4	Derivatisation and dilution	27	Marker technique for cobalaim defiency, 0.1 μM/L LOD with LIF
Steroids in serum	MECC with SDS or DTAB	Ultrafiltration	28	Steroid profiling to diagnose Cushing's syndrome, Addisons disease and congenital adrenal hyperplasia
Metabolism				
S-carboxymethyl-L-cysteine and metabolites in urine	pH 9 phosphate	SPE	29	Coated 25 μm capillaries used
Cefixime and metabolites in urine	Phosphate pH 6.8	Direct injection	30	Separation from a range of metabolites
Cefotaxime and metabolites in plasma	MECC, SDS	Direct injection	31	Validation exercise, theobromide internal standard
Coumarin metabolites in urine and serum	Phosphate pH 7.5	Liquid extraction	32	HPLC and CE results comparable – CE analysis time 1.5 min compared to 12 min for HPLC
Famotidine	PEO and dextran, pH 4.5	Centrifugation and dilution	33	Novel mixed polymer electrolyte system used – also used for chiral separation

Haloperidol and 10 synthetic metabolites	pH 4.5 with 10 % MeOH	Centrifugation	34	Metabolism studies of holperidol with guinea pig hepatic microsomes
Oxprenolol and its metabolites in human urine	Phosphate pH 2.5 with HP-β-CD	Incubation with β-gluronidase, liquid extraction	35	Enantioselective metabolism studies performed, metabolites and oxprenolol all chiral resolved by same method
Pyrazolacridine metabolites in urine	NH$_4$ acetate, MeOH, acetic acid	Solvent extraction	36	Non-aqueous CE, improved detection with LIF
Theophylline and its metabolites in rat liver microsomes and human urine	MECC,SDS	SPE	37	Novel separation of all theophylline metabolites
Thalidomide and metabolites	CM-β-CD pH 6	Liquid extraction	38	Enantioselective metabolism studies of thalidomide following incubation with liver microsomes
Zolpidem and metabolites in urine	Phosphate pH 5.6	Incubation with β-gluronidase	39	LOD of 2 ng/ml with LIF detection, metabolism studies performed in human volunteers
Zopiclone and its metabolites in urine and saliva	Phosphate pH 2.8 with β-CD	Liquid extraction	40	S-enantiomer of zopiclone metabolised faster than r-enantiomer

Abbreviations:

CM-β-CD = carboxymethyl-beta-cyclodextrin

CAPSO = zwitterionic buffer

DTAB = dodecyltrimethylammonium bromide, a cationic surfactant

HP-β-CD = hydroxpropyl-beta-cyclodextrin

LIF = laser induced fluorescence

PEO = polyethylene oxide

SPE = solid phase extraction

TTAB = tetradecyltrimethylammonium bromide, a cationic surfactant

ment. The strategies and approaches for monitoring drugs in biofluids have been reviewed (46). Sample pre-treatments are similar to those involved in HPLC. These procedures include ultracentrifugation (46), and deproteinisation by the addition of solvents such as acetonitrile (47, 48). Alternatively the drug can be retained on a chromatographic support to remove the extraneous sample components and to concentrate the solute(s) of interest - typically this involves use of solid phase extraction cartridges (46). However analyte specific columns can be used to preconcentrate and extract components of interest, for example a dexamethasone specific affinity column has been used in the determination of dexamethasone levels in equine urine (49).

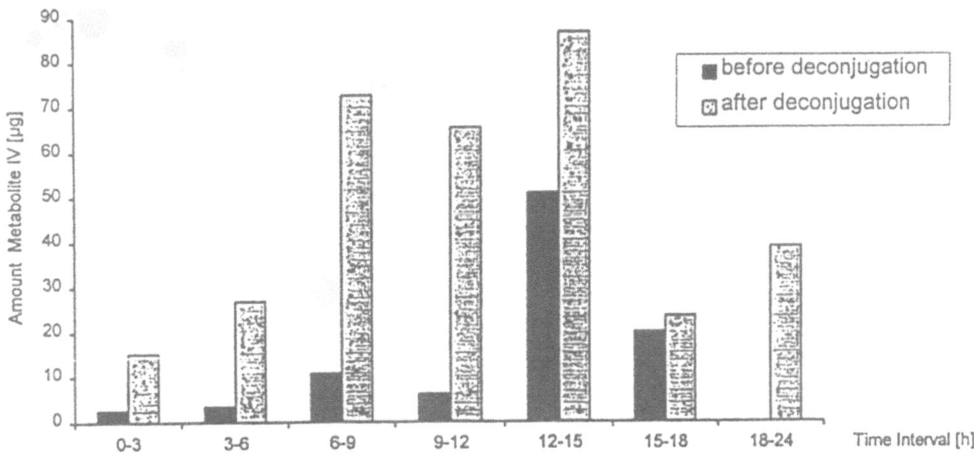

Figure 10.1   Total amount of zolpidem metabolite in volunteer urine before and after deconjugation with beta-glucuronidase. Reproduced with permission from reference 39.

In drug metabolism studies conjugates can become formed and incubation with appropriate enzyme may be needed (35, 39, 50) to release all the parent drug. For example (50) serum samples were incubated with beta-glucuronidase for 1 h prior to the determination of lamotrigine – this incubation period increased the peak height of lamotrigine by about 24 %. Beta-glucuronidase was also used (39) to deconjuate zolpidem and metabolites present in urine sample. Figure 10.1 shows the increased amount of a zolpidem metabolite found following deconjugation.

Liquid-liquid extraction of the drug into a water-immiscible phase such as hexane or ether is another popular means of preconcentration and can be effective in removal of salt in the sample as the salt will remain in the aqueous layer. The drug may be extracted (15, 22) into several mls of solvent which is then totally evaporated and the residue redissolved in a small volume to increase the sample solution concentration.

10.3 Direct Sample Injection

Direct injection of untreated biofluid samples solutions such as serum and urine (Table 10.1) can be made into CE capillaries. This can result in considerable savings in analysis time and cost of consumables. Direct injection of the sample solution also guarantees a 100 % recovery as the intact sample solution is injected into the capillary. Whenever a sample pretreatment procedure is adopted suitable recovery data must be demonstrated in method validation studies.

The principal difficulty with direct injection of plasma or serum is that sample contain large amounts of proteins which can generate interfering peaks. Therefore the majority of direct injection analyses are performed by MECC as the SDS micelles strongly interact with the sample proteins causing the proteins to be eluted after the drug peaks of interest. Figure

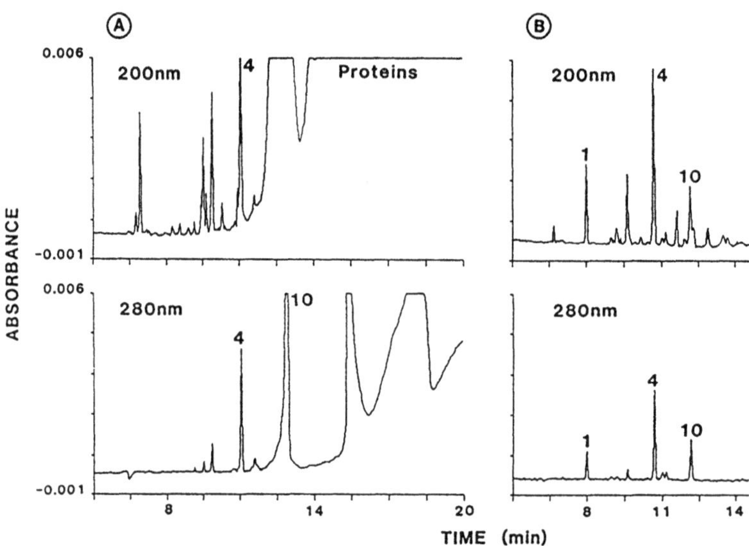

Figure 10.2   Separation of a plasma sample with direct injection (A) or following SPE (B) with detection at 200 nm and 280 nm. Operating conditions : borate-phosphate pH 9 containing
75 mM SDS, 90 cm × 75 μm, 20 kV. Peak identities 1 = theobromide, 4 = theophylline
and 10 = uric acid. Reproduced with permission from reference 18.

10.2 shows the determination (18) of purines in plasma when directly injected or injected
following extraction through a SPE column. All components of interest were sufficiently resolved to be detected at 280 nm with no pretreatment but detection of uric acid at 200 nm required SPE pretreatment. Figure 10.2 shows that the proteins are strongly bound to the SDS
in the MECC electrolyte and are well retained.

When direct sample injection is performed the inter-analysis rinses used becomes of particular importance to remove all the proteineous components which have a high tendency to
stick to the capillary walls resulting in variations in the rate of EOF and migration times. A
variety of rinsing procedures have been used in (51) including inter-analysis rinses with concentrated SDS solutions (52). Concentrated SDS solutions are particularly effective as proteins strongly interact with SDS and can be removed from the capillary wall when standard
rinses with NaOH or phosphoric acid may be ineffective (52).

Quantitation of a number of drugs in human plasma have been shown (53) on uncoated
fused-silica capillaries with direct sample pretreatment using acetonitrile as a between-run
rinsing reagent. Intra- and inter-day precision values of about 1–2 % R.S.D. (n = 20) and 2–
3 % R.S.D. (n > 80) respectively are obtained using a sodium dodecyl sulfate-containing be-
rate buffer, pH 10 . This method was highly robust with uninterrupted routine operations for
several weeks. The efficiency of the acetonitrile rinse procedure was compared to rinsing
procedures using enzyme-containing solutions, different organic solvents and hydrofluoric
acid.

The different approaches to quantifying drug in human serum following direct sample
injection have been compared (54).

10.4 Sample Matrix Effects

The matrix of the sample can have a pronounced effect upon the quality of the separation achieved (19) when compared to that of standards prepared in water or pure solvents. In particular the presence of high salt contents in the sample solution can result in deformation of the peak shape and/or shifts in migration time. Therefore the use of high ionic strength buffers is advocated (19) for use with direct serum analysis by CE. Solvent extraction and use of solid phase extraction (SPE) procedures also serves to remove high salt content from samples as the drug is retained on the SPE column whilst the salt is washed through. The drug is collected from the column by a short flush with an appropriate organic solvent. The use of SPE procedures is most suitable as SPE-based sample clean-up procedures can be readily automated for high sample throughput activities. The other major complication is the presence of high concentrations of proteins in plasma. Deproteinisation of the sample solution by the addition of acetonitrile effectively removes protein interferences and also beneficially (48) promotes "stacking effects" leading to improved sensitivity.

10.5 Sensitivity Enhancement

Typically the sensitivity requirements in drug bioassay monitoring activities is extremely low and approaches have been developed to improve the sensitivity of CE methods to meet these requirements. These approaches have include the use of highly sensitive/selective detectors such as laser-induced fluorescence (LIF) detectors which may give 3 orders of magnitude better detection limits than use of UV absorbance detectors.

The sample injection procedure can also be optimised to give improved sensitivity. For example large volume sample stacking capillary electrophoresis (LVSS-CE) allows (55) injection of much larger sample volumes than normally permitted. The LVSS-CE procedure involves injection of a large sample volume and then the sample solvent is removed from the capillary by establishing a reversed EOF direction by the preanalysis addition of a cationic surfactant such as cetyltrimethylammonium bromide to the sample diluent. When the sample solvent is effectively removed from the capillary the voltage polarity is switched and the analyte peak migrates through the detector. Use of LVSS-CE can improve sensitivity several fold compared to normal pressure injections.

Electrokinetic injection of sample ions can also be used to generate improved sensitivity. The use of injections of pure solvent into the capillary end prior to application of the voltage used for electrokinetic injection can amplify the electric field at the end of the capillary and increase the sample loading for electrokinetic injections. This procedure is known as field amplified sample injection and has been used (56) to determine amiodarone and desethylamiodarone in 20 µL serum samples with LOD values of 80 nmol/L. The method could be optimised to give sample volume requirements of as little as 2 µl and a LOD value of <1 nmol/L of for amiodarone. Electrokinetic injections from samples dissolved in pure organic solvents such as acetonitrile is also another means of increasing sensitivity as the viscosity of these solvents is lower than water and therefore the sample ions are more mobile and more ions enter the capillary per second.

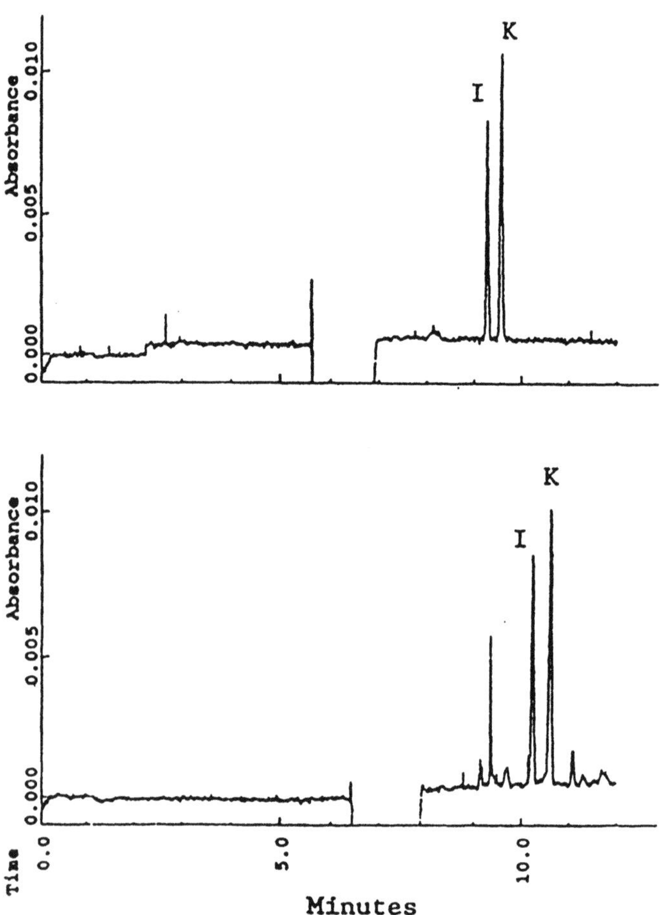

Figure 10.3   Separation of ketoprofen standard and serum sample containing ketoprofen at 10 mg/L. Operating conditions : 250 mM boric acid pH adjusted to pH 8.9 with 2 M NaOH containing 1 % ACN and 1 g/L β-cyclodextrin, 254 nm detection, 50 cm × 50 μm capillary. Peak identity : K = ketoprofen, I = isobutyl methyxanthine (internal standard). Reproduced with permission from reference 48.

The low viscosity of acetonitrile sample solutions promotes stacking effects and allows longer injection times to be used to improve sensitivity. For example Figure 10.3 shows (48) a 99 second injection of standard dissolved in 66 % ACN 34 % water containing 0.7 % NaCl and a 99 second injection of a acetonitrile deproteinised serum sample containing 10 mg/L.

The use of solid phase extraction and solvent-solvent extraction can be a highly effective means of improving method sensitivity. This two approaches act as concentration enhancers as several mls of biofluid may be extracted into a small volume of organic liquid. This organic solvent solution can be further evaporated and the sample reconstituted in a few μl ready for injection. This can produce increases of sensitivity in the order of 100–1000 fold.

Other approaches to increasing sensitivity include use of wider bore capillaries or capillaries with modified geometry's. These modified capillaries include bubble cell capillaries, z-cell capillary and the flow-cell recently introduced by Hewlett-Packard. Sensitivity is also often improved by performing the analysis using low UV wavelengths such as 200 nm (17) where many compounds have increased UV activity.

10.6 Quantitative Precision

Invariably internal standards are used in bioassays performed by CE as the injection volumes can be variable and the use of an internal standard compensates effectively for injection volume changes. Section 2.5 of Chapter 2 which discusses these aspects in greater detail. Many factors affect the volumes injected in a pressure injection in CE including the viscosity and surface tension. For example (21) internal standards were needed in the analysis of thiopental in serum as the standards and sample solutions had considerably different viscosity values.

Peak areas are usually used for quantitation purposes in bioassays as they offer greater linearity ranges. For example considerable improvements in detector linearity range were noted for suramin determinations in serum using (57) peak areas instead of peak heights. Use of an internal standard also assisting obtaining better detector linearity coefficients as it reduces (58) the errors involved in measuring each point along the linearity calibration line.

10.7 Applications

The performance of CE methods in clinical assays has been assessed by many workers and general comments indicate that CE is not so sensitive as HPLC but has benefits in terms of simplicity and possible sample pre-treatment reductions. Validation parameters such as linearity, recoveries and precision show acceptable performance (10, 24, 41, 42). Cross-validation of CE results with other techniques such as HPLC (24) and immunoassays (59), show that CE is capable of generating accurate results. Table 10.2 compares levels of creatinine and uric acid as determined (5) by both an enzymatic method and by MECC.

MECC using a phosphate buffer, pH 8.00, containing 165 mM SDS and CZE using a pH 9.2 borate buffer have been applied (31) to determine therapeutic levels of cefotaxime (C) and its deacetyl metabolite (DA) in human plasma. Plasma samples required deproteinization with acetonitrile prior to CZE analysis. Direct injection of the plasma samples was possible with MECC. Theobromine as internal standard in MECC. Both methods gave satisfactory interday precision with respect to migration times (RSD < 1 %) and gave linear responses over the concentration ranges investigated (5–100 mg L^{-1} C and 5–20 mg L^{-1} DA). Interday precision (n = 4 days) was 1.49 % for MECC when theobromine was used as internal standard. Poorer precision data was obtained for the CZE method or when MECC was performed without use of the internal standard. Detection limits (S / N = 3) of 2 mg L^{-1} (CZE) and 1 mg L^{-1} in plasma (MECC) were reported.

The number of biomedical applications of CE continues to expand rapidly, some particular examples are discussed covering both applications and methodology approaches. Theo-

Table 10.2 Comparison of levels as determined by MECC and enzymatic methods

Plasma sample	Creatinine (µg/ml)		Uric acid (µg/ml)	
	Enzymatic	MECC	Enzymatic	MECC
1	5	5	28	32
2	7	7	50	51
3	4	4	38	n. r.
4	6	5	27	32
5	7	7	66	66
6	1	n. r.	12	11
7	8	8	56	58
8	4	4	60	60

Reprinted with permission from reference 5.

n. r. = no result obtained

phylline and metabolites have been determined in urine (37) using solid phase extraction pretreatment, a variety of ephedrine alkaloids were determined in urine with direct sample injection (60), and levels of free and total 7-hydroxycoumarin were determined (32) in both urine and serum samples.

Naproxen was determined in serum extracts following extraction (15) with an hexane/ether mixture and drying down. Separation was achieved in 10 mM TRIS – 10 mM Tricine pH 8 using a 50 µm × 70 (50) cm capillary. Using 254 nm the detection limit was 100 fmol and with He-Cd laser (native fluorescence ex 325 nm, em 375 nm) LOD was 3 fmol. With UV detection the naproxen levels by CE in 10 subjects compared very well to an HPLC method. CE has been used (48) to determine serum levels of non-steroidal anti-inflammatory, ketoprofen. The 10 minute analysis time method has been applied to emergency toxicology and therapeutic monitoring. Serum was diluted with acetonitrile (containing an internal standard) to deproteinized serum proteins and to induce sample stacking. The assay was linear between 1–10 mg/l without any interferences. The method compared well to an HPLC assay. The HPLC method afforded a better detection limit, but the CE was less expensive to operate. This method demonstrates that capillary electrophoresis is a simple and effective method for determination of ketoprofen as well as other drugs in human serum at levels close to 1 mg/l.

Serum levels of lamotrigine, a new antiepileptic drug have been determined (50). Samples were deproteinized with acetonitrile containing an internal standard, acidified with dilute acetic acid and injected into the capillary. The method generated a lamotrigine peak at 3.5 min free from interferences. Linearity was shown between 0.5–10 mg/l and a correlation of r = 0.97 was obtained for 35 sample results obtained by both CE and HPLC.

Levels of cicletanine in human plasma were determined (6) using a MECC method of 100 mM borate containing 25 mM SDS and 10 % acetonitrile. Samples were solvent extracted and analysed by both MECC and HPLC. Table 10.3 shows the recovery data obtained for spiked plasma samples for both MECC and HPLC. An internal standard was used in both methods and gave good injection precision.

Table 10.3 Assay results for plasma samples spiked with cicletanine obtained by MECC and HPLC.

Spiking concentration (ng/ml)	HPLC result (ng/ml)	MECC result (ng/ml)
20	19.97	19.88
50	50.13	50.11
100	100.02	102.50
500	512.34	485.51
1000	967.48	993.23

Reproduced with permission from reference 6

Piracetam has been determined in human plasma using (17) a borax buffer containing alpha-cyclodextrin with detection at 200 nm. The detection limit of the authentic samples was 1 µg/ml. The calibration curve was linear over a range of 4 to 24 mu g/ml ($r = 0.997$). Inter-assay R.S.D. was below 9.3 %. The method was suitable for use in clinical and bioavailability studies. MECC has been used to determine (13) levels of heparinoid mimetics which are spaced persulfated carbohydrates designed to increase the success rate of angioplasty and bypass surgery. CE was selected as these compounds are highly charged and only minute volumes of rat plasma were available. Plasma was directly injected following a 1:1 dilution with 100 mM SDS. A LOD of 3 mu g/ml was sufficient to allow analysis of more than 350 plasma samples. Serum levels of pentobarbital assay have been determined (16) using 300 mM boric acid pH 8.5. 3-isobutyl-1-methyl-xanthine was used an internal standard to give a 4 min assay

Levels of antipyrine were determined (1) in saliva by both HPLC and MECC. The MECC method employed a short capillary and high voltage and gave a sub 4 minute analysis time. Recoveries of approximately 94-97% were obtained for spiked samples. Figure 10.4 shows the excellent correlation obtained between HPLC and MECC results for 75 samples.

In certain cases when detergent is present in the separation buffer it has been shown (61) that between-run capillary rinsing/regeneration procedures can be avoided. The reproducibility for migration time, peak area, and peak height with separation in a borate/phosphate buffer containing 75 mM sodium cholate was found to be acceptable in the absence of between-run capillary cleansing. Continuous-sequential injection of hypoglycemic drug standards over the course of 39 consecutive runs without between-run capillary regeneration showed acceptable reproducibility. In excess of 100 continuous-sequential injections were performed for hypoglycemic drug standards, urinary estrogen and the human urine with no significant effects on electroosmotic flow or reproducibility.

In complex separations it may be difficult to positively identify peaks of interest and confirm peak purity. The use of chemometric software to analyse diode array data and (62) has been used to determine amphetamine from common interferences in human urine. Principle Component Analysis and Iterative Target Transform Factor Analysis were used to inspect each electropherogram for spectral homogeneity of the peaks and to deconvolute comigrations and to confirm the assay results.

CE has also been applied to monitor the metabolism of administered drugs through analysis of a range of biofluids. For example a CE method has been used (63) to determine the

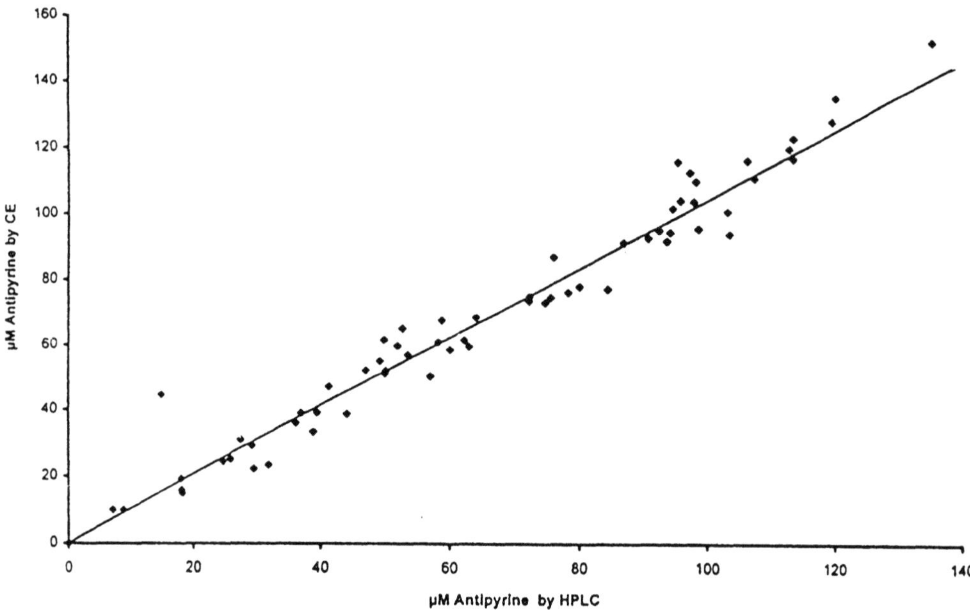

Figure 10.4 Comparison of saliva antipyrine levels as determined by MECC and HPLC. Reproduced with permission from reference 1.

concentration-time metabolism profiles of dexamethasone by the analysis of tears following drug application. Indoprofen was used as an internal standard. The metabolism of caffeine has been followed (23) in 2–4 h urine following ingestion of 140 mg of caffeine in coffee. A range of caffeine metabolites were measured using a MECC method with a 70 mM SDS in 16 mM sodium tetraborate electrolyte.

Capillary electrophoresis has been used (64) for the determination of serum iohexol levels in order to monitor glomerular filtration rate (GFR). Capillary electrophoresis was considered to be a simple, rapid method that can be used to calculate GFR and provides results at least as accurate as those obtained by HPLC and x-ray fluorescence

Diseases such as adenolysuccinase deficiency, 5-oxoprolinuria, propionic acidemia and disorders having erotic acid as diagnostic metabolite (e.g. the HHH-syndrome) can be (65) diagnosed by CE. The CE method could resolve over 50 metabolites within 15 minutes with direct injection of urine samples. Confirmation of metabolite peaks was established by reference to their diode-array spectra and spectra obtained from standards.

Capillary electrophoresis equipped with a diode-array detector and GC-MS have been used to determine diagnostic metabolites occurring in urine of patients with various metabolic disorders. Direct injection of urine samples was possible (66) with pretreatment whilst GC-MS samples required extraction and derivatisation. Identification of abnormal metabolites was based on migration times and characteristic diode-array spectra, or mass spectral library search when GC-MS was used. CE readily diagnosed glyceric aciduria and the secondary metabolite in lysinuric protein intolerance, erotic acid. Methylmalonic aciduria required pressure elution in addition to high voltage to accomplish diagnosis.

Familial defective apolipoprotein B-100 (FDB) is a dominantly inherited disorder. Capillary gel electrophoresis in combination with molecular biology techniques has been used (67) to allow diagnosis of FDB. Mutation screening for FDB is performed (67) by an allele-specific amplification followed by capillary gel electrophoresis (CGE). For the combined polymerase chain reaction (PCR)-CGE method, a total analysis time of only 3 h is needed. In a pilot study 4 of 43 hypercholesterolemic patients were reported to have the predominant apoB 3500 codon mutation.

10.8 Chiral Clinical Applications

The use of CE for chiral separations is attractive as the methods can be simple to develop and operate. Often the chiral resolution is obtained by the addition of a chiral selector such as cyclodextrin into the buffer. Chapter 4 covers a range of applications and method development approaches. The use of chiral CE to monitor enantioselective processes such as metabolism studies offers considerable advantages. The columns used in HPLC can become fouled by endogenous compounds in biofluids. Therefore extensive sample pretreatment may be required prior to chiral HPLC analysis. This problem does not occur in chiral CE as the chiral separation is obtained through interactions with the electrolyte additives.

A number of studies have shown CE methods to be of useful for the resolution of chiral compounds in biofluids. The subject has been recently reviewed (68). A number of examples of enantiomeric methods include oxprenolol and its metabolites (35), thalidomide and its neutral metabolites (38), zopiclone and its metabolites in urine (40), diastereoisomers of l-buthionine-(r,s)-sulfoximine in human plasma (69), local anaesthetic drugs (70), 3,4-methylenedioxymethamphetamine and two of its metabolites in human urine (71), and ondansetron in human serum (72).

The simultaneous CE chiral separation of 3, 4-methylenedioxymethamphetamine (MDMA or Ecstasy) and its two metabolites 4-hydroxy-3-methoxymethamphetamine (HMMA) and 3, 4-methylenedioxyamphetamine (MDA) in human urine was obtained (71) using capillary zone electrophoresis with a pH 2.5 phosphate buffer containing 30 mM (2-hydroxypropyl)-beta-cyclodextrin. The method involved enzymatic hydrolysis of conjugated HMMA (major urinary metabolite) and solid phase extraction with improved sensitivity obtained at 195 nm. The detection limit was in the 20–50 ng/mL range. Higher urinary amounts of R-(-)-MDMA were excreted compared to S-(+)-MDMA. Within 72 h after drug administration one patient excreted 42.28 and 10.16 % of the racemic MDMA dose (1.5 mg/kg body weight) as R-(-) and S-(+)-MDMA enantiomers, respectively.

S-(+)- and R-(-)-ondansetron enantiomers were determined (72) in human serum using 100 mM phosphate buffer (pH 2.5) containing 15 mM heptakis-(2, 6-di-O-methyl)-beta-cyclodextrin (DM-beta-CD). Figure 10.5 shows the analysis of a serum blank and a serum sample spiked with S-(+)- and R-(-)-ondansetron and procainamide which was used as an internal standard. Recoveries of 85 % for both S-(+)- and R-(-)-ondansetron from the cyanopropyl solid phase cartridge were reported. The limit of quantitation was 15 ng/ml which were similar to those obtained in a HPLC procedure. The CE method was considered a useful alternative to existing chiral high-performance liquid chromatographic methods.

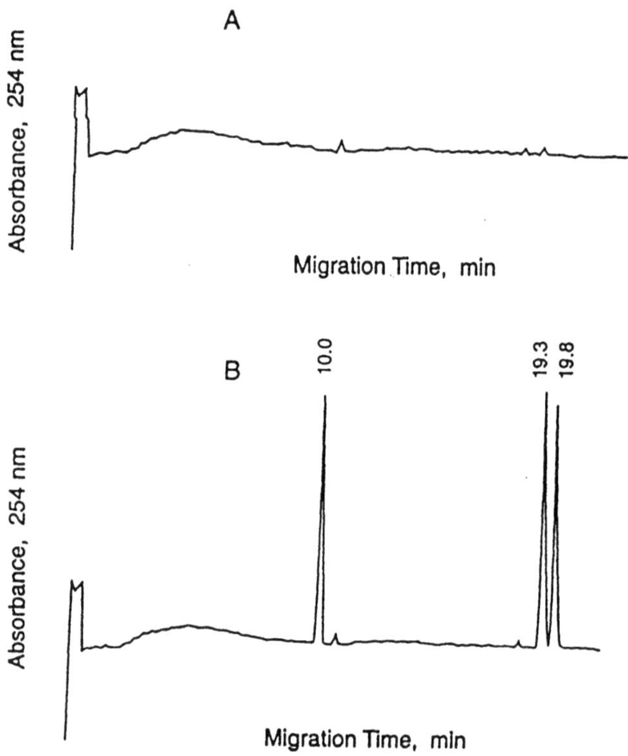

Figure 10.5 Analysis of a serum blank and a serum sample spiked with S-(+)- and R-(-)-ondansetron and internal standard procainamide. Separation conditions : 100 mM phosphate buffer (pH 2.5) containing 15 mM heptakis-(2, 6-di-O-methyl)-beta-cyclodextrin (DM-beta-CD), 50 cm × 50 µm, 20 kV, detection at 254 nm. Reproduced with permission from reference 72.

10.9 Non-Aqueous CE and CEC Clinical Applications

Both Capillary electrochromatography (CEC) and non-aqueous CE (NACE) are rapidly developing areas and are covered in greater details in Chapters 12 and 13 respectively.

Gradient elution (CEC) was used with UV detection at 240 nm (73) to determine corticosteroids (adrenosterone, hydrocortisone, dexamethasone, fluocortolone) in extracts of equine urine and plasma. Urine samples were first purified using solid phase extraction, Two purification steps were necessary to prevent contamination of the CEC column, Plasma was purified using automated dialysis. Capillaries were packed with 3 µm Apex ODS, a mobile phase of ammonium acetate (5 mM) in water/acetonitrile was used. Acetonitrile levels were varied from 9 to 80 % (v/v) using a gradient formation device. Long periods of unattended operation were possible although a broadening of peaks was observed after analysis of 200 urine extracts.

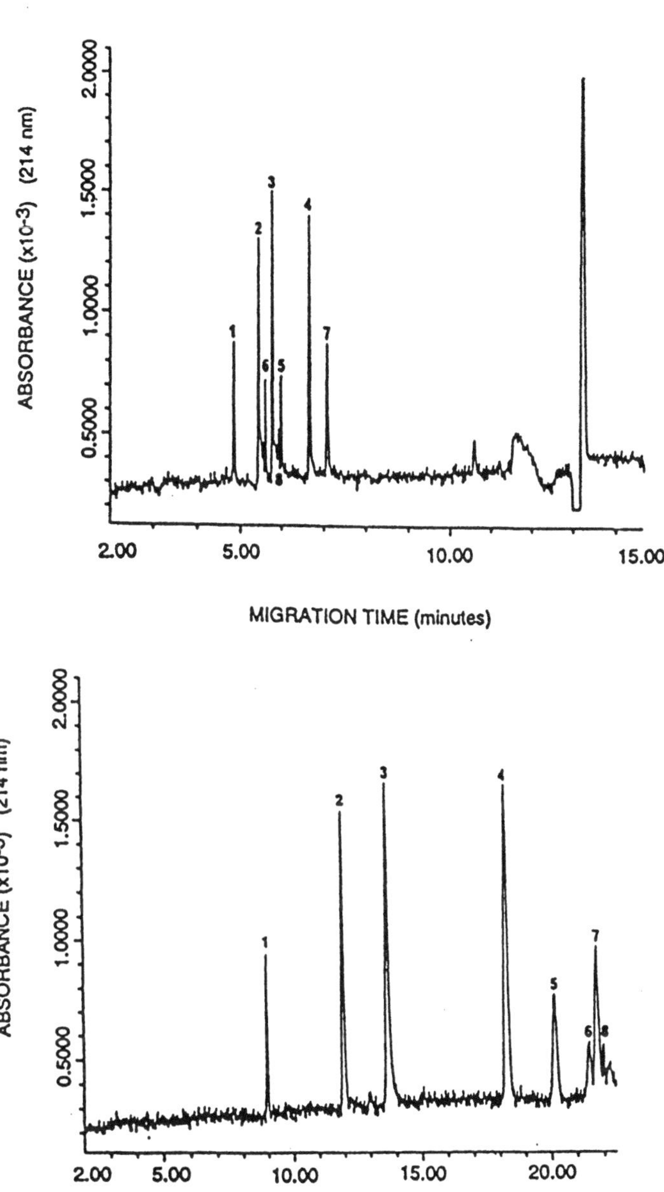

Figure 10.6 Separation of mifentidine and impurities/metabolites using electrolytes containing different % v/v levels of methanol. Operating conditions: 57 cm (50 cm) × 75 mm, 214 nm, (A) 20 mM NH₄OAc in 30 % aqueous methanol containing 1 % acetic acid (B) 20 mM NH₄OAc in 100 % methanol containing 1% acetic acid. Peak 1 mifentidine, peaks 2–8 mifentidine related compounds. Reproduced with permission from reference 76.

A non-aqueous buffer system of 1:1 methanol-acetonitrile containing 50 mM ammonium acetate and 1 % acetic acid has been used (74) to determine levels of the antiestrogen compound Tamoxifen (TAM) and metabolites in rodent serum. 4-Dimethylaminopyridine was used as internal standard with UV detection at 214 nm. Tamoxifen was successfully separated from N-desmethyltamoxifen (DMT), and 4-hydroxytamoxifen (4-HT). TAM and DMT were determined in serum of female rats 4 h following a single oral dose of 120 mg/kg. Results from another group of workers on the NACE separation of Tamoxifen (TAM) and metabolites showed (75) that use of methanolic electrolyte provided better resolution and detection sensitivity compared to aqueous systems. The CE-MS signal was increased using methanolic electrolytes due to the ease of methanol solvent evaporation giving an improved yield of protonated analytes compared to aqueous electrolytes. The NACE method was used to monitor metabolites formed after incubation of tamoxifen with mouse hepatocytes

Non-aqueous CE separation conditions have been used in (76) the *in vitro* metabolism of the H-2-antagonist mifentidine with MS and UV detection. Figure 10.6 shows that the separation of mifentidine and its metabolites could be manipulated by varying the methanol in the solution used to prepare the electrolyte. Improved resolution was obtained with 100% methanol at the expense of a longer analysis time. The results from the studies allowed a metabolic pathway scheme to be postulated.

References

1. Perrett D and Ross GA, Rapid determination of drugs in biofluids by capillary electrophoresis – Measurement of antipyrine in saliva for pharmacokinetic studies, *J. Chromatogr. A* 700 (**1995**) 179-186.

2. Nishi H, Fukayama T, and Matsuo M, Separation and determination of aspoxicillin in human plasma by micellar electrokinetic chromatography with direct sample injection, *J. Chromatogr.*, 515 (**1990**) 245-255.

3. Nakagawa T, Oda Y, Shibukawa A, and Tanaka H, Separation and determination of cefpiramide in plasma by electrokinetic chromatography with micellar solution and an open tubular-fused silica capillary, *Chem. Pharm. Bull.*, 36 (**1988**) 1622-1625.

4. Choi O-K and Song Y-S, Determination of cefuroxim levels in human serum by micellar electrokinetic capillary chromatography with direct sample injection, *J. Pharm. Biomed. Anal.*, 15 (**1997**) 1265-1270.

5. Miyake M, Shiukawa A, and Nakagawa T, Simultaneous determination of creatinine and uric acid in human plasma and urine by micellar electrokinetic chromatography, *J. High Resol. Chromatogr.*, 14 (**1991**) 180-185.

6. Prunonosa J, Obach R, Diezcascon A, and Gouesclou L, Comparison of high-performance liquid chromatography and high-performance capillary electrophoresis for the determination of cicletanine in plasma, *J. Chromatogr.*, 581 (**1992**) 219-226.

7. Luksa J and Josic D, Determination of cimetidine in human plasma by free capillary zone electrophoresis, *J. Chromatogr. B*, 667 (**1995**) 321-327.

8. Soini H, Tsuda T, and Novotny M V, Electrochromatographic solid-phase extraction for determination of cimetidine in serum by micellar electrokinetic capillary chromatography, *J. Chromatogr.*, 559 (**1991**) 547-558.

9. Wang C J, Tian Z, Watkins R, Cook J, and Lin C, Determination of a cyclic guanine monophosphate phosphodiesterase inhibitor (SCH 51866) in rat serum using capillary zone electrophoresis, *J. Chromatogr. B*, 669 (**1995**) 141-148.

10. Lloyd D K, Cypess A M, and Wainer I W, Determination of cytosine-ß-D-arabinoside in plasma using capillary electrophoresis, *J. Chromatogr.*, 568 (**1991**) 117-124.

11. Schmutz A and Thormann W, Determination of phenobarbital, ethosuximide, and primidone in human serum by micellar electrokinetic capillary chromatography with direct sample injection, *Therapeutic Drug Monitoring*, 15 (**1993**) 310-316.

12. Baillet A, Pianetti G A, Taverna M, Mahuzier G, and Baylocq-Ferrier D, Fosfomycin determination in serum by capillary zone electrophoresis with indirect ultraviolet detection, *J. Chromatogr.*, 616 (**1993**) 311-316.

13. Mayer S and Schleimer M, Quantitative determination of heparinoid mimetics in human and rat plasma by micellar electrokinetic chromatography, *J. Chromatogr. A*, 730 (**1996**) 297-303.

14. Shihabi Z K and Constantinescu M S, Iohexol in serum determined by capillary electrophoresis, *Clin. Chem.*, 38 (**1992**) 2117-2120.

15. Soini H, Novotny M V, and Riekkola M L Determination of naproxen in serum by capillary electrophoresis with ultraviolet absorbance and laser-induced fluorescence detection, *J. Microcol. Sep.*, 4 (**1992**) 313-318.

16. Shihabi Z K, Serum pentobarbital assay by capillary electrophoresis, *J. Liq. Chromatogr.*, 16 (**1993**) 2059-2068.

17. Lamparczyk H, Kowalski P, Rajzer D, and Nowakowska J, Determination of piracetam in human plasma by capillary electrophoresis, *J. Chromatogr. B*, 692 (**1997**) 483-487.

18. Thormann W, Minger A, Molteni S, Caslavska J, and Gebauer P, Determination of substituted purines in body fluids by micellar electrokinetic capillary chromatography with direct sample injection, *J. Chromatogr.*, 593 (**1992**) 275-288.

19. Garcia L L and Shihabi Z K, Sample matrix effects in capillary electrophoresis. I. Basic considerations, *J. Chromatogr.*, 652 (**1993**) 465-469.

20. Linhares MC and Kissinger PT, Pharmacokinetic studies using micellar electrokinetic capillary chromatography with in-vivo capillary ultrafiltration probes, *J. Chromatogr. B*, 615 (**1993**) 327-333.

21. Meier P and Thormann W, Determination of thiopental in human serum and plasma by high-performance capillary electrophoresis – micellar electrokinetic chromatography, *J. Chromatogr.*, 559 (**1991**) 505-513.

22. Lee K J, Lee J J, and Moon D C, Determination of tricyclic antidepressants in human plasma by micellar electrokinetic capillary chromatography, *J. Chromatogr. B*, 616 (**1993**) 135-143.

23. Guo R and Thormann W, Acetylator phenotyping via analysis of 4 caffeine metabolites in human urine by micellar electrokinetic capillary chromatography with multiwavelength detection, *Electrophoresis*, 14 (**1993**) 547-553.

24. Li S, Fried K, Wainer I W, and Lloyd D K, Determination of dextromethorphan and dextrorphan in urine by capillary zone electrophoresis – Application to the determination of debrisoquin-oxidation metabolic phenotype, *Chromatographia*, 35 (**1993**) 216-222.

25. Hufschmid E, Theurillat R, Martin U, and Thormann W, Exploration of the metabolism of dihydrocodeine via determination of its metabolites in human urine using micellar electrokinetic capillary chromatography, *J.Chromatogr.B*, 668 (**1995**) 159-170.

26. Bory C, Chantin C, and Boulieu R, Comparison of capillary electrophoretic and liquid chromatographic determination of hypoxanthine and xanthine for the diagnosis of xanthinuria, *J. Chromatogr. A,* 730 (**1996**) 329-331.

27. Schneede J and Ueland PM, Application of capillary electrophoresis with laser-induced fluorescence detection for routine determination of methymalonic acid in human serum, *Anal. Chem.,* 67 (**1995**) 812-819.

28. Abubaker M A, Petersen J R, and Bissell, M G, Micellar electrokinetic capillary chromatographic separation of steroids in urine by trioctylphosphine oxide and cationic surfactant, *J. Chromatogr. B,* 674 (**1995**) 31-38.

29. Tanaka Y and Thormann W, Capillary electrophoretic determination of S-carboxymethyl-L-cysteine and its major metabolites in human urine: Feasibility investigation using on-column detection of non-derivatized solutes in capillaries with minimal electroosmosis, *Electrophoresis,* 11 (**1990**) 760-764.

30. Honda S, Taga A, Kakehi K, Koda S, and Okamoto Y, Determination of cefixime and its metabolites by high-performance capillary electrophoresis, *J.Chromatogr.,* 590 (**1992**) 364-368.

31. Penalvo G C, Kelly M, Maillols H, and Fabre H, Evaluation of capillary zone electrophoresis and micellar electrokinetic capillary chromatography with direct injection of plasma for the determination of cefotaxime and its metabolite, *Anal. Chem.,* 69 (**1997**) 1364-1369.

32. Bogan D P, Deasy B, O'Kennedy R, Smyth M R, and Fuhr U, Determination of free and total 7-hydroxycoumarin in urine and serum by capillary electrophoresis, *J. Chromatogr. B,* 663 (**1995**) 371-378.

33. Soini H, Riekkola M L, and Novotny M V, Mixed polymer networks in the direct analysis of pharmaceuticals in urine by capillary electrophoresis, *J. Chromatogr. A,* 680 (**1994**) 623-634.

34. Tomlinson A J, Benson L M, Landers J P, Scanlan G F, Fang J, Gorrod J W, and Naylor S, Investigation of the metabolism of the neuroleptic drug haloperidol by capillary electrophoresis, *J. Chromatogr.,* 652 (**1993**) 417-426.

35. Li F, Cooper S F, and Mikkelsen S R, Enantioselective determination of oxprenolol and its metabolites in human urine by cyclodextrin-modified capillary zone electrophoresis, *J. Chromatogr. B,* 674 (**1995**) 277-285.

36. Tomlinson A J, Benson L M, Gorrod J W, and Naylor S, Investigation of the *in-vitro* metabolism of the H-2-antagonist mifentidine by on-line capillary electrophoresis mass spectrometry using non-aqueous separation conditions, *J. Chromatogr. B,* 657 (**1994**) 373-381.

37. Zhang ZY, Fasco MJ, and Kaminsky LS, Determination of theophylline and its metabolites in rat liver microsomes and human urine by capillary electrophoresis, *J. Chromatogr. B,* 665 (**1995**) 201-208.

38. Weinz C and Blaschke G, Investigation of the *in-vitro* biotransformation and simultaneous enantioselective separation of thalidomide and its neutral metabolites by capillary electrophoresis, *J. Chromatogr. B,* 674 (**1995**) 287-292.

39. Hempel G and Blaschke G, Direct determination of zolpidem and its main metabolites in urine using capillary electrophoresis with laser-induced fluorescence detection, *J. Chromatogr. B,* 675 (**1996**) 131-137.

40. Hempel G and Blaschke G, Enantioselective determination of zopiclone and its metabolites in urine by capillary electrophoresis, *J. Chromatogr. B,* 675 (**1996**) 139-144.

41. Thormann W, Molteni S, Caslavska J, and Schmutz, Clinical and forensic applications of capillary electrophoresis, *Electrophoresis,* 15 (**1994**) 3-12.

42. Deyl Z, Tagliaro F, and Miksik I, Biomedical applications of capillary electrophoresis, *J. Chromatogr.,* 656 (**1994**) 3-27.

43. Guzman N A, Moschera J, Iqbal K, and Malick A N, Effect of buffer constituents on the determination of therapeutic proteins by capillary electrophoresis, *J. Chromatogr.*, 608 (**1992**) 197-204.

44. Iversen HH, Celsing F, Leone AM, Gustafsson LE, and Wiklund NP, Nerve-induced release of nitric oxide in the rabbit gastrointestinal tract as measured by in vivo microdialysis, *British J. of Pharmacology*, 120 (**1997**) 702-706.

45. Tagliaro F, Moretto S, Valentini R, Gambaro G, Antonioli C, Moffa M, and Tato L, Capillary zone electrophoresis determination of phenylalanine in serum - A rapid, inexpensive and simple method for the diagnosis of phenylketonuria, *Electrophoresis*, 15 (**1994**) 94-97.

46. Thormann W, Lienhard S, and Wernly P, Strategies for the monitoring of drugs in body fluids by micellar electrokinetic capillary chromatography, *J. Chromatogr.*, 636 (**1993**) 137-148.

47. Shihabi Z K, Serum phenobarbital assay by capillary electrophoresis, *J. Liq. Chromatogr.*, 16 (**1993**) 2059-2068.

48. Friedberg M and Shihabi ZK, Ketoprofen analysis in serum by capillary electrophoresis, *J. Chromatogr. B*, 695 (**1997**) 193-198.

49. Gu X, Meleka-Boules M, and Chen C-L, Micellar electrokinetic capillary chromatography with immunoaffinity chromatography for identification and determination of dexamethasone and flumethasone in equine urine, *J. Cap. Elect.*, 3 (**1996**) 43-49.

50. Shihabi ZK and Oles KS, Serum lamotrigine analysis by capillary electrophoresis, *J. Chromatogr. B*, 683 (**1996**) 119-123.

51. Smith S C, Strasters J K, and Khaledi M G, Influence of operating parameters on reproducibility in capillary electrophoresis, *J. Chromatogr.*, 559 (**1991**) 57-68.

52. Lloyd DK and Watzig H, Sodium dodecyl sulphate is an effective between-run rinse for capillary electrophoresis of samples in biological matrices, *J. Chromatogr. B*, 663, (**1995**) 400-405.

53. Kunkel A, Gunter S, and Watzig H, Quantitation of acetaminophen and salicylic acid in plasma using capillary electrophoresis without sample pretreatment – improvement of precision, *J. Chromatogr. A*, 768 (**1997**) 125-133.

54. Zhang C-X and Thormann W, Determination of drug levels in human serum by micellar electrokinetic capillary chromatography with direct sample injection using different quantitation strategies, *J. Cap. Elect.*, 1 (**1994**) 208-218.

55. Mcgrath G and Smyth WF, Large-volume sample stacking of selected drugs of forensic significance by capillary electrophoresis, *J. Chromatogr. B*, 681 (**1996**) 125-131.

56. Zhang CX, Aebi Y, and Thormann W, Microassay of amiodarone and desethylamiodarone in serum by capillary electrophoresis with head-column field-amplified sample stacking, *Clin. Chem.*, 42 (**1996**) 1805-1811.

57. Garcia L L and Shihabi Z K, Suramin determination by capillary electrophoresis, *J. Liq. Chromatogr.*, 16 (**1993**) 2049-2057.

58. Altria K D and J Bestford, Main component assay of pharmaceuticals by capillary electrophoresis - considerations regarding precision, accuracy and linearity data, *J.Cap. Elect.*, 3 (**1996**) 13-23.

59. Caslavska J, Lienhard S, and Thormann W, Comparative use of three electrokinetic capillary methods for the determination of drugs in body fluids. Prospects for rapid determination of intoxications, *J. Chromatogr.*, 638 (**1993**) 335-342.

60. Chicharro M, Zapardiel A, Bermejo E, Perex-Lopex JA, and Hernandez L, Direct determination of ephedrine alkaloids and epinephrine in human urine by capillary zone electrophoresis, *J. Liq. Chromatogr.*, 18 (**1995**) 1363-1381.

61. Roche M E, Oda R P, Machacek D, Lawson G M, and Landers J P, Enhanced throughput with capillary electrophoresis via continuous-sequential sample injection, *Anal. Chem.*, 69 (**1997**) 99-104.

62. Lilley K A and Wheat T E, Drug identification in biological matrices using capillary electrophoresis and chemometric software, *J. Chromatogr. B*, 683 (**1996**) 67-76.

63. Baeyens V, Varesio E, Veuthey J L, and Gurny R, Determination of dexamethasone in tears by capillary electrophoresis, J. *Chromatogr. B*, 692 (**1997**) 222-226.

64. Rocco M V, Buckalew V M, Moore L C, and Shihabi Z K, Capillary electrophoresis for the determination of glomerular filtration rate using nonradioactive iohexol, *American J. of Kidney Diseases*, 28 (**1996**) 173-177.

65. Jellum E, Dollekamp H, and Blessum C, Capillary electrophoresis for clinical problem solving – analysis of urinary diagnostic metabolites and serum proteins, *J. Chromatogr. B*, 683 (**1996**) 55-65.

66. Jellum E, Dollekamp H, Brunsvig A, and Gislefoss R Diagnostic applications of chromatography and capillary electrophoresis, *J. Chromatogr. B*, 689 (**1997**) 155-164.

67. Lehmann R, Koch M, Pfohl M, Voelter W, Haring H U, and Liebich H M, Screening and identification of familial defective apolipoprotein B-100 in clinical samples by capillary gel electrophoresis, *J. Chromatogr. A*, 744 (**1996**) 187-194.

68. Bojarski J and Bboulenein HY, Application of capillary electrophoresis for the analysis of chiral drugs in biological fluids, *Electrophoresis*, 18 (**1997**) 965-969.

69. Sandor V, Flarakos T, Batist G, Wainer I W, and Lloyd D K Quantitation of the diastereoisomers of l-buthionine-(r,s)-sulfoximine in human plasma, : A validated assay by capillary electrophoresis, *J. Chromatogr. B*, 673 (**1995**) 123-131.

70. Amini A and Paulsensorman U, Enantioseparation of local anaesthetic drugs by capillary zone electrophoresis with cyclodextrins as chiral selectors using a partial filling technique, *Electrophoresis*, 18 (**1997**) 1019-1025.

71. Lanz M, Brenneisen R, and Thormann W, Enantioselective determination of 3,4-methylenedioxymethamphetamine and two of its metabolites in human urine by cyclodextrin-modified capillary zone electrophoresis, *Electrophoresis*, 18 (**1997**) 1035-1043.

72. Siluveru M and Stewart J T, Enantioselective determination of s-(+)- and r-(-)-ondansetron in human serum using derivatized cyclodextrin-modified capillary electrophoresis and solid-phase extraction, *J.Chromatogr. B*, 691 (**1997**) 217-222.

73. Taylor M R, Teale P, Westwood S A, and Perrett D, Analysis of corticosteroids in biofluids by capillary electrochromatography with gradient elution, *Anal.Chem.*, 69 (**1997**) 2554-2558.

74. Sanders J M, Burka L T, Shelby M D, Newbold R R, and Cunningham M L, Determination of tamoxifen and metabolites in serum by capillary electrophoresis using a nonaqueous buffer system, *J. Chromatogr. B*, 695 (**1997**) 181-185.

75. Lu W Z, Poon G K, Carmichael P L, and Cole R B, Analysis of tamoxifen and its metabolites by on-line capillary electrophoresis-electrospray ionization mass spectrometry employing nonaqueous media containing surfactants, *Anal. Chem.*, 68 (**1996**) 668-674.

76. Tomlinson A J, Benson L M, Gorrod J W, and Naylor S, Investigation of the in vitro metabolism of the H-2-antagonist mifentidine by on-line capillary electrophoresis mass spectrometry using non-aqueous separation conditions, *J Chromatogr B*, 657 (**1994**) 373-381.

11 Method Validation

11.1 Introduction

Method validation is a necessary exercise in which the method under assessment is proven to be "fit for purpose". The experiments follow a predetermined protocol and the results must meet predetermined criteria of performance. The exact experiments conducted are dependent on the type of testing being undertaken and the extent of the method's usage. Table 11.1 gives some examples of validation studies in CE, detailing the validation aspects studied in the individual studies (1–20). Each of these validation aspects are then discussed in this chapter.

The general format and procedure for the validation of a CE method is similar (21) to that of other analytical separative techniques. However there are a number of CE-specific validation issues which are illustrated in this chapter.

11.2 Specific method validation aspects

11.2.1 Specificity (selectivity)

This aspect demonstrates that the method is capable of resolving all the likely compounds of interest. The complexity of demonstrating selectivity is highly dependent on the type of testing being undertaken. For example if the method was to be used to quantitatively determine drug-related impurities then demonstrating appropriate selectivity would be an extensive study. It would be necessary to demonstrate resolution of all likely synthetic and degradative impurities from each other and from the main component. This would normally involve preparation of test mixture solutions containing the main component at the method concentration and the impurities present at their likely levels. For example if an 0.5 mg/ml drug solution is used and typical impurity levels are 0.1 % then impurity compounds should be added to the test solution at the 0.5 µg/ml level. If the method is to be employed as a stability indicating method then the method is often challenged with deliberately degraded solutions. These solutions may have been treated with acid, alkali, heated or stored in the presence of an intense light source. If the method is to be used to assess the purity of drug substance then the method is challenged with a range of likely side-products and intermediates.

For example both solid and solution samples of the metal chelator TMT-NCS were exposed to either heat (70°C) and light (900 ft candles) for several days. Solutions of the samples were then analysed by CE and HPLC to confirm that both methods had the required se-

Table 11.1 Examples of validated CE methods

Application	Ref.	1	2	3	4	5	6	7	8	9	Solute(s)
				Validation aspects included							
Impurities	1	√	√	√	√	√	√		√	√	Cefotaxime
	2	√	√	√	√	√	√	√	√	√	Ranitidine
	3	√	√	√	√	√	√	√	√	√	p-toluene sulphonic acid
	4	√	√	√	√	√	√		√	√	Metal chelator TMT-NCS
	5	√	√	√	√	√	√	√	√	√	Atenolol
Assay	6	√	√	√	√	√	√	√	√	√	Acidic drugs
	7	√	√	√	√	√	√	√	√	√	Mirtazarpine
	8	√	√	√	√	√	√	√	√		Hydrochlorothiazide and chlorothiazide
	9	√	√	√	√	√	√	√	√		Non-steroidal anti-inflammatories
	10	√	√	√	√	√	√	√	√	√	Losartan
	11	√	√	√	√	√	√	√	√	√	Basic drugs
Chiral	12	√	√	√	√	√	√	√	√		Ropivacaine
	13	√	√	√	√	√	√	√	√		LY231514
	14	√	√	√	√	√	√		√		BMS-180431-90
	15	√	√	√	√	√	√		√		Tryptophan
Drug counter-ions	16	√	√	√	√	√	√	√	√	√	Organic acids
	17	√	√	√	√	√	√		√		Acetate
	18	√	√	√	√	√	√	√	√		Potassium
Bioassay	19	√	√	√	√	√	√		√	√	Diltiazem
	20	√	√	√	√	√	√		√	√	L-buthionine-(R,S)-sulfoximine

Validation aspects

1. Specificity (selectivity)
2. Linearity
3. Sensitivity
4. Accuracy/recovery
5. Injection Repeatability

6. Method Repeatability
7. Method Robustness
8. Cross-Validation
9. Solution Stability

lectivity. Figure 11.1 shows separation of a stressed solution by both CE and HPLC, individually impurity peaks are denoted in Figure 11.1 which clearly shows the different selectivity obtained by the two methods.

This exercise is complicated in chiral separations where the impurity enantiomer should be separated from the main enantiomer as well as all other potential impurities. A number of studies (12, 13) have demonstrated that no synthetic or degradative impurities would interfere with the determination of either enantiomer at the required levels.

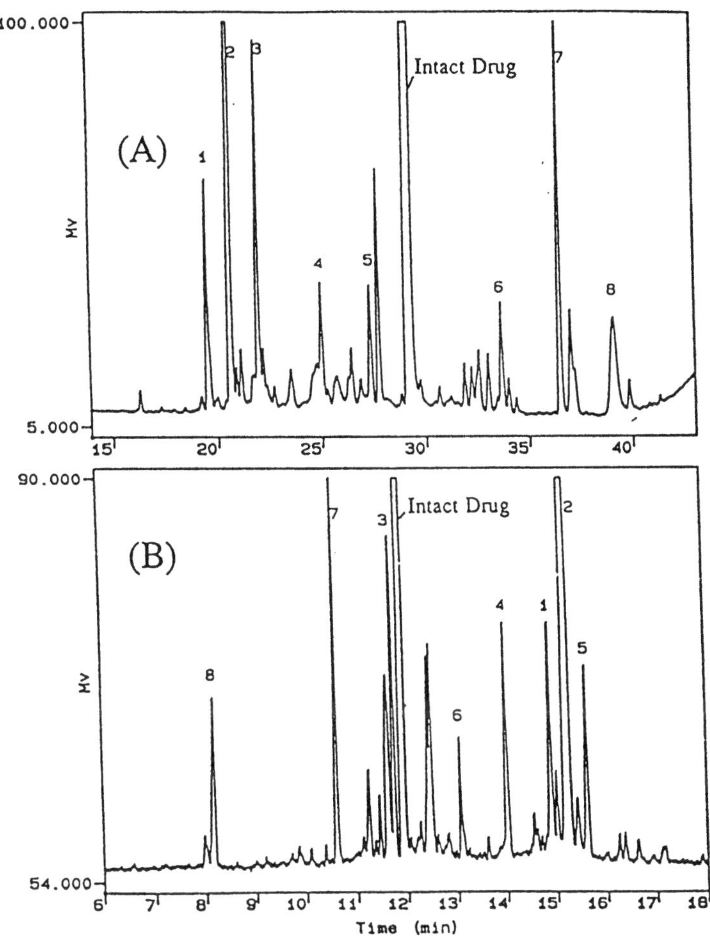

Figure 11.1 Separation of a stressed drug sample solution by A) HPLC and B) CE. Reproduced with permission from reference 4. Separation conditions: A) Hypersil BDS C-18, Solvent A NaH_2PO_4 (50 mM) – $Na_2EDTA \cdot 2\ H_2O$ (10 mM) adjusted to pH 7, Solvent B was methanol, gradient 10 %B to 70 % B in 30 minutes and 100 % B at 40 min., 1 ml/min, 287 nm. B) Boric acid (100 mM) – tricine (25 mM) – EDTA (0.5 mM) (pH 8.05): acetonitrile (67:33 % v/v), 280 nm, 30 kV.

11.2.2 Linearity

This exercise is conducted for quantitative CE methods where a change in detector output should be linearity related to the concentration of the test solute. The linear dynamic range of CE is significantly better for peak areas rather than peak heights (22) due to unavoidable peak broadening that occurs at higher concentrations. Therefore peak areas are generally used in CE (22). For a main component assay for the active drug content in a formulated product the linearity is often assessed in a range covering 80–120 % of the expected sample concentration which may reflect the likely variations in sample concentrations in routine manufacture. If the purpose of the method is to determine drug-related impurities or trace level enantiomeric impurities it is necessary to demonstrate suitable detector linearity is obtained for variation in solute concentration in the expected range. For example if impurities are present at levels from 0.01–0.2 then detector linearity for the impurities should be demonstrated covering this range with the main peak present at the method concentration.

For example (3) levels of p-toluenesulfonic acid were spiked into drug substance at levels of 0.02–5.0% and a correlation coefficient of 0.9999 was obtained for peak area plotted against impurity concentration. Correlation coefficients of >0.99 have been reported (22) for fluparoxan enantiomer levels of 1–10 % of the desired enantiomer content. Typically linearity is assessed by analysing 5 standards covering concentrations over the range 50–150 % of the nominal assay concentration. Many reports (6–11) have shown that this performance is possible with standard CE instruments.

11.2.3 Sensitivity

This parameter is of importance in trace level determinations such as the determination of drug-related impurities and undesired enantiomeric impurities. The method may be to determine trace level contamination/residues of drugs or to demonstrate absence of the drug such as in the testing of placebo formulations. For drug-related and enantiomeric impurity determinations sensitivity (limit of detection, LOD) is defined as the lowest level of material that can be detected in the presence of the main component. Typical LOD values for impurities are in the region of 0.01–0.1 %. The LOD is often defined as the smallest peak detected with a signal height three times that of the baseline noise (signal-to-noise ratio, S/N). The limit of quantitation (LOQ) refers to the lowest level of material can be determined with an acceptable degree of confidence/accuracy. The LOQ value is often calculated as 10 times the S / N, alternatively the signal obtained from repeated injections may be measured and the variability of signal should meet some predetermined measure of precision such as less than 10 % RSD.

These LOD values will be dependent upon the UV activity of the solute, the capillary bore and the sample loading. Limits of detection of achiral impurities may be lower than 0.1%; often values of 0.01 % are required. For example (2) levels of ranitidine hydrochloride impurities have been determined at 0.01 % of the main peak. Figure 11.2 shows separation of ranitidine spiked with 0.1 % of a range of impurities. Limits of detection of 0.1 % and lower are possible for trace level enantiomer contents (12–15).

11.2.4 Accuracy/recovery

This validation criteria is necessary to show analytically that all the material in the test sample is being correctly quantified by the method. If there are issues of insufficient solubility or binding/interaction of the solute with components in the sample matrix then low recoveries and incorrect results will be obtained. This involves recovery studies such as addition/spiking of known amounts of the drug substance into mixtures of the tablet excipients or

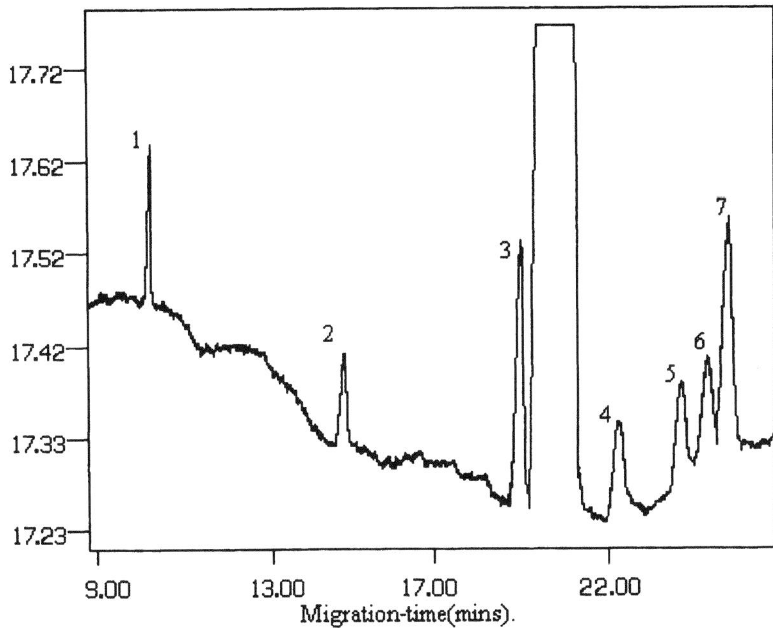

Figure 11.2   Separation of ranitidine (10 mg/ml) and seven related substances at the 0.1 % area/area level (see Table 11.1 for related substances). Separation conditions: Fused silica capillary 27 cm × 50 µm i.d., Buffer; 25 mM sodium di-hydrogen orthophosphate adj. to pH 2.6 with conc. phosphoric acid. Voltage 8 kV, wavelength 230 nm and an operating temperature of 25°C.

spiking of known amounts of impurities into the drug substance. The method is then applied to analyse the spiked sample to determine the concentration of the added component. The determined concentration of the added component is then compared to the concentration added. It is necessary to compensate for any residual level of the compound already present in the sample. For example when 0.1 mg of an impurity is added to 100 ml of solution containing 10 mg of drug substance which contains a residual 0.1 % of the impurity. If the level of the impurity is then measured and found to be 1.1 % then an acceptable recovery has been demonstrated.

This aspect is of particular importance in the area of bioassays where drug-binding may generate low recoveries. Table 11.2 shows the results from recovery experiments for desactyldiltiazem and diltiazem from plasma samples.

11.2.5 Injection Repeatability

Multiple injections from a single sample solution are performed to demonstrate that the CE instrument is functioning correctly. If the temperature and voltage controls are performing correctly then consistent migration times will be achieved. If the sample injection device is

Table 11.2   Recovery results for spiked desactyldiltiazem and diltiazem in plasma

Desacteyldiltiazem				
Spiked content (ng/ml)	5	10	50	250
Average CE result (ng/ml)	4.87	10.1	50.7	252.9
Diltiazem				
Spiked content (ng/ml)	5	10	50	250
Average CE result (ng/ml)	4.77	10.4	47.2	250.8

Reproduced with permission from reference 19

operating correctly then an acceptable level of variation for peak areas will be obtained. The level of variability will be highly dependent upon the type of analysis being conducted; if the method is for trace level determinations peak area precision of 5–20 %RSD may be typical. For assay of the main component a figure of 1–2 %RSD is more typical. Use of internal standards and high sample concentrations are generally used to meet this acceptance criteria. The number of repeated injections used in this assessment is in the order of 6–10. Reported precision in CE is typically of the order of 0.5–2% RSD for main peak assay (6–11).

The long-term performance of a method should also be assessed if the method is to be applied to large numbers of samples in extensive injection sequences. For example in the determination of desactyldiltiazem and diltiazem in plasma (19) 600 injections were performed over a 3 months period. This may be necessary if long-term problems such as sample evaporation, sample absorption and buffer depletion occur. Many of the autosampler vials used on commercial CE instruments are unsealed to allow the capillary to enter the vial. Evaporation problems can occur which will be magnified in long injection sequences or if volatile organic solvents are used to prepare electrolyte or sample solutions. Use of an internal standard will minimise evaporation effects on long-term injection precision. Absorption of sample matrix components, such as proteins from bio-fluid samples, of excipients from tablet formulations, can cause long-term variability in migration times and peak areas (as the area of the peak obtained in CE is directly related to the migration time (24).

Buffer depletion is an electrolytic process that occurs (25) when the separation voltage is applied. When the voltage is applied cations, such as the protons and metal ions from the electrolyte (ie Na^+ from a sodium phosphate buffer), migrate to the cathode electrolyte reservoir and gradually cause a decrease in the pH of the solution in the buffer reservoir. This gradual alteration of the pH and ionic composition of the solutions in the electrolyte reservoirs can cause problems with repeatability and separation selectivity (25). Routine approaches towards eliminating this problem (25) include use of high concentration buffers, zwitterionic buffers, narrow bore capillaries (to limit current), automatically replacing the contents of the electrolyte vials with fresh buffer or programming use of vials of fresh electrolyte during a long injection sequence (for example every 5 or 10 injections).

11.2.6 Method Repeatability

This assessment is sometimes described as "intermediate precision" and is performed to demonstrate that the method can be successfully applied on different occasions and under different circumstances. The analysis is repeated by a number of analysts using different instruments and capillaries in various laboratories. Buffer and sample/standard solutions are prepared by the different analysts. The primary objective of this exercise is to demonstrate that the method can be transferred between laboratories/analysts etc. and that method performance is not specific to the successful performance of a specific analyst/instrument etc. Demonstration of method repeatability may be especially important if coated capillaries or chemically impure electrolyte additives are employed in a specific method. For example [22] use of a chemically derivatised cyclodextrin for a chiral CE method highlighted differences between batches of the cyclodextrin from a single supplier and large differences between individual suppliers of the cyclodextrin.

For example (18) in the validation of a method for determination of potassium levels ten individual calibration solutions were prepared and analysed in duplicate, the calculated response factors for the calibrations produced a precision of 0.8 % RSD. Ten samples were also prepared and analysed (Table 11.3) and this gave an RSD of 0.65 %.

As the method may be employed on CE instruments from various manufacturers it is important to demonstrate that acceptable performance can be attained using different sample introduction modes and capillary dimensions. Examples of such studies include assessments of CE methods for the determination of hydrochlorothiazide (8) and a range of basic (11) and acidic (6) drugs. This testing indicates (21) the ability of the method to be transferred between operators and instrumentation.

Table 11.3 Precision of sample preparation. Results as %w/w potassium (calculated using peak area ratios)

Sample No.	Injection 1	Injection 2	Average result
1	5.69	5.71	5.70
2	5.74	5.73	5.74
3	5.80	5.78	5.79
4	5.73	5.73	5.73
5	5.72	5.72	5.72
6	5.66	5.67	5.67
7	5.74	5.75	5.75
8	5.71	5.79	5.75
9	5.70	5.72	5.71
10	5.77	5.76	5.77
Mean			5.73 (0.65 % RSD)

(Reproduced with permission from reference 18)

11.2.7 Method Robustness

This study involves assessing whether the method is able to tolerate small changes in the operating conditions. The results from testing allow limits to be assigned to the various operating parameters. The assessment involves monitoring the effects on the separation of small, deliberate variations from the preset method operating conditions. For example if two components are acceptably separated at pH 7.1 with a 55 mM phosphate buffer will they remain resolved with a 60 mM pH 6.9 buffer? The use of robustness testing to determine the operating limits that method can be operated over introduces a very useful validated flexibility to the use of the method. Without this data it is necessary to scientifically justify any slight deviation to the method. For instance if the method robustness studies confirm the method successfully operates over a range of 45-55 mM buffer concentration then the operator has a simple task of weighing the required amount of buffer salt.

Robustness testing traditionally involves evaluating individual operating parameters separately over the predetermined ranges. This "one-by-one" approach has been performed (8) for the determination of hydrochlorothiazide content in tablets by CE. An alternative approach is to use a statistically developed experimental designed robustness evaluation (6, 11, 18) to simultaneously monitor the effect of changing several factors. In this approach an sequence of injections is performed during which each of operating parameter is randomly varied to cover the range under investigation. Specific performance measurements of concern for the method such as resolution, sensitivity, peak efficiency etc. are recorded for each combination of operating conditions tested. The performance data results can be statistically processed to allow confidence intervals to be set to the limits for each operating parameter. An early evaluation of the robustness of a method generates a confidence in the method and a higher likelihood that the method will be reliable in long-term routine usage and that will be possible to transfer the method to other laboratories or sites. The results from these robustness testing exercises allow the tolerance limits to be prescribed on each parameter in the method.

11.2.8 Cross-Validation

This aspect is generally performed to demonstrate method accuracy by comparison of the CE results to those obtained by an established analytical method. Alternatively the CE data may be compared to the expected, or label claim result. Comparisons of CE and HPLC are frequently used (1, 10, 11) and can act as a core element in the validation of either, or, both the CE and HPLC methods. Often the CE data is statistically correlated to other data sets to demonstrate equivalence.

11.2.9 Solution Stability

Experiments should be conducted to support the shelf-life assigned to sample, calibration and reagent solutions. The shelf-life should normally specify storage conditions ie. shelf-life of 10 days, protected from light, stored in a fridge in a glass container. Determination of shelf-life is essential to demonstrate that the sample or standard do not degrade whilst awaiting analysis or that buffer solutions do not deteriorate on storage. In CE the buffer volumes used are relatively small and stock solutions may often have shelf-lives of 3 months. Support data is obtained by preparing buffer and then storing this for the required length of time under the prescribed storage conditions. Following the storage period fresh buffer is prepared and a sample solution is analysed using the fresh and stored buffer. There should be no significant difference between the results generated using the different buffers. This is important as buffers can deteriorate on storage, for example cyclodextrin solutions used for chiral CE can encourage microbial growth. A similar approach can be used for sample/calibration solutions. Alternatively the sample/calibration may be stored and re-injected later to observe if any increase in impurities has occurred. It is useful to demonstrate solution stability of the samples/calibration solutions in the vials on the CE autosampler as this would then support re-injection of the solutions if necessary. For example a shelf-life of 8 days (stored in a refrigerator) has been assigned (11) for aqueous solutions of various basic drugs tested by CE.

11.3 Response factors

The response factors for each quantified components of a mixture which is quantified in a method should be established to ensure that they are being accurately determined. This is essential if impurity data is to be calculated as % area/area of the total peak area. Impurities may have markedly different UV characteristics from each other and from the main component and may therefore be considerably over- or under-estimated. Data handling is different in CE to that in HPLC as the separated peaks in CE move through the detector at different speeds. Therefore, to compare the areas of peaks within a single separation all peak areas must (25) be divided by their corresponding migration times. The calculated areas are often referred to as "normalised areas". All impurity data (and response factors) should be calculated using normalised areas which compensate for the different migration speed of peaks through the detector.

In the determination of impurities it is necessary to ascertain if the impurities have a similar UV response compared to the main component. The detection wavelength is typically a maximum for the main component but may not necessarily be the case for the impurities. If the UV absorbance of both main component and impurity are significantly different then the quoted % area / area may not be a true indication of the level of impurity present. Therefore, response factors (RF's) are measured for all available isolated impurities and for the main component. Typically, if the Rf of an impurity is 80–120 % that of the main component, it may be considered to be equivalent to the main component and no correction would be necessary. However, if the response factor difference exceeds these limits it may be necessary to use a response factor to calculate the impurity content. If response factors differ greatly from 1 then useful comparisons between different techniques will be impossible.

Response factors obtained from HPLC measurements cannot be directly applied to peaks obtained in CE separations. The ionic form of the solute may be different in CE, compared to HPLC, as it may be charged. Changes in solute charge can cause UV absorbance shifts and alter response factors. Inclusion of solutes into commonly used CE additives such as surfactant micelles and cyclodextrins may also cause spectral shifts.

11.4 Peak Homogeneity

Another issue of concern is the possibility of co-migration of peaks is possible in CE as in any other separative technique, therefore it is useful to investigate the purity of separated CE peaks. Approaches to peak purity determination in CE are similar to those employed in HPLC. The principal methods are fraction collection (26) and spectral characterisation (4). Figure 11.3 shows an assessment of the purity of a peak in a stressed sample solution using a diode array. Common approaches would be used in related impurity determinations, main component assay and enantiomeric separations.

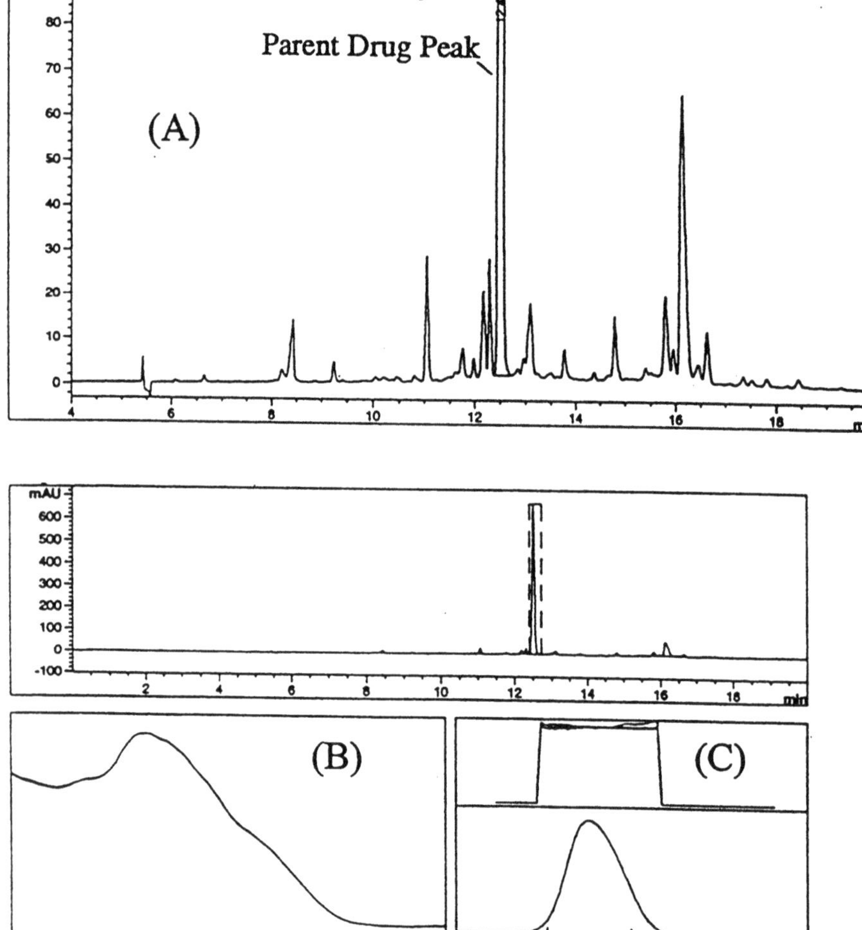

Figure 11.3   Assessment of the peak purity of the TMT-NCS peak in a stressed sample using a photodiode array detector. A) Sample electropherogram, B) overlay of 7 spectra across the main peak (purity level = 999.7), C) wavelength ratio plot.

11.5 Method Transfer

The formal transfer of methods between laboratories usually follows a number of predetermined protocols. For example if the method is a quantitative assay the protocol may include acceptable criteria such as performance limits for injection precision, sample preparation repeatability and statistical comparison of the results by the laboratories involved in the transfer. Usually subsamples from particular batches are analysed by both laboratories involved in the transfer following the exact procedures described in the method.

A series of inter-company cross-validation exercises in which methods have been repeated in seven independent pharmaceutical companies for a chiral separation (27), assay of the main component in a formulation (28) and drug stoichiometry determination (29).

11.6 System Suitability

This testing is performed usually on a daily basis to demonstrate that, on the occasion of testing, that the particular instrumentation and method set-up was capable of meeting predetermined acceptance criteria. These criteria must be met prior to use of the method to obtain analytical results. The exact criteria applied depends on the application for the method. For example if the method is to be used to determine trace enantiomeric impurity levels then a test mixture containing low levels of the enantiomeric impurity will be separated. In this case the detection sensitivity for the impurity and the resolution of the enantiomers would be specified. If the method were to be used for quantitative analysis an assessment of injection precision may be performed by repeated injections. Alternatively performance limits for aspects such as this may be listed in the method, for example that the method should give an injection precision of less than 1.5 % RSD for calibration response factors for standard solutions injected throughout a sequence.

Examples of system suitability requirements for CE methods include the requirement for acceptable resolution and detection of 0.1 % of the R-enantiomer in the S-enantiomer of the Eli-Lilly development LY231514 (13). These aspects have been discussed (30) with regard to determination of the trace enantiomer in naproxen. The main naproxen peak was severely asymmetric (peak asymmetry at 10 % of peak width, $A_{10\%}$ was greater than 10) however the system suitability criteria of Rs greater than 2.0 was maintained despite the relatively poor peak shape. System suitability criteria for the separation of mirtazaprine and related impurities (31) included resolution between specific impurities, migration time position and a minimum separation efficiency value.

11.7 Conclusions

The various validation aspects for a CE method are broadly similar to those used in HPLC method validation. Fundamentally the validation is performed to demonstrate that the method is "fit-for-purpose" and that the results obtained are accurate and precise to the required level. There are a number of CE specific issues, most notably transfer between instrument types, robustness study requirements, long-term injection repeatability, and extended shelf-life assessments.

References

1. Castaneda G Penalvo E J, and Fabre H, Cross validation of capillary electrophoresis and high-performance liquid chromatography for cefotaxime and related impurities, *Chromatographia*, 42 (**1996**) 159-164.

2. Kelly M A, Altria K D, Grace C, and Clark B J, Optimisation, validation and application of a capillary electrophoresis method for the determination of ranitidinehydrochloride and related substances, *J. Chromatogr. A*, in press **1998**.

3. Shah PA and Quinones L, Validation of a micellar electrokinetic capillary chromatography (MECC) method for the determination of p-toluenesulfonic acid impurity in a pharmaceutical intermediate, *J. Liq. Chromatogr.*, 18 (**1995**) 1349-1362.

4. Bullock J, Capillary Electrophoresis purity method for the novel metal chelator TMT-NCS, *J. Pharm. Biomed. Anal.*, 14 (**1996**) 845-854.

5. Shaafati A and Clark B J, Development and validation of a capillary zone electrophoretic method for the determination of atenolol in presence of its related substances in bulk and tablet dosage form, *J. Pharm. Biomed. Anal.*, 14 (**1996**) 1547-1554.

6. Altria K D, Bryant S M, and Hadgett T, Validated capillary electrophoresis method for the assay of a range of acidic drugs and excipients, *J. Pharm. Biomed. Anal.*, 15 (**1997**) 1091-1101.

7. Wynia G S, Windhorst G, Post P C, and Maris F A, Development and validation of a capillary electrophoresis method within a pharmaceutical quality control environment and comparison with high-performance liquid chromatography, *J. Chromatogr. A*, 773 (**1997**) 339-350.

8. Thomas B R, Fang X G, Chen X, Tyrell R J, and Ghodbane S, Validated micellar electrokinetic capillary chromatography method for the quality control of the drug substances hydrochlorothiazide and chlorothiazide, *J. Chromatogr.*, 657 (**1994**) 383-394.

9. Bechet I, Fillet M, Hubert Ph, and Crommen J, Quantitative analysis of non-steroidal anti-inflammatory drugs by capillary zone electrophoresis, *J. Pharm. Biomed. Anal.*, 13 (**1995**) 497-503.

10. Williams R C, Alasandro M S, Fasone V L, Boucher R J, and Edwards J F, Comparison of liquid chromatography, capillary electrophoresis and super-critical fluid chromatography in the determination of Losartan Potassium drug substance in Cozaar tablets, *J. Pharm. Biomed. Anal.*, 14 (**1996**) 1539-1546.

11. Altria KD, Frake P, Gill I, Hadgett T, Kelly MA, and Rudd D R, Validated capillary electrophoresis method for the assay of a range of basic drugs and excipients, *J. Pharm. Biomed. Anal.*, 13 (**1995**) 951-957.

12. Sanger-van de Griend C E, Wahlstrom H, Groningsson K, and Widahl-Nasman M, A chiral capillary electrophoresis method for ropivacaine hydrochloride in pharmaceutical formulations: validation and comparison with chiral liquid chromatography, *J. Pharm. Biomed. Anal.*, 15 (**1997**) 1051-1061.

13. Liu L, Osborne LM, and Nussbaum MA, Development and validation of a combined assay and enantiomeric purity for a chiral pharmaceutical compound using capillary electrophoresis, *J.Chromatogr.A*, 745 (**1996**) 45-52.

14. Noroski J E, Mayo D J, and Moran M, Determination of the enantiomer of a cholesterol-lowering drug by cyclodextrin-modified micellar electrokinetic chromatography, *J. Pharm. Biomed. Anal.*, 13 (**1995**) 54-52.

15. Altria K D, Harkin P, and Hindson M, Validation of a CE method for the quantitative determination of trytophan enantiomers, *J. Chromatogr. B*, 686 (**1996**) 103-110.

16. Altria K D, Assi K, Bryant S, and Clarke B J, Quantitative analysis of organic acid drug counter-ions by capillary electrophoresis, *Chromatographia*, 44 (**1997**) 367-371.

17. Zhou L and Dovletoglou A, Practical capillary electrophoresis method for the quantitation of the acetate counter-ion , *J. Chromatogr.* A, 763 (**1997**) 279-284.

18. Altria K D, Wood T, Kitscha R, and Roberts-McIntosh A, Validation of a capillary electrophoresis method for the determination of potassium counter-ion levels in an acidic drug salt, *J. Pharm. Biomed. Analysis*, 13 (**1995**) 33-38.

19. Coors C, Schulz H-G, and Stache F, Development and validation of bioanalytical method for the quantitation of diltiazem and desacetyldiltiazem in plasma by capillary zone electrophoresis, *J. Chromatogr.* A, 717 (**1995**) 235-243.

20. Sandor V, Flarakos T, Batist G, Wainer I W, and Lloyd D K, Quantitation of the diastereoisomers of l-buthionine-(r,s)-sulfoximine in human plasma: A validated assay by capillary electrophoresis, *J. Chromatogr.* B, 673 (**1995**) 123-131.

21. Altria K D and Rudd D R, An overview of method validation and system suitability aspects in capillary electrophoresis, *Chromatographia*, 41 (**1995**) 325-331.

22. Watzig H, Appropriate calibration functions for capillary electrophoresis. 1. Precision and sensitivity using peak areas and heights, *J. Chromatogr.* A, 700 (**1995**) 1-7.

23. Altria K D, Walsh A R, and Smith N W, Validation of a capillary electrophoresis method for the enantiomeric purity testing of fluparoxan, *J. Chromatogr.*, 645 (**1993**) 193-196.

24. Altria K D, Essential peak area normalisation in Capillary Electrophoresis, *Chromatographia*, 35 (**1993**) 177-182.

25. Kelly M A, Altria KD, and Clark B J, Approaches used in the reduction of buffer electrolysis effects in routine capillary electrophoresis procedures, *J.Chromatogr.* A, 768 (**1997**) 73-80.

26. Altria K D and Dave K, Peak homogeniety determination and micro-preparative fraction collection by capillary electrophoresis in pharmaceutical analysis, *J. Chromatogr.*, 633 (**1993**) 221-225.

27. Altria K D, Harden R C, Hart M, Hevizi J, Hailey P A, Makwana J, and Portsmouth M J, An inter-company cross-validation exercise on capillary electrophoresis. 1. Chiral analysis of clenbuterol, *J. Chromatogr.*, 641 (**1993**) 147-153.

28. Altria K D, Clayton N G, Harden R C, Hart M, Hevizi J, Makwana J, and Portsmouth M J, An inter-company cross-validation exercise on capillary electrophoresis testing of dose uniformity of paracetamol content in formulations, *Chromatographia*, 39 (**1994**) 180-184.

29. Altria K D, Clayton N G, Harden R C, Hart M, Hevizi J, Makwana J, and Portsmouth M J, Inter-company cross-validation exercise on capillary electrophoresis. Quantitative determination of drug counter-ion level, *Chromatographia*, 40 (**1995**) 47-50.

30. Guttman A and Cooke N, Practical aspects in chiral separation of pharmaceuticals by capillary electrophoresis. 2. Quantitative separation of naproxen enantiomers *J. Chromatogr.* A, 685 (**1994**) 155-159.

31. Wynia G S, Windhorst G, Post P C, and Mars F A, Development and validation of a capillary electrophoresis method within a pharmaceutical quality control environment and comparison with high-performance liquid chromatography, *J.Chromatogr.A*, 773 (**1997**) 339-350.

12 Capillary Electrochromatography

12.1 Introduction

Capillary Electrochromatography (CEC) is an analytical technique in which capillaries are either packed or internally coated with stationary phase. A voltage is applied across the capillary to generate an electro-osmotic flow (EOF). The EOF sweeps solutes along the capillary towards the detector. The solutes chromatographichally interact with the stationary phase resulting in separations. Ionic solutes also separate by virtue of differences in their electrophoretic mobilities.

In 1981 Jorgenson and Lukacs (1) performed CEC separations of 9-methylanthracene and perylene using a 170 µm ID column packed with 10 µm Partisil ODS-2, using acetonitrile as the mobile phase and on-column fluorescence detection. They achieved reasonable efficiencies for their separations with relatively long retention times. In 1987 Knox and Grant (2) demonstrated the use of packed CEC capillaries to obtain highly efficient separations of neutral aromatics with fluorescence detection. Current work is predominantly focused on the effective operation of CEC systems and reliable construction of packed CEC capillaries.

CEC is highly applicable to the separation of pharmaceuticals and Table 12.1 shows that CEC has been used to separate a range of compounds including drug related impurities and chiral species. Although the number of pharmaceutical applications are currently somewhat limited it is widely recognised that CEC is currently undergoing extensive investigation and that this, coupled with the recent availability of CEC compatible CE instrumentation and increasing commercial availability of CEC packed capillaries, is leading to a rapid increase in the number, and range, of applications.

12.2 CEC Instrumentation

Figure 12.1 shows a schematic of a packed capillary used in CEC. Typically 50 or 75 µm fused silica capillaries are partially filled with HPLC packing material.

Commercial CE instruments can be employed for CEC purposes with little or no modifications. The major modification that may be required enables the system to be pressurised whilst applying the separation voltage. HPLC column suppliers have now begun to manufacture pre-packed columns for CEC which will greatly accelerate the development of CEC as a viable alternative to HPLC and CE.

CEC separations are often performed on CE equipment which has been modified to apply pressure to the inlet and outlet buffer vials to prevent bubble formation. Formation of a solvent gradient is possible in CEC using external pumps to mix and supply solvent of varying composition to the high voltage electrolyte reservoir.

Table 12.1 CEC literature reports

Analytes	Stationary Phase	Comments	Ref.
Alkylbenzoates	ODS2	Evaluation of performance parameters	3
DNA adducts	ODS	Qualitative separations	4
Drug intermediates	ODS	Diastereoisomer separation, buffer depletion	5
Highly basic drugs including nortryptyline	SCX	Ion-exchange CEC	6
Israolypine and by-products	ODS	Industrial application	7
Nicotinamide	ODS	Main component assay	8
Oligonucleotides	Various	Gradient CEC	9
Pharmaceuticals	ODS1	Resolution of impurities, routine operation evaluation.	10
Polyaromatic hydrocarbons	ODS	Laser induced fluorescence (10^{-11} M sensitivity)	11
Polyaromatic hydrocarbons	1.5 μm Non-porous ODS	Routine use of CEC, rapid separations	12
Steroids, prostaglandins, cephalosporins and intermediates	Zorbax	Resolution of related impurities	13
Steroids and aromatics	ODS	On-line CEC-MS detection	14
Textile dyes	ODS	On-line CEC-MS detection	15
Chiral CEC Separations			
1,1' binapthyl-2,2'-diyl-hydrogenphosphate	pH 7 Chirasil-dex coated capillary	20 minutes separation with 20 kV	16
Chlorthalidone enantiomer, Mianserin cationic enantiomer	HP-β-CD bonded silica stationary phase	Comparison with CE	17
Hexobarbital	Cyclobond	Flow reversal with TEAA	18
Mephobarbitol	CD coated capillary	pH 8, phosphate	19
Racemic pharmaceuticals	Cellulose coated capillaries	Cellulose coated phase	20
Various including hexobarbital and cicloprofen	Chirasil-dex	Chirasil-dex as coating or as additive	21
Various including carproxen and cicloprofen	CD coated capillaries and CD additive	Range of compounds separated by FSCE and coated capillary EC.	22
Various including hexobarbital and propanolol	AGP stationary phase	10 out of 19 compounds resolved	23

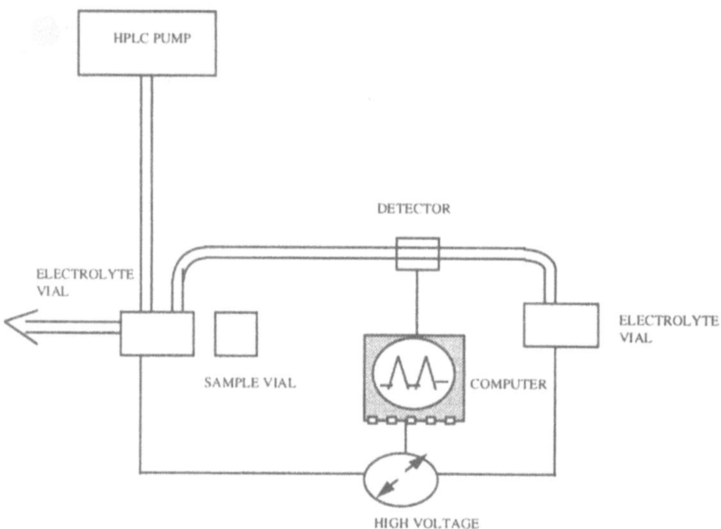

Figure 12.1   Schematic of gradient CEC system. Reproduced with permission from reference 8.

The packing material is held within the separation capillary by use of 2 frits prepared in the capillary. Initially the capillaries were self-packed and a variety of approaches to packing have been described. However a range of packed capillaries are available commercially (see Table 12.2). The early work of Knox and Grant (2) involved packing relatively wide bore capillaries and then drawing them down to the smaller diameters required. The procedures for directly packing 50 or 75 µm capillaries are more difficult and have been detailed by both Boughtflower et al (5) and Smith and Evans (13). Retaining frits can be prepared by careful heating of the stationary phase packing. The production of good quality frits is paramount in achieving acceptable CEC capillaries. Use of ultrasound during the packing to keep the packing slurry mobile has been shown (24) to improve the packing procedure. Alternatively the silica packing can be electrokinetically loaded by application of an applied field (25). In all instances the production of packed capillaries is an expert practice and it is therefore advised to purchase commercially available capillaries. Table 12.2 shows details for a number of companies supplying CEC capillaries.

12.3 CEC Operation

The principal operating consideration in CEC is to avoid air bubbles being formed within the packed portion of the capillary during the application of the separation voltage. These bubbles can occur due to joule heating inside the capillary. Alternatively bubbles can be formed due to pressure differences occurring at the frit separating the packed and empty portions of the capillary as the EOF rate is different in these two portions. The presence of these air bubbles prevents passage of the voltage and the packed capillary must be flushed with mobi-

Table 12.2 Selection of commercial suppliers of packed CEC capillaries

Supplier	Stationary phases	Contact details
Capital HPLC (Electrokinetics Division)	Range of standard and customer specific phases to fit all instrument types.	Capital HPLC, East Mains Ind. Estate, Broxburn, EH52 5NN, UK, phone +44 1506 858505, fax +44 1506 858506, www.capital-hplc.co.uk
Electropak	Capillary columns for CEC and micro-HPLC	Unimicro Technologies Inc. 4713 First Street, Suite 225, Pleasanton, CA 94566 Tel: +1 510 846-8638 Fax: +1 510 846-3687
Hewlett-Packard	ODS packed columns to fit HP instrumentation	National HP offices world-wide.
Hypersil	C_{18} columns, diameters 50 or 100 μm $\times$ 250 mm. Compatible with most CE equipment. Available in cartridges for HP CEC system.	Hypersil, Chadwick Road, Astmoor, Runcorn, Cheshire, WA7 1PR Tel: +44 1928 562633 Fax: +44 1928 581078
Innovatech.	Range of standard and specialist phases to fit all instrument types	UK based phone/fax +44 1438 724727
LC Packings	Range of customer specific phases to fit all instrument types	Dufourstrasse 30, 8008 Zürich, Switzerland, phone +41 1 252 2158 fax +41 1 361 3602

le phase before it can be used again. The likelihood of bubble formation increases with the level of current generated within the capillary and therefore low buffer concentrations such as 5 mM phosphate are often used to restrict routine currents to below 5 μA. Pressurisation of the packed capillary during application of the voltage (10) reduces the tendency for bubbles to occur and this is the main modification required of a CE instrument to make it CEC compatible. Specially designed instruments offer the possibility of capillary pressurisation or alternatively the air-lines in some existing CE instruments can be replumbed to generate the pressurisation. Thorough de-gassing of mobile phases is also strongly recommended (12).

These low buffer concentrations can give problems of reproducibility for repeated separations due to buffer electrolysis (depletion) which alters the pH and ionic strength of the solution in the electrolyte vials. Depletion effects are magnified using low concentration buffers. Buffer depletion problems can be restricted by use of zwitterionic electrolytes such as 20 mM MES (5) or 50 mM Tris (10) which generate very low operating currents, or by automatically refilling the vial buffer replenishment (26).

Generally 50 or 75 micron inner diameter packed capillaries are employed. However, use of shorter columns in combination with lower applied voltages can permit wider bore, such as 150 micron (25), columns to be used. Increased diameters allow improved sensitivity and increased column capacity.

Sample is generally injected by electrokinetic loading in which the capillary tip and the high voltage electrode is introduced into the sample solution. Application of the appropriate polarity voltage causes sample components to enter the capillary by virtue of EOF and their charge. Pressure based CEC injections are not easily attainable on standard commercial CE instruments as the back pressure of the packed capillary is extremely high. However it has also been reported (26) that it is possible to perform hydrodynamic injections in CEC using a modified CE system capable of high pressure injections. Alternatively 30–60 second injection times can be used on unmodified CE instruments (8).

Detection is generally performed on-capillary at a position immediately after the second retaining frit (see Figure 12.1). The standard detector used has been UV absorbance, although other options such as laser induced fluorescence (11), fluorescence and mass-spectrometric detection (4,14,27) have also been reported. It has been shown (12) that it is possible to detect within the packed area of the capillary and this reduces peak broadening effects (400,000 plates compared to 150,000) – however lower sensitivities were obtained due to light scattering within the packing.

A variety of standard HPLC stationary phase packing materials have been employed (Table 12.1) including ODS (3-5), Zorbax SBC 8 (13), Cyclobond (18) ABZ (5), and Spherisorb SCX (6). The EOF generated in a CEC capillary produces effectively no back pressure across the system and this allows small diameter (sub 5 µm) silica particles to be employed. The high back pressures associated with these small particles generally prohibits their use in conventional HPLC. Since separation efficiencies are significantly improved the use of these small particles is a major advantage of CEC.

Basic compounds can tail appreciably when using conventional phases such as ODS as the basic compounds interact strongly (6) with the silanols. SCX phase has been specifically prepared for CEC and involved silica bonded with a propyl sulphonate function. The sulphonate functionality changed both the level of EOF and also allowed ion-pair interactions. Figure 12.2 shows separation of 4 highly basic drugs using the SCX column which generated up to 8 million plates per metre. As CEC is currently undergoing a rapid and widespread development it is anticipated that an expanded range of conventional HPLC and CEC specific stationary phase packings will become commercially available.

Gradient CEC has been reported (28). Sixteen PAH's were separated in 100 minutes by use of a gradient ranging from 55–80 % acetonitrile. The gradient was produced by attaching two feed capillaries to the separation capillary. These feed capillaries were inserted into reservoirs containing electrolytes containing different levels of acetonitrile. The level of voltage applied to the individual feed capillaries was varied to alter the solvent composition entering the CEC separation capillary. The magnitude and direction of the EOF in the capillary can be modified by addition of ion-pair reagents such as triethylammonium acetate (18).

12.4 Detection Options in CEC

A variety of detectors have been employed, most commonly UV detection, in-column fluorescence detection and mass spectrometry. UV detection is achieved by viewing through a portion of capillary immediately after the retaining frit in the capillary. In-packing detection (25) prevents possible loss of integrity of peak elution order and decrease in separation efficiency, which can arise with other detection methods due to the increased flow rate in the unpacked sections of the column causing charged molecules to migrate at differing speeds to those in the packed section of the column. However, sensitivity can be significantly de-

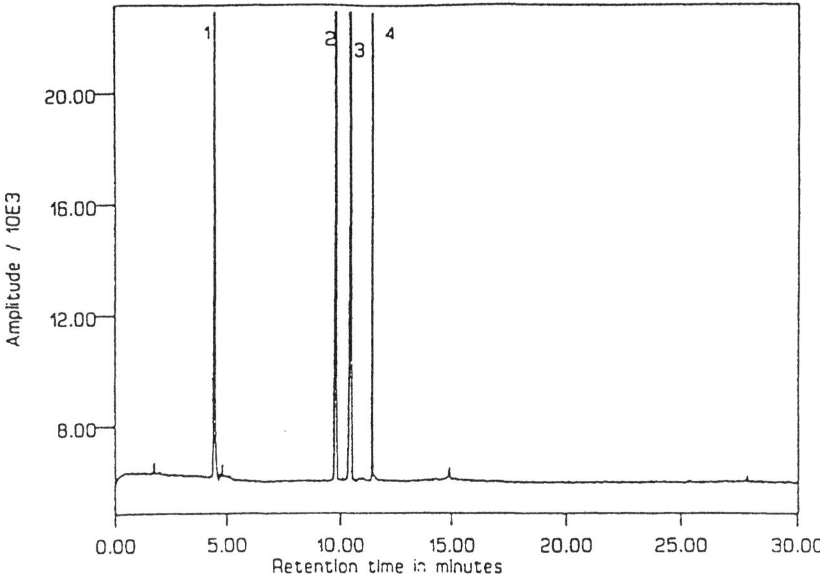

Figure 12.2 Highly efficient separation of basic drugs using a SCX packed column. Instrument Modified ABI 270A; Capillary: 26 cm packed, 50 cm total × 50 μm ID; Packing: 3 μm Spherisorb SCX; Detection:λ = 210 nm, 0.04 absorbance units full scale; Rise time: 2 s; Applied Voltage: 30 kV; Temperature: 30°C; Mobile phase: 70 % acetonitrile / 30´% 0.05 mol/L NaH_2PO_4 pH = 2.3; Injection: 2 kV for 0.5 mins. Peaks: 1 = bendroflume-thiazide, 2 = nortriptyline, 3 = clomipramine, 4 = imipramine, 5 = methdilazine. Reproduced with permission from reference 6.

creased with in-packing detection. CEC-MS is commonly performed with the column being packed right up to the point where the sample is injected into the mass spectrometer and the voltage is earthed in the mass spectrometer, thus avoiding the aforementioned problem.

CEC-LIF is also a popular choice giving high sensitivity in-column detection, however not all compounds are suitable for this method of analysis, as not all compounds fluoresce and background fluorescence of the buffer can also limit sensitivity. Figure 12.3 shows use of LIF to detect (29) a range of polyaromatic hydrocarbons (PAHs).

12.5 Analytical Performance of CEC

To-date the major focus of the reports has centred on developing applications and optimising operating conditions. Therefore there has been little emphasis on the analytical performance of CEC. The most extensive quantitative report is that of Yan et al who studied the CEC separation of 16 polyaromatic hydrocarbons (PAH'S). Repeated separations were performed on 4 separate days within a week. The precision for the retention times of these peaks were less than 2.6 % RSD. The authors mentioned that they had obtained similar results from other columns. Detection limits of 10^{-9}–10^{-11} M were reported for LIF detection of the PAH's.

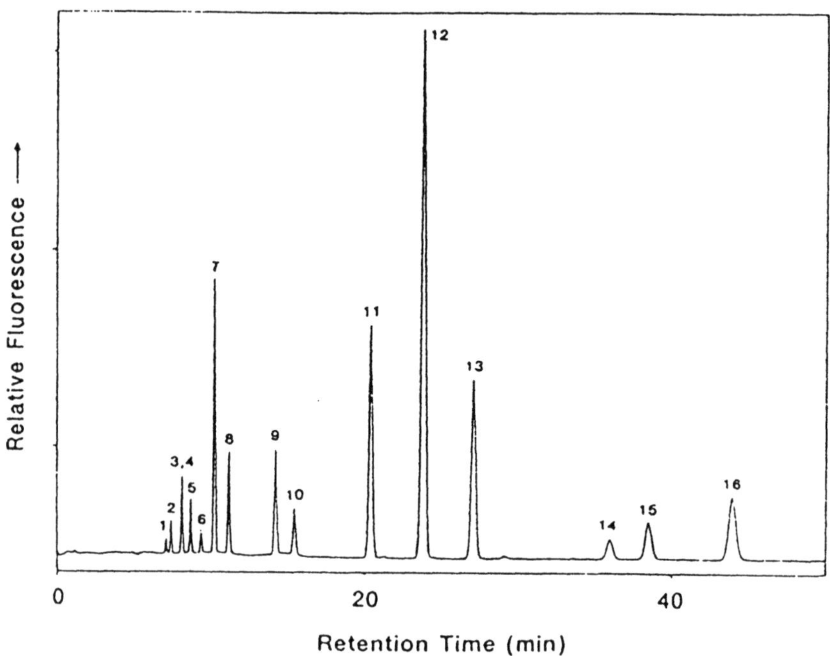

Figure 12.3   Electrochromatogram (CEC-LIF) of separation of 16 polyaromatic hydrocarbons. Column dimensions: 75 µm ID × 365 µm OD (33 cm packed length); Mobile phase: 80 % acetonitrile in a 4 mM sodium borate solution; Applied voltage: 15kV; Injection: Electrokinetic 5 kV for 5 s; Peaks: 1 = naphthalene 2 = acenaphthylene 3 = acenaphthene 4 = fluorene 5 = phenanthrene 6 = anthracene 7 = fluoranthene 8 = pyrene 9 = benz[a]anthracene 10 = chrysene 11 = benzo[b]fluoranthene 12 = benzo[k]fluoranthene 13 = benzo[a]pyrene 14 = dibenz[a,h]anthracene 15 = benzo[ghi]perylene 16 = indeno[1,2,3-cd]pyrene. Reproduced with permission from reference 29.

The long-term stability of CEC columns was reported by Boughtflower et al (5) who described several hundred repeat analyses of a neutral solute test mixture using a ODS filled capillary. Rebscher and Pyell reported (30) a RSD of 0.2 % for peak area ratios for 5 replicate injections of thiourea (which was used a CEC dead volume marker) on a capillary filled with ODS-Li Chromspher. The same workers (30) reported a 2.2 % RSD for six injections for the peak area of p-nitrotoluene using 2,4-dinitro-toluene as an internal standard on a ODS-Polygosil packed capillary. The use of higher concentration zwitterionic MES electrolyte allowed Boughtflower et al (5) to obtain RSD values of 1–1.5 % for repeated injections of uracil, benzamide and anisole. Dittmann et al described (26) the consistency of migration times obtained for 300 repeated injections of a test mixture of various aromatics on a ODS-Hypersil column. This consistency was only possible by automated replacement (replenishment) of the contents of the electrolyte reservoirs after every five separations.

Use of zwitterionic buffers and capillary cooling to minimise the current generated allowed good precision to be obtained (Table 12.3) for both electrokinetic and pressure injection on unpressurised CEC systems.

Table 12.3 Reproducibility of migration times and peak areas for electrokinetic and pressure injection in CEC.

| Sample | Electrokinetic injection (5 kV / 5 sec) | | | |
	Average Migration time (min)	%RSD (n = 10)	Average Peak height	%RSD (n = 10)
Thiourea	2.70	0.17	293924.5	0.87
Benzamide	2.94	0.17	453229.5	0.74
Naphthalene	4.04	0.13	318967.2	0.92
Sample	Pressure injection (0.5 mins)			
	Average Migration time (min)	%RSD (n = 10)	Average Peak height	%RSD (n = 10)
Thiourea	2.66	0.33	1010056.1	3.84
Benzamide	2.91	0.29	1438958.6	3.97
Naphthalene	3.98	0.24	951747.6	3.97

Reproduced with permission from reference 8.

12.6 Applications

Table 12.1 shows that the current number of applications is somewhat limited although it is anticipated that the improved availability of commercial CEC packed capillaries, increased awareness of CEC operating procedures and the availability of CE instruments modified for CEC operation, will all serve to expand this number exponentially within the next 3–5 years.

12.6.1 Pharmaceutical applications

Eurby et al (10) separated a range of pharmaceuticals by CEC. These included baseline separation of 2 diastereoisomers within 10 minutes – this separation had been impossible to achieve by HPLC using similar stationary phases. The separation was achieved with 180,000 plates per meter on the CEC column. Impurities in a tetrapeptide were also resolved (10) by CEC with detection at 210 nm. The impurities were separated using a 25cm CEC column and gave a similar profile to a 30 minute HPLC separation. Sample diluents effects were observed for the CEC separation. Good separation selectivity and efficiencies were obtained when the sample was dissolved in acetonitrile but this was heavily diminished when the sample was diluted with methanol. Detection limits of 0.05 % were calculated for related impurities.

Smith and Evans (13) have performed an extensive evaluation of the application of CEC to the analysis of pharmaceuticals. They have reported highly efficient separations of test solution containing steroids, prostaglandins, separation of a polyaromatic synthetic intermediate from 13 related impurities (Figure 12.4), and the simultaneous resolution of the diastereoisomers of both E- and Z-isomers of the antibiotic cefuroxime axetil.

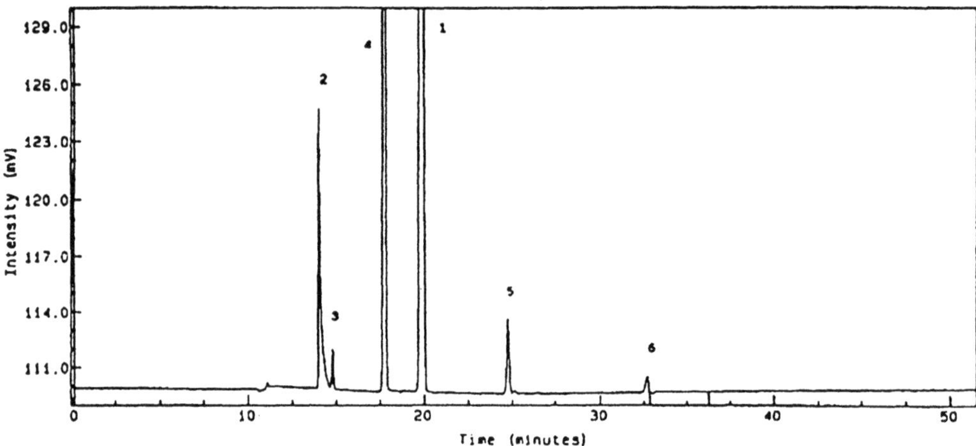

Figure 12.4   Separation of a prostaglandin from 6 related impurities by electrochromatography. Conditions : 70 % ACN : 30 % 10 mM Na_2HPO_4 (pH 9.9), 40 cm × 50 μm capillary filled with 1.8 μm Zorbax SBC8, 270 nm, 30 kV. Reproduced with permission from ref.13.

In a further report (6) the same workers reported the highly efficient separations of a test solution containing 4 development compounds and a solution containing 7 highly basic antidepressants which are very difficult to analyse using conventional HPLC due to peak tailing difficulties. Boughtflower et al reported (5) separation of the diastereoisomers of an intermediate compound used in the synthesis of a serine protease inhibitor.

Gordon et al (14) showed the CEC separation of the 3 steroids aldosterone, hydrocortisone and testosterone with both UV and MS detection.

Nicotinamide (Vitamin PP) has been determined (8) in multivitamin drops using benzamide as an internal standard. Five consecutive injections of each of the calibration and the vitamin drop sample and using a 50:50 $CH_3CN:H_2O$ with 10 mM TRIS on a 20 cm (packed length) column. Table 12.4 shows data for injection repeatability in this study (8).

It can be seen that the RSDs for the peak areas are significantly lowered by the use of an internal standard. Such a method is routinely employed in CE to ensure that RSDs are below

Table 12.4   Reproducibility of injection for nicotinamide analysis with benzamide as an internal standard.

	%RSD Calibration ($n = 5$)	% RSD Sample ($n = 5$)
Benzamide peak area	1.41	1.24
Benzamide retention time	0.73	0.62
Nicotinamide peak area	1.41	1.79
Nicotinamide retention time	0.67	0.63
Peak area ratio	*0.63*	*0.74*
Relative retention time	*0.14*	*0.10*

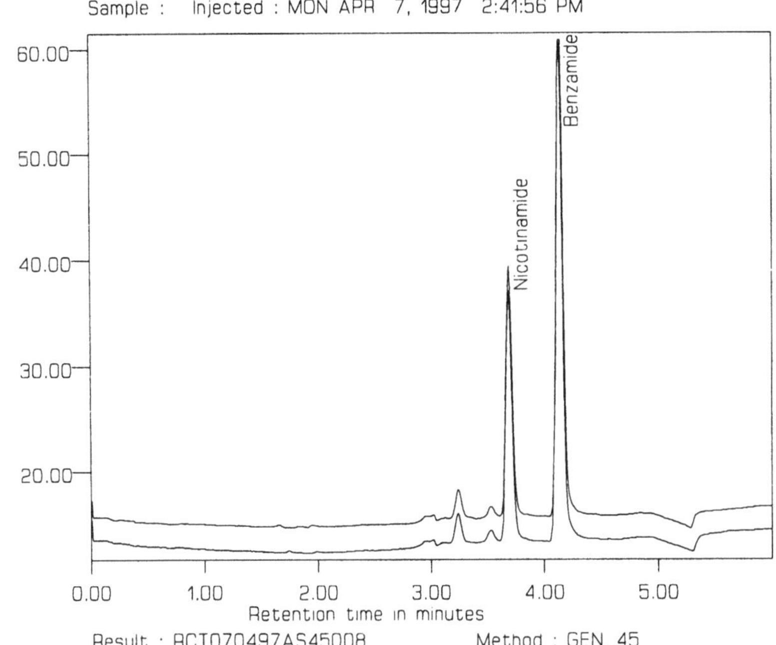

Figure 12.5 Overlaid electropherograms for multivitamin separation using acetonitrile. Separation conditions: Tris buffer (50:50, 10 mM overall), 15 kV, 200 nm. Separation performed using a Beckman P/ACE system. No pressure applied to inlet or outlet vials. Reproduced with permission from reference 8.

1 % and this is particularly important when an electrokinetic injections are used. The reproducibility of the retention times is shown clearly Figure 12.5 which is an overlay of two injections of a vitamin formulation sample.

12.7 Chiral CEC

The use of chiral stationary phases in CEC is a logical extension of chiral HPLC as a multitude of well characterised chiral LC phases are commercially available. A number of research groups have shown chiral CEC separations (Table 12.1). The resolutions have been achieved using three approaches:

1) Packed columns filled with chirally selective stationary phase.
2) CEC columns coated with chirally selective stationary phase material.
3) A combination of chiral electrolyte additives and packed columns filled with achiral stationary phase material

Mayer and Schurig have recently reported chiral separations using electrochromatography (16) by coating the inner surface of a 50 μm ID capillary with an immobilised dimethylpolysiloxane containing chemically-bonded permethylated β- or γ-CD. Enantioselective partition from a pH 7 phosphate was used (16) to separate enantiomers of ibuprofen, flurbiprofen, cicloprofen, etodolac, 1-phenylethanol and 1,1'-binaphthyl-2,2'-diylhydrogenphosphate. Efficiencies as high as $N = 3 \times 10^5$ theoretical plates were obtained for films of 0.2 μm thickness. CEC capillaries packed with cyclobond material have been used to chirally resolve (18) the enantiomers of benzoin, barbital and a range of dansylated amino acids. The separations showed a similar degree of selectivity to that obtained in CE with cyclodextrin addition to the electrolyte. However, the CE separation efficiencies were better (18) in this report possibly due to non-optimised packing of the CEC column.

α_1-acid glycoprotein (AGP) is an acidic protein with an isoelectric point of 2.7 which is widely used as a chiral stationary phase in HPLC. Li and Lloyd packed a 50 μm capillary with 5 μm AGP coated silica (23) and separated 10 chiral compounds (benzoin, hexobarbital, pentobarbital, ifosfamide, cyclophosphamide, metoprolol, alprenolol, oxprenolol, propranolol and disopyramide). Increasing phosphate concentration from 0.5 mM to 10 mM improved enantioselectivity.

A range of chiral pharmaceutical have been resolved using CEC capillaries coated with derivatised cellulose phases (20).

12.8 Benefits and Disadvantages of CEC Compared to CE and HPLC

CEC in its current state of infancy offers a tremendous blend of the experience database of HPLC and the miniaturisation benefits that CE offers. There are a large number of well characterised and commercially available HPLC packings that would be suitable for a range of application types. The considerable cost gains associated with use of CE or CEC are largely related to the reduced solvent purchase and disposal costs.

12.8.1 Advantages of CEC

The major advantages are listed below:

- *Unique selectivity* – since the separation occurs by both chromatographic and electrophoretic principles it is possible to obtain unique selectivity's. For example Figure 12.6 shows separation of fluticasone propionate and related impurities (peaks 1–4), peak 5 represents a component previously unseparated by the routine HPLC method.

- *Reduced particle sizes* – the EOF generates no significant back pressure which allows routine use of 3 μm or lower particles which generate highly efficient separations. Significantly improved peak efficiencies are obtained with smaller particle which permits improved separations or decreased analysis times for the same degree of resolutions.

- *Instrument usage* – CEC, CE and microbore LC can be performed on a single instrument. This ability would maximise equipment usage and reduced capital costs.

- *Solvents costs* – it is estimated that routine overnight operation of a CEC system with electrolyte replenishment would represent a 90 % + saving on organic solvent costs compared to HPLC.

- *Water insoluble* solutes – these separations have largely been considered difficult in CE. However, the use of high organic solvent content electrolytes in CEC allows high efficiency separations of water insoluble species such as steroids.

- *Packing material/additive costs* – the reduced amounts of packing material or electrolyte additives allow the cost effective use of elaborate stationary phases or electrolyte additives that would be considered prohibitive in conventional HPLC.

- *Low wavelength detection* – the short detection pathlength across the CEC capillary allows detection at low UV wavelengths such as 200 nm (8) where many solutes have enhanced UV activity. This would allow compounds to be separated by CEC that have limited chromophores and would be undetected by HPLC which is forced to use higher detection wavelengths due to the background absorbance of the solutes.

- *Mass-spectroscopy compatible* – the separation of neutral species in CE is generally performed by MECC using ionic micelles such as SDS. These MECC electrolytes are not MS compatible and MECC-SDS is not routinely considered. However, neutral solutes can be directly resolved using CEC which is MS-compatible.

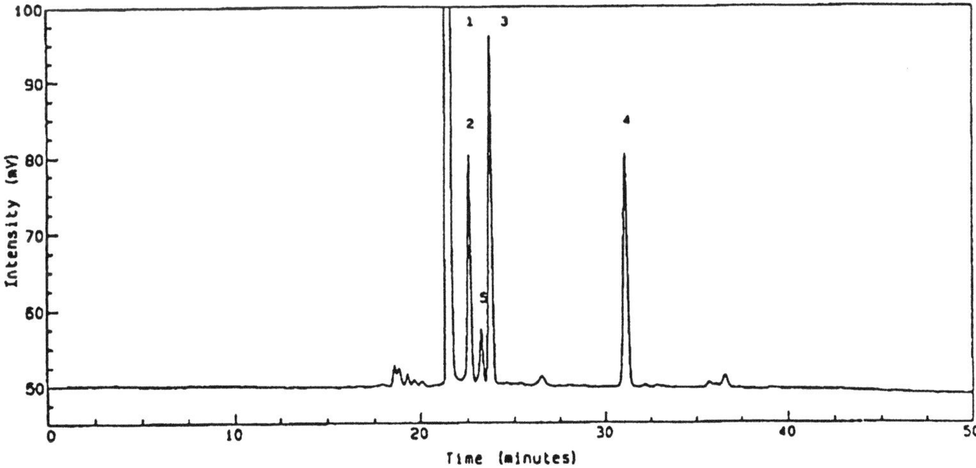

Figure 12.6 Fluticasone Propionate separation from related impurities. Instrument: Modified ABI 270 A; Capillary: ~60 cm × 50 μm ID; Packing: 3 μm ODS-1; Detector: λ = 238 nm, 0.05 aufs; Rise time: 1.0 s; Applied Voltage: 30 kV; Temperature: 30°C; Mobile phase: 80 % CH₃CN / 20 % 5 mM Borate, Injection: 0.4 mins at 20 kV; Peaks 1 = Fluticasone Propionate; 2, 3 and 4 = related impurities, 5 = previously undetected peak: Reproduced with permission from reference 13.

- *Separation speed* – the production of high plate counts allows reductions in analysis time. High voltages can be applied across short packed capillaries to produce extremely fast separations. For example Figure 12.7 shows (25) separation of 5 polyaromatic hydrocarbons within 5 seconds. The separation was achieved using a 6.5 cm packed area in a 10 cm long capillary.

12.8.2 Current disadvantages of CEC

The current disadvantages of CEC are largely technological issues which are likely to be overcome in the foreseeable future.

- *Analyte range* – current the vast majority of CEC research has been performed on the optimisation of systems using test mixtures of neutral compounds. Tailing of basic drugs onto the silanol groups present on the packing can be appreciable. Acidic compounds can be only marginally retained as their negative charge repels them from the negatively charged silanols on the packing at the high pH values generally needed to generate sufficient EOF.

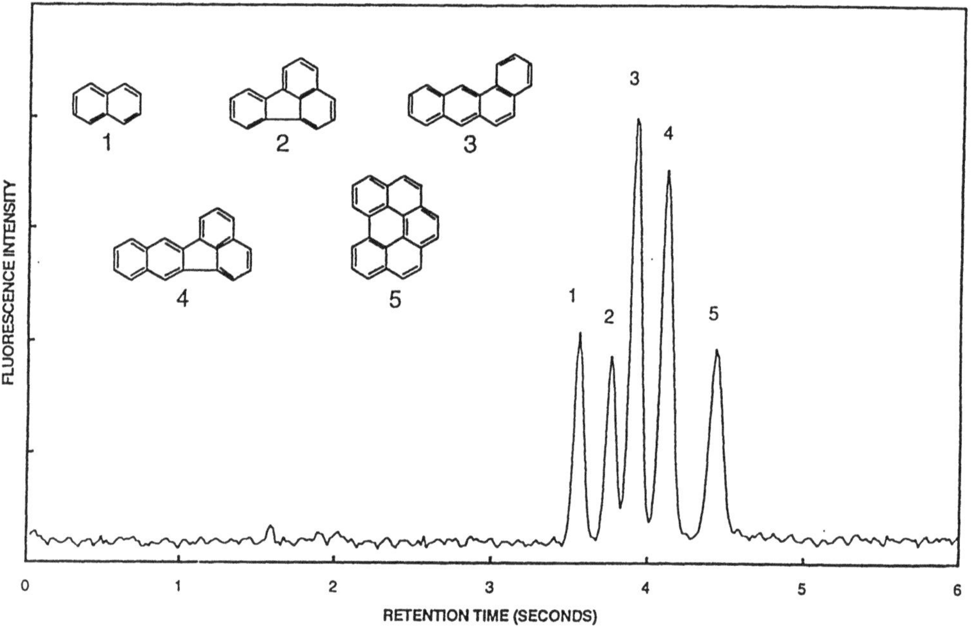

Figure 12.7   Rapid separation of 5 PAH's on non-porous ODS packing. Separation conditions: 100 µm ID × 6.5 cm packing of 1.5 µm ODS (10cm total length), 70:30 Acetonitrile:4 mM borate, +28 kV, injection 2 sec at 5 kV. Fluorescence detection. Peaks: 1= naphthalene, 2 = fluoranthene, 3 = benz[a]anthracene, 4 = benzo[k]fluoranthrene and 5 = benzo[ghi]perylene. Reproduced with permission from reference 25.

- *Column-to-column repeatability* – currently there are a variety of means of packing capillaries and no extensive data is yet available to demonstrate consistency of capillary packing.

- *Practical experience* – there are presently only a limited number of groups who have the levels of experience necessary to perform routine CEC.

- *Chiral CEC* – the column capacity in CEC appears to currently limit the usage of CEC in this area. However, the use of CEC specific chiral phases and the use of wider bore CEC capillaries may alter this situation.

- *Commercial availability of capillaries* – the majority of workers employ self-packed capillaries as supplies of commercially available packed capillaries are currently relatively scarce and expensive.

- *Preparative CEC* – the micro-preparative ability of CEC is likely to be similar to CE which represents only a small fraction of the amounts that can be collected by HPLC.

12.9 Conclusions

CEC is currently at the same position as CE was in around 1988. A number of workers have generated valuable contributions showing the potential advantages and applications possible. The real driving force to the current widespread application of CE was the advent of commercially available instrumentation and increased awareness of operating procedures. It is anticipated that CEC will emerge over the next 3–5 years to become a viable and routine alternative and compliment to both CE and HPLC.

Stop press update

CEC is a rapidly evolving technique and new developments are reported on a weekly basis. During the editing period of this book a number of references have appeared and are briefly described below. the subject has been recently reviewed (32).

Packing developments

Microparticulate cellulose based packing has been used (33) for separation of a range of compounds including basic compounds using non-aqueous electrolyte systems. Stationary phase has been packed (34) into continous beds which assists in production and ease of use. Dextran was added (34) to promote EOF. Advantages of the continous bed approach include elimination of the need for a frit.

A suspension of packing material has been used to separate PAH's (35). The capillary was simply flushed with a suspension of particles in a high pH buffer containing cyclodextrin and urea as modifiers. Other workers (36, 37) have investigated the use of new particle types with some success.

Pharmaceutical applications

A gradient CEC instrument coupled to a MS detector has been used (38) to separate mixtures of steroids, thiazide diuretics and benzodiazepines. A modified CEC-MS interface has been used (39) to analysis steroid mixtures.

The separation of 4 related impurities of an antibacterial agent have been separated by CEC within 8 minutes (40). The separation was optimised using an experimental design.

Chiral separation

Molecularly imprinted polymer has been used as a stationary phase in CEC (41). These imprinted polymers have been used by other group of workers (42) who also explored the possibility of cyclodextrin addition to the CEC mobile phase and a protein coated stationary phase.

Ion-exchange CEC

Nucleosil 5 µm SB has been used (43) to achieve separations of anions such as iodate and iodide. The solutes were detected at 190 nm.

References

1. Jorgenson J W and Lukacs K D, High-resolution separations based on electrophoresis and electroosmosis, *J. Chromatogr.*, 218 (**1981**) 209-219.

2. Knox J H and Grant I H, Miniaturisation in pressure and electroendosmotically driven liquid chromatography: some theoretical considerations, *Chromatographia*, 24 (**1987**) 135-143.

3. Behnke B, Grom E, and Bayer E, Evaluation of the parameters determining the performance of electrochromatography in packed capillary columns, *J. Chromatogr. A*, 716 (**1995**) 207-213.

4. Ding J and Vouros P, Capillary electrochromatography and capillary electrochromatography-mass spectrometry for the analysis of DNA adduct mixtures, *Anal. Chem.*, 69 (**1997**) 379-384.

5. Boughtflower R J, Paterson C J, and Underwood T, Capillary Electrochromatography – some important considerations in the preparation of packed capillaries and the choice of mobile phase buffers, *Chromatographia*, 40 (**1995**) 329-336.

6. Smith N W and Evans M B, The efficient analysis of neutral and highly polar pharmaceutical compounds using reversed – phase and ion-exchange electrochromatography, *Chromatographia*, 41 (**1995**) 197-203.

7. Yamamoto H, Baumann J, and Erni F, Electrokinetic reversed-phase chromatography with packed capillaries, *J. Chromatogr.*, 593 (**1992**) 313-319.

8. Altria K D, Smith N W, and Turnbull C H, A review of the current status of capillary electrochromatography technology and applications, *Chromatographia*, 46 (**1997**) 664-674.

9. Behnke B and Bayer E, Pressurised gradient electro-high-performance liquid chromatography, *J. Chromatogr. A*, 680 (**1994**) 93-98.

10. Euerby M R, Johnson C M, Bartle K D, Myers P, and Roulin C P, Capillary Electrochromatography in the pharmaceutical industry. Practical reality or fantasy? *Anal. Comm.*, 33 (**1996**) 403-405.

11. Yan C, Schafelberger D, and Erni F, Electrochromatography and micro high-performance liquid chromatography with 320μm I.D. packed columns, *J. Chromatogr. A*, 670 (**1994**) 15-23.

12. Dadoo R, Behnke B, Grom E, and Bayer E, Evaluation of the parameters determining the performance of electrochromatography in packed capillary columns, *J. Chromatogr. A*, 716 (**1995**) 207-213.

13. Smith N W and Evans M B, The analysis of pharmaceutical compounds using electrochromatography, *Chromatographia*, 38 (**1994**) 649-657.

14. Gordon D B, Lord G A, and Jones D S, Development of packed capillary column electrochromatography/mass spectrometry, *Rapid Comm. in Mass Spec.*, 8 (**1994**) 544-548.

15. Lord G A, Gordon D B, Tetler L W, and Carr C M, Electrochromatography – electrospray mass spectrometry of textile dyes, *J. Chromatogr. A*, 700 (**1995**) 27-33.

16. Mayer S and Schurig V, Enantiomer separation by electrochromatography on capillaries coated with chirasil-dex, *J. High Res. Chromatogr.*, 15 (**1992**) 129-131.

17. Lelievre F, Yan C, Zare R N, and Gareil P, Capillary electrochromatography: operating characteristics and enantiomeric separations, *J. Chromatogr. A*, 723 (**1996**) 145-156.

18. Li S and Lloyd DK, Packed-capillary electrochromatographic separation of the enantiomers of neutral and anionic compounds using β-cyclodextrin as a chiral selector. Effect of operating parameters and comparison with free-solution capillary electrophoresis, *J. Chromatogr. A*, 666 (**1994**) 321-335.

19. Armstrong D W, Tang Y, Ward T, and Nichols M, Derivatised cyclodextrins immobilised on fused-silica capillaries for enantiomeric separations via capillary electrophoresis, gas chromatography, or supercritical fluid chromatography *Anal. Chem.*, 65 (**1993**) 1114-1117.

20. Francotte E and Jung M, Enantiomer aeparation by open-tubular liquid chromatgraphy and electrochromatography in cellulose-coated capillaries, *Chromatographia*, 42 (**1996**) 521.

21. Mayer S and Schurig V, Enantiomer separation by electrochromatography in open tubular columns coated with chirasil-dex, *J. Liq. Chromatogr.*, 16 (**1993**) 915-931.

22. Mayer S and Schurig V, Enantiomer separation using mobile and immobile cyclodextrin derivatives with electromigration, *Electrophoresis*, 15 (**1994**) 835-841.

23. Li S and Lloyd D K, Direct chiral separations by capillary electrophoresis using capillaris packed with α_1 acid gylcoprotein chiral stationary phase, *Anal. Chem.*, 65 (**1993**) 3684-3690.

24. Boughtflower R J, Underwood T, and Maddin J, The production of packed capillaries using a novel pressurised ultrasound device, *Chromatographia*, 41 (**1995**) 398-402.

25. Yan C, Dadoo R, Yan C, Zare R N, Anex D S, and Rakestraw D J, Advances towards the routine use of CEC, *LC GC Int.*, March **1997** 164-174.

26. Dittmann M, Weinand K, Bek F, and Rozing G, Theory and practice of capillary electrochromatography, *LC-GC*, 13 (**1995**) 800-814.

27. Lane S J, Boughtflower R, Paterson C J, and Underwood T, Capillary electrochromatography / mass spectrometry: principles and potential for application in the pharmaceutical industry, *Rapid Comm. in Mass Spec.*, 9 (**1995**) 1283-1287.

28. Yan C, Dadoo R, Zare R N, Rakestraw D J, and Anex D S, Gradient elution in Capillary Electrochromatography, *Anal. Chem.*, 68 (**1996**) 2726-2730.

29. Yan C, Dadoo R, Zare R N, and Rakestraw D J, Capillary electrochromatography : analysis of polycyclic aromatic hydrocabons, *Anal. Chem.*, 67 (**1995**) 2026-2029.

30. Rebscher H and Pyell U Instrumental developments in capillary electrochromatography, *Chromatographia*, 42 (**1996**) 171-176.

31. Zimina T M, Smith R M, and Myers P, Electrochromatography in packed capillaries, *J. Chromatogr. A*, 758 (**1997**) 191-197.

32. Colon L A, Guo Y, and Fermier A, Capillary Electrochromatography, *Analytical Chemistry News and Features* A, **1997** (Aug 1), 461-467.

33. Maruska A and Pyell U, The development of microparticulate cellulose based packing materials and evaluation of potential use in capillary electrochromatography, *Chromatographia*, 45 (**1997**) 229-234.

34. Ericson C, Liao J L, Nakazato K, and Hjerten S, Preparation of continuous beds for electrochromatography and reversed-phase liquid chromatography of low-molecular-mass compounds, *J. Chromatogr. A*, 767 (**1997**) 33-41.

35. Gottlicher B and Bachmann K, Investigation of the separation efficiency of hydrophobic compounds in suspension electrokinetic chromatography, *J. Chromatogr. A*, 768 (**1997**) 320-324.

36. Fujimoto C and Muranaka Y, Electrokinetic chromatography using nanometer sized particles as the dynamic stationary phase, *J. High Resol. Chromatogr.*, 20 (**1997**) 400-402.

37. Bachmann K and Gottlicher B, New particles as pseudostationary phase for electrokinetic chromatography, *Chromatographia*, 45 (**1997**) 249-254.

38. Taylor M R and Teale P, Gradient capillary electrochromatography of drug mixtures with UV and electrospray ionisation mass spectrometric detection, *J. Chromatog. A*, 768 (**1997**) 89-95.

39. Lord G A, Gordon D B, Myers P, and King B W, Tapers and restrictors for capillary electrochromatography and capillary electrochromatography-mass spectrometry, *J. Chromatogr. A*, 768 (**1997**) 9-16.

40. Miyawa J H, Alasandro M S, and Riley C M, Application of a modified central composite design to optimise the capillary electrochromatographic separation of related S-oxidation compounds, *J. Chromatogr. A*, 769 (**1997**) 145-153.

41. Lin J-M, Nakagama T, Uchiyama K, and Hobo T, Capillary electrochromatographic separation of amino acid enantiomers using on-column prepared molecularly imprinted polymer, *J. Pharm. Biomed. Anal.*, 15 (**1997**) 1351-1358.

42. Nilsson S Schweitz L, and Petersson M, Three approaches to enantiomer separation of β-adrenergic antagonists by capillary electrochromatography, *Electrophoresis*, 18 (**1997**) 884-890.

43. Li D, Knobel H H, and Remcho V T, Capillary zone electrophoresis and ion-exchange capillary electrochromatography: analytical tools for probing the Hanford nuclear site environment, *J. Chromatogr. B*, 695 (**1997**) 169-174.

13 Use of non-aqueous electrolytes in pharmaceutical analysis

13.1 Introduction

There are many advantages to the use of organic solvents in CE and a number of research groups have demonstrated the possibility of operating CE using non-aqueous electrolyte solutions. Table 13.1 shows a number of applications of pharmaceutical interest (1–17). Recent reviews has covered a general overview of the use of non-aqueous solvents in CE (18, 19).

The addition of organic solvents induces shifts the dissociation constants for both acidic and basic compounds away from the pKa values obtained in water. Generally pKa values for acids become greater with increased organic solvent content (18, 19). The ionisation of basic compounds generally occurs at higher pH values in organic solvent-water mixtures. The pKa of solutes is shifted to varying extents by the different solvents. The solvation of the sample ions is also changed differentially when various solvents or solvent ratios are used as the solvents have different dielectric constants and viscosities (18,19). The viscosity and dielectric constants of the solvents govern both the sample ion mobility and the level of EOF generated. The selectivity can be therefore be manipulated in non-aqueous CE (NACE) by varying the ratio of components of the solvent mixture. This has been demonstrated by Bjornsdottir and Hansen (11) in the change in migration order of basic opium alkaloids when varying organic solvent compositions. Further selectivity manipulation can be achieved in aqueous FSCE by use of additives such as ion-pair reagents and cyclodextrins. Organic solvent-soluble forms of these additives used such as hydroxide salts of ion-pair reagents or acetylated-β-cyclodextrin must be used in non-aqueous CE (NACE).

In NACE the choice of electrolyte involves selection of the different organic solvent(s) and an electrolyte soluble in the organic solvent(s). Mixture of MeOH/ACN have been widely used (1,2,4) as this combination offers the possibility of general use and allow use of allowed detection at 200 nm. Other solvents such as Formamide (FA) N-methylformamide (NMF) and N,N-dimethylformamide (DMF) are useful alternatives as these have appropriate dielectric constants and viscosity values, but these solvents suffer from high UV background absorbance which reduces sensitivity and they may also attack the components of some CE instruments. The typical electrolyte salts used for high pH aqueous CE (such as phosphate and borate) cannot be used in non-aqueous CE as they are insoluble in pure organic solvents. Therefore non-aqueous soluble electrolytes such as sodium acetate or tris acetate are used. Organic soluble ion-pair reagents such as tetra-n-butylammonium tetrafluoroborate (2, 7) can also be used.

When organic solvents are added to aqueous buffers, pH readings are no longer accurate and are replaced by pH* which is indicative of the approximate pH of the solution. To analyse basic drugs low pH* electrolytes are used typically produced by the addition of glacial acetic acid into the electrolyte. Acidic compounds are analysed using high pH* electrolytes prepared by the addition for example of sodium acetate buffer or sodium hydroxide.

A limited amount of work has been performed on the use of indirect UV detection in NACE. A range of inorganic anions (9), metal ions (20,21) and small cations (20) have been determined.

The currents generated in NACE are considerably lower (18, 19) than in aqueous CE. However the upper current limits possible are higher in aqueous CE as non-aqueous media have lower thermal conductivity than aqueous solutions which reduces the heat dissipated at the capillary wall. The organic solvents have low heat capacities and can boil and/or outgas at lower temperatures than water. For example methanol and acetonitrile boil at 65°C and 82°C respectively (2). In practice electrolyte ionic strengths and applied voltages were adjusted to produce currents no greater than ~45 µA.

These non-aqueous solvents are ideal for use in CE-MS studies (10, 16) as the electrolytes are volatile and improved solute ionisation can occur due to the ease of solvent evaporation from the solute ion. For example (16) the sensitivity for Tamoxifen and metabolites was ~5 times higher using a methanolic electrolyte compared to an aqueous buffer.

13.2 Basic Drugs

The main attention of the use of NACE in pharmaceutical analysis has focused on the use of low pH* electrolytes to separate basic drugs (Table 13.1). Typically ammonium acetate and acetic acid have been added to solvents, or solvent mixtures, using methanol, acetonitrile, FA, NMF or DMF. These studies have confirmed that improved and different separation selectivities for basic compounds can be obtained in organic solvents when compared to the separations obtained using aqueous electrolytes. Figure 13.1 shows separation of a range of opiates using an aqueous electrolyte or a MeOH-ACN electrolyte. Similar electrolyte conditions have been used (17) to achieve the separation of tetracyclineoxytetracycline, doxycycline and chlortetracycline by non-aqueous CE.

20 mM ammonium acetate in acetonitrile-methanol-acetic acid (49:50:1) has been used (2) to obtain the baseline separation of nine morphine analogues, eleven antihistamines, eleven antipsychotics and ten stimulants within 6 minutes. The migration order observed was very different from one observed for an aqueous medium. Improved speed, efficiency and precision were achieved when using tetra-n-butylammonium tetrafluoroborate as the electrolyte.

In binary solvent systems the resolution can be greatly influenced by switching the ratios of solvents. For example several opium alkaloids were resolved (3) using a 25/75 ACN/MeOH. The alkaloids were unresolved using a range of other ratios of the solvents.

Table 13.1 Range of non-aqueous CE applications

Compound (s)	Solvent	Ref. No.	Comments
Acidic drugs, surfactants and dyes	ACN and MeOH	1	Quantitative analysis of drug content, impurities determinations
Basic drugs including 9 morphine analogues, 11 antihistamines, 10 stimulants and 11 antipsychotics	ACN-MeOH mixtures	2	Addition of ion-pair reagent improved separation, RSD of 1% for migration times
Basic drugs	FA, NMF, ACN	3	Comparisons of aqueous and non-aqueous electrolytes
Basic opium alkaloids	ACN-MeOH mixes	4	Quantitative analysis in pharmaceutical formulation
Chiral amino acids	FA, NMF	5, 6	β-CD added into electrolyte
Chiral compounds including 11 pharmaceutical amines	FA, NMF, DMF	7	Separation with β-CD, improved with ion-pair reagent addition
Chiral primary amine	FA with crown ether	8	Aromatic amines, amino acids and amino alcohols chirally separated
Inorganic anions – including chloride, nitrate, sulphate	MeOH	9	Indirect detection with phthalate as UV absorber, also electrochemical detection
Mifentidine	MeOH	10	Metabolism studies with CE-MS detection
Opium alkaloids	MeOH and ACN	11	Determination of opium alkaloids in crude opium
Peptides	FA	12	Higher separation efficiencies than aqueous buffer
Racemic drugs including cisapride and salbutamol	MeOH-ACN mixes	13	(R) or (S)-camphorsulphonate added to electrolyte for chiral resolutions
Racemic pharmaceuticals	FA, NMF DMF	14	Addition of cyclodextrins into buffers
Surfactants	ACN-MeOH mixes	15	Indirect UV with toluenesulphonate as UV absorber
Tamoxifen and its metabolites	MeOH with 7 mM SDS	16	On-Line CE – Electrospray Ionization MS
Tetracycline and impurities	MeOH-DMF-ACN mix	17	LOD of 0.02 % for impurities

Key:

ACN = Acetonitrile
DMF = N,N-dimethylformamide
FA = formamide

NMF = N-methylformamide
MeOH = Methanol

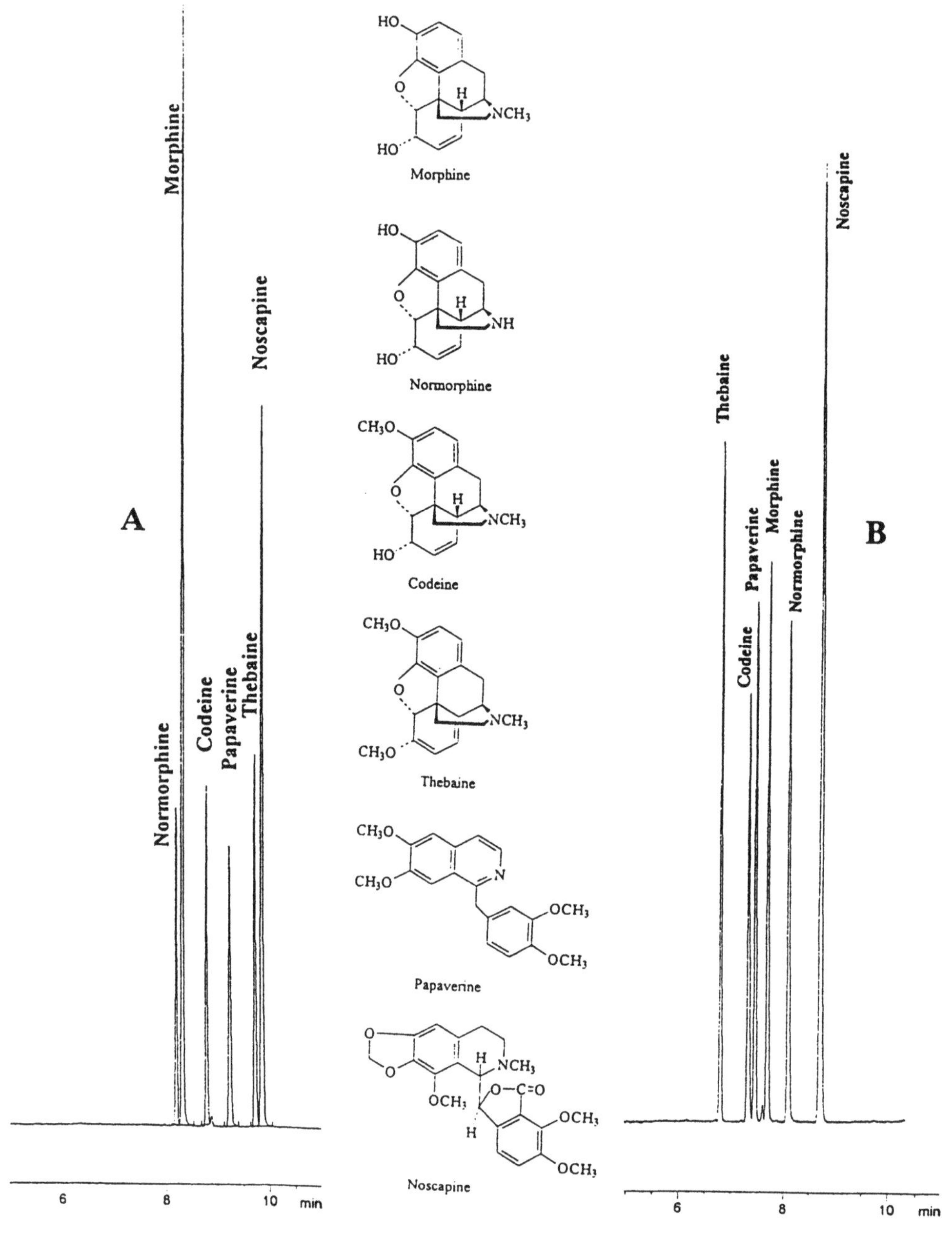

Figure 13.1 Separation of 6 opium alkaloids using aqueous (A) or non-aqueous (B) electrolytes. Separation conditions: A) 56 cm to detector × 75 micron capillary, 30°C, 214 nm, 30 kV, 50 mM 6-aminocaproic acid pH 4 with 30 mM dimethyl-β-CD B) 55 cm to detector × 50 micron capillary, 30°C, 214 nm, 25 kV, 25 mM ammonium acetate and 1 M acetic acid in ACN. Reproduced with permission from reference 3.

13.3 Acidic Drugs

The addition of sodium acetate to 50/50% v/v ACN/MeOH produces a solution with a pH* value of 9.3. This electrolyte has been used to resolve a large number of acidic drugs (1). Figure 13.2 shows the effects of the ionic strength variation on the separations achieved (1) for a range of acidic drugs. The EOF flow was significantly increased at lower ionic strengths which reduced the resolution of the solutes. At higher sodium acetate concentrations the symmetry and resolution of early migrating peaks improved. Use of lower ionic strengths improved (1) the symmetry for the late migrating peaks.

13.4 Impurity/Metabolites Determinations

A highly selective method for quantitative determination of impurities in tetracycline antibiotics based on non-aqueous capillary electrophoresis has been developed (17). The degradation products 4-epitetracycline, anhydrotetracycline and 4-epianhydrotetracycline were determined in tetracycline hydrochloride drug substance with limits of detection corresponding to 0.06 %, 0.04 % and 0.02 % respectively. The relative standard deviations were about 4 % at the 0.1 % level of impurity in tetracycline hydrochloride. Desmethyltetracycline and a number of unknown impurities were also separated within a 10 minute analysis.

A low pH* ACN electrolyte was used (3) in the purity testing of Imipramine-N-oxide. The limit of detection for the impurities was 0.02 % when detected at 214 nm. The low current generated using the non-aqueous electrolyte allowed use of a 100 micron bore capillary to achieve an improved sensitivity.

Tamoxifen and a range of metabolites were separated (16) using low pH* methanol electrolyte with CE-MS detection. Solutions of Tamoxifen that had been incubated with rat microsomes were analysed by CE-MS to determine metabolic pathways. A similar exercise involving CE-MS and a low pH* electrolyte was used by another group of workers (10) to monitor the *in vitro* metabolism of Mifentidine following incubation with guinea pig hepatic microsomes. A metabolic pathway for Mifentidine was devised (10) from the CE-MS data produced.

A partially degraded solution of phenoxymethoxypenicillin has been separated (1) using various mixtures of ACN/MeOH containing 10 mM Na acetate. The separation selectivity was altered dramatically by varying the solvent ratios (Figure 13.3). The detection limits for the impurities were found to be 0.05 % area/area. Penicillin's are prone to degradation in aqueous solution. In these studies it was observed that the rate of degradation was appreciably reduced when the samples were prepared in non-aqueous solvents.

13.5 Chiral Separations

The range of chiral selectors employed in aqueous CE (Chapter 4) can also be employed in NACE. Underivatised cyclodextrins (CD's) are readily soluble (5, 6, 14) in solvents such as FA and NMF. For example 100 mM β-CD dissolved in FA has been used for the chiral separation of racemic basic drugs. This compares with aqueous solubility of β-CD which is

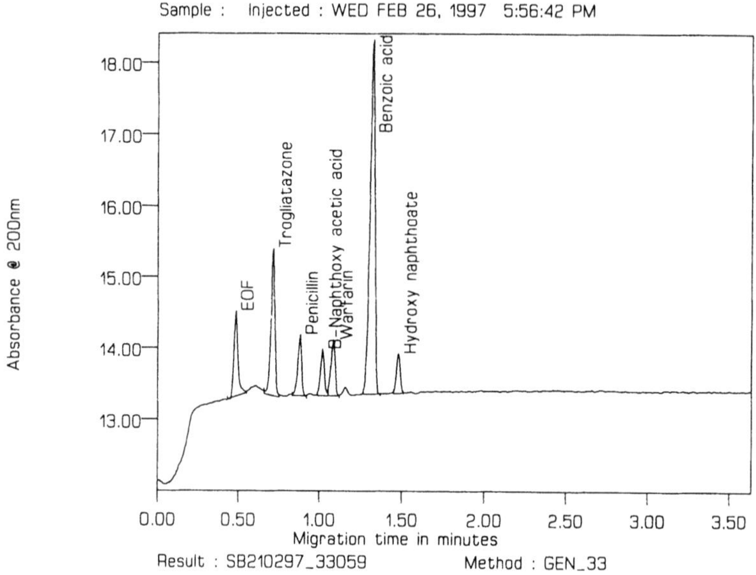

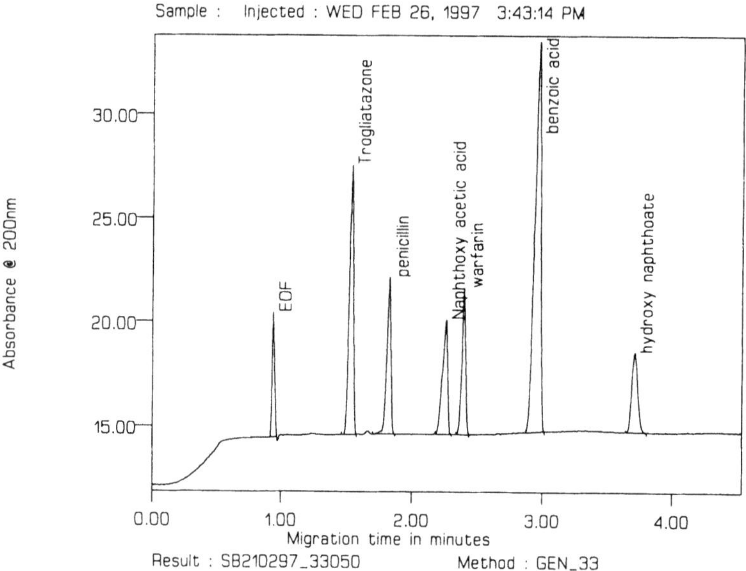

Figure 13.2 Effect of buffer ionic strength on separation performance for a range of acidic compounds. a) 1 mM Na acetate, b) 10 mM Na acetate. Separation conditions : 27 cm × 50 μm capillary (20 cm to detector), 200 nm, 25 kV, 1 second injection, using 50/50% v/v ACN/MeOH containing either 1mm or 10 mM sodium acetate pH* 9.3 (unadjusted). Reproduced with permission from reference 1.

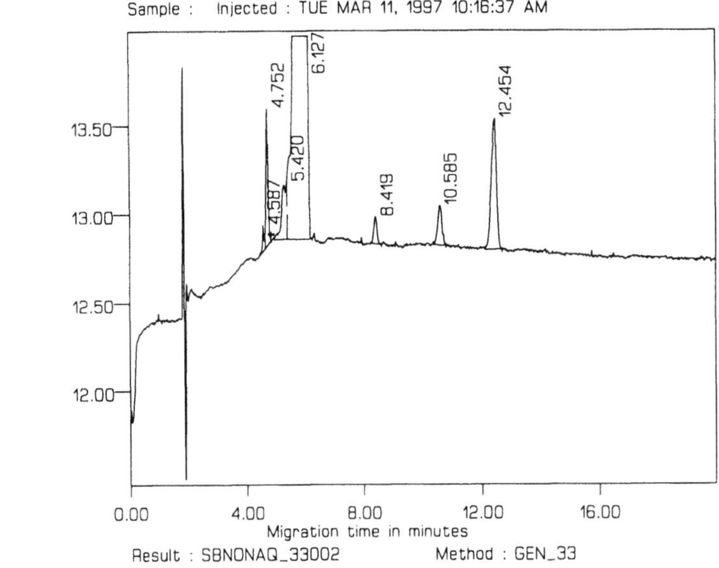

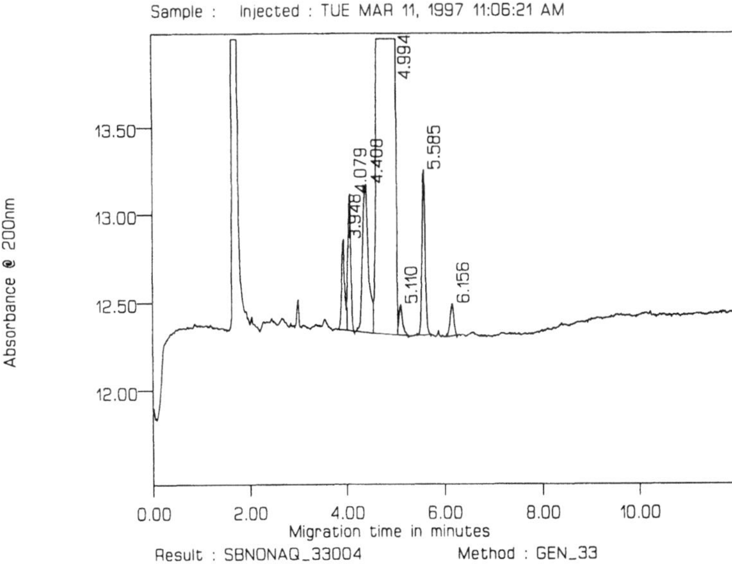

Figure 13.3 Separation of partially degraded penicillin sample solution with different organic solvent ratios, a) Separation using 75:25 MeOH:ACN, b) Separation using 25:75 MeOH:ACN. Separation conditions : 27 cm × 50 μm capillary (20 cm to detector), 200 nm, 25 kV, 1 second injection, using 10 mM sodium acetate dissolved in solvent. Reproduced with permission from reference 1.

only 16 mM. However, underivatised CD's have only limited solubility in solvents such as ACN or MeOH. Chemically derivatised CD's such as acetylated-β-CD are readily soluble in ACN/MeOH.

The enantiomeric separations of eight primary amino compounds including aromatic amines, amino acids and amino alcohols have been achieved (8) by addition of a chiral crown ether, (+)-18-crown-6 tetracarboxylic acid, to FA. The addition of tetra-n-butylammonium perchlorate as a supporting electrolyte to the electrophoretic solution improved the separation efficiency and gave the baseline enantiomeric separation of D,L-1-phenylethylamine which had not been separated (8) by any other separation mode of CE .

(R) or (S)-camphorsulphonate has been added (13) into MeOH-ACN mixtures to chirally resolve a range of basic pharmaceuticals. The camphorsulphonate enantiomer differentially forms ion-pairs with the positively charged drug enantiomers resulting in chiral separations.

13.6 Excipients and Raw Materials

The range of excipients and raw materials used in pharmaceutical manufacture are considerable and include dyes, preservatives, dyes and surfactants. Aqueous based CE has been used to analyse many of these materials (Chapter 7). NACE offers distinct advantages in this type of application as many of the solutes have limited water solubility.

The highly efficient resolution (Figure 13.4) of three acidic dyes has been obtained (1) with 2 mM NaAc in 50/50 ACN/MeOH. Detection was performed at 200 nm and the sample was dissolved in 50/50 % v/v ACN/MeOH.

The separation of the homologues of dodecylbenzenesulphonate surfactants has been achieved using different NACE conditions (1, 15). Linear alkyl benzenesulphonates (LAS) were separated using methanol containing acetic acid and tetramethylammonium hydroxide (15) and the optimise method was applied to separation of LAS homologues in commercial detergent solutions. Indirect UV detection was also used (15) to determine alkylsulphate contents in detergents – the sensitivity and selectivity in the NACE was considerably better than the aqueous based buffers.

Parahydroxybenzoate preservatives are widely used within the pharmaceutical industry and require use of MECC to obtain separation of a mixture of parahydroxybenzoates. Figure 13.5 shows separation of methyl-,ethyl-,butyl-and propyl-parahydroxybenzoate using a pH*13 electrolyte and their separation has been attempted by aqueous FSCE (19). Incomplete resolution by aqueous CE necessitated the use of MECC.

13.7 Main Component Assay

Leung et al (2) reported migration time repeatability of 0.8–3.7 % RSD for the analysis of basic drugs in non-aqueous CE. Acceptable within-day precision data has been reported (8) and RSD values of 10 % were shown for day-to-day EOF repeatability.

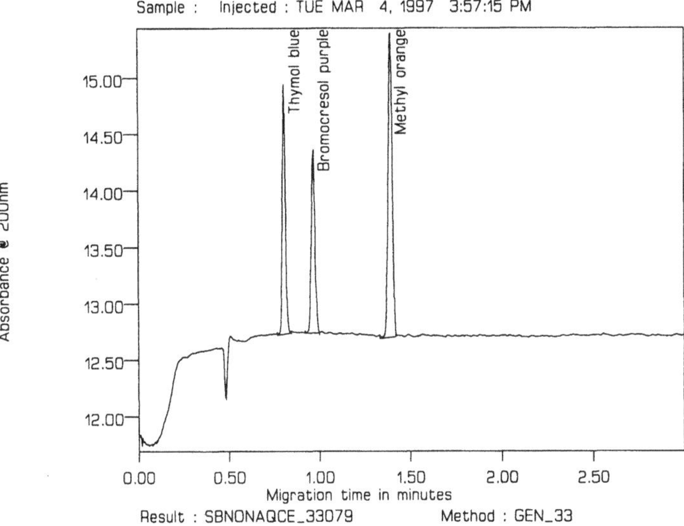

Figure 13.4   Separation of acidic dyes using 2 mM Na Acetate in 50/50 ACN/MeOH. Separation conditions: 27 cm × 50 µm capillary (20 cm to detector), 200 nm, 30 kV, 1 second injection, using 50/50 % v/v ACN/MeOH containing 2mM sodium acetate pH* 9.3 (unadjusted). Reproduced with permission from reference 1.

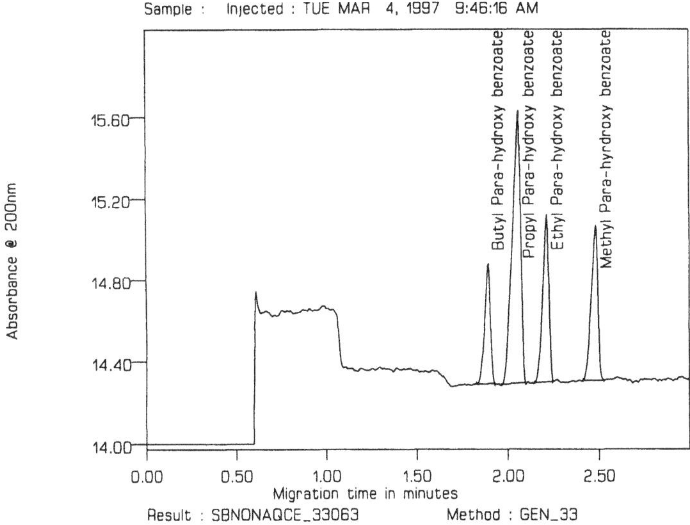

Figure 13.5   Separation of parahydroxybenzoate preservatives using a pH*13 electrolyte. Separation conditions: 27 cm × 50 µm capillary (20 cm to detector), 200 nm, 25 kV, 1 second injection, using 50/50% v/v ACN/MeOH containing 2mM sodium hydroxide pH*13 (unadjusted). Reproduced with permission from reference 1.

Table 13.2 Repeatability results ($n = 10$) for quantitation of an acidic drug

	Analyte	Migration time	Relative migration time	Peak area ratio
Study 1	Troglitazone RSD	0.54	0.52	1.0
Study 2	Butyl PHB RSD	2.1	0.4	1.2

PHB = Para Hydroxybenzoate
Relative migration times and peak area ratios compared to β-naphthoxyacetic acid internal standard.
Study 1: 50 μm × 27 cm, 2 mM Na acetate in 50 % v/v ACN/MeOH pH* 9.3 (unadjusted) +30 kV.
Study 2: 50 μm × 27 cm, 2 mM NaOH 50 % v/v ACN/MeOH pH* 13 (unadjusted) +30 kV
Reproduced with permission from reference 1

A number of injection repeatability studies have been performed (1) using high pH* electrolyte. The separations were found to be stable in terms of EOF, migration time and peak area ratios. Table 13.2 gives data obtained for 10 injections of two test mixtures on different instruments and capillaries on different days with different preparations of electrolyte. Internal standards were necessary to obtain acceptable injection precision.

Levels of an acidic drug, troglitazone, in a tablet formulation were assayed (1) using non-aqueous CE. β-naphthoxyacetic acid was used as an internal standard to obtain acceptable precision data. An average assay result of 300.2 mg/tablet was obtained for a 300 mg label claim tablet formulation. This result was well within the manufacturing specification range which confirmed that the CE method was capable of generating accurate and precise data.

13.8 Inorganic Ions and Small Organic Ions

The use of non-aqueous solvents combined with indirect UV detection allows useful separations of inorganic ions or organic ions with no chromophore. There have been a limited number of reports in this area to-date (9, 20) which have demonstrated that different selectivities can be obtained compared to use of aqueous electrolyte systems. The use of NACE to determine these small ions is of particular importance if they are present as contaminants in water-insoluble compounds.

Figure 13.6 shows the NACE separation of a range of metal ions and small cations (20) with indirect UV detection. The methanolic electrolyte contained imidazole as the UV absorbing species and a low pH* was obtained by the addition of glacial acetic acid. The selectivity of the method was different from that of an equivalent aqueous system. A similar set of operating conditions were used to separate a range of quaternary ion-pair reagents which can be present as contaminants in drug substance materials (21). Selectivity can be manipulated in these NACE separations by the additions of additives such as crown ethers to selectively chelate with the ions being separated. For example (20) tromethamine (tris, Trizma) was resolved from hydroxylamine by the addition of 2 mM 18-6 crown ether to the electrolyte. This method was validated for quantitation of tromethamine which is a widely used buffering agent in pharmaceutical preparations. The linearity correlation coefficient of $R = 0.9996$ was obtained for Trizma (dissolved in 100 ppm internal standard, Dodecyltrimethylammonium

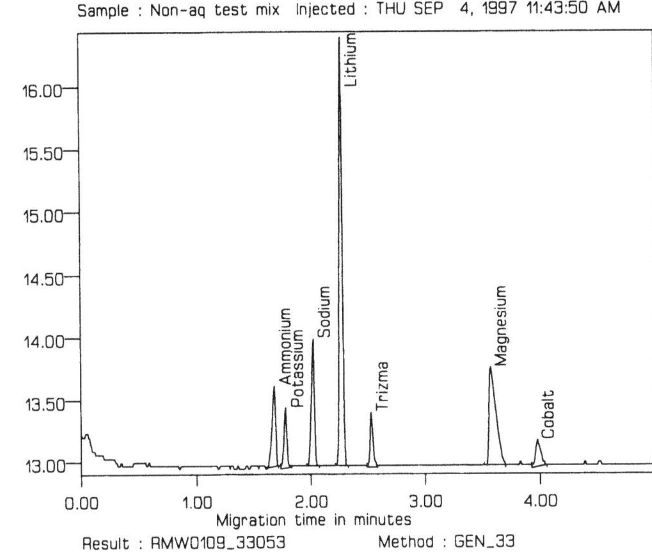

Figure 13.6   Separation of a range of metal ions and cations using non-aqueous CE. Separation conditions: 27 cm × 50 µm capillary (20 cm to detector), indirect UV detection at 214 nm, 10 kV, 5 mM imidazole in 99 % v/v MeOH with 1 % v/v glacial acetic acid

bromide), over the concentration range 50–150 ppm. The precision for 10 injections of the solution was 1.1 % RSD for peak area ratios and less than 0.1 %RSD for relative migration times. The response factor for repeated injections of 5 individual calibrations gave an RSD of 1.2 %

A range of inorganic anions such as nitrate, azide, chloride and sulphate have been separated in NACE (9) using phthalate as the UV absorbing species dissolved in methanol:water 98:2 % v/v.

13.9 Practicalities of Routine NACE Operation

Methanol and acetonitrile are more volatile than water and has been demonstrated (17) that evaporation of the solvent from sample as well as from the electrophoresis medium may cause severe problems in some capillary electrophoresis instruments. However this was not found to be as significant an issue as expected if analysis times were short. The use of cooled and sealed electrolyte vials, or a replenishment system, reduces these effects.

Attention should be paid to the solvent resistant properties of the instruments used. Instrument-type specific problems with corrosion of seals or component dissolution may occur and it is recommended to check solvent compatibility with particular instrument suppliers.

The generation of low currents leads to relatively quiet baselines as increased baseline noise is evident in CE at higher operating currents. This may be of particular importance when utilising indirect UV detection in non-aqueous CE, as this form of detection is particularly prone to increased signal noise at currents greater than 15–20 μA. Increased sensitivity in non-aqueous CE may be obtained using specially purified (Far UV) organic solvents.

Capillaries should be regenerated prior to their initial use by rinsing with 0.1 M aqueous NaOH solution followed by 10 minutes with pure methanol.

13.10 Comparison of the Advantages of NACE and Aqueous CE

There are a number of major advantages gained when operating NACE compared to aqueous CE. These advantages include increased selectivity, improved solubility for water insoluble compounds, reduced currents which allow application of high field strengths to achieve rapid separation. The principal advantage is the improved ability to manipulate selectivity as the pKa and mobilities of solutes can be altered dramatically by varying the organic solvent choices and ratios and the pH* value. This improved selectivity options provides significantly more method development possibilities in comparison to the use of electrolyte additives. Other advantages include the high MS compatibility of NACE electrolytes and the ability to operate wider bore capillaries as operating currents are lower than in aqueous CE.

The principal disadvantage of NACE is the increased volatility of the solvents causing evaporation problems. Although the currents generated in NACE are lower than in aqueous CE, the poorer thermal properties of the organic solvents give a lower maximum current level prior to out-gassing/boiling problems. Another factor to consider is that commonly employed reagents are not always soluble in NACE electrolytes, but this apparent difficulty may offer new opportunities as previously unusable water insoluble reagents can be employed.

References

1. Altria K D and Bryant S M, Highly selective and efficient separations of a wide range of acidic species in capillary electrophoresis employing non-aqueous media, *Chromatographia*, 46 (**1997**) 122-130.

2. Leung G N W, Tang H P O, Tso T S C, and Wan T S M, Separation of basic drugs with non-aqueous capillary electrophoresis, *J. Chromatogr. A*, 738 (**1996**) 141-154.

3. Bjornsdottir I and Hansen S H, Comparison of separation selectivity in aqueous and non-aqueous capillary electrophoresis, *J. Chromatogr. A*, 711 (**1995**) 313-322.

4. Bjornsdottir I and Hansen S H, Determination of opium alkaloids in crude opium using non-aqueous capillary electrophoresis, *J.Pharm. Biomed. Anal.*, 13 (**1995**) 1473-1481.

5. Valko I E, Siren H, and Riekkola M L, Chiral separation of dansyl-amino acids in non aqueous medium of capillary electrophoresis, *J. Chromatogr. A*, 737 (**1996**) 263-272.

6. Valko I E, Siren H, and Riekkola M L, Chiral separation of dansyl amino acids by capillary electrophoresis: comparison of formamide and n-methylformamide as background electrolytes, *Chromatographia*, 43 (**1996**) 242-246.

7. Wang F and Khaledi M G, Chiral separations by nonaqueous capillary electrophoresis, *Anal. Chem.*, 68 (**1996**) 3460-3467.

8. Mori Y, Ueno K and Umeda T, Enantiomeric separations of primary amino compounds by nonaqueous capillary zone electrophoresis with a chiral crown ether, *J. Chromatogr.A*, 757 (**1997**) 328-332.

9. Salimi-Moosavi N and Cassidy R M, Capillary electrophoresis of inorganic anions in nonaqueous media with electrochemical and direct UV detection, *Anal. Chem.*, 67 (**1995**) 1067-1073.

10. Tomlinson A J, Benson L M, Gorrod J W, and Naylor S, Investigation of the *in vitro* metabolism of the H$_2$-antagonist mifentidine by on-line capillary electrophoresis-mass spectrometry using non-aqueous separation conditions, *J. Chromatogr. B*, 657 (**1994**) 373-381.

11. Bjornsdottir I and Hansen S H, Determination of opium alkaloids in crude opium using nonaqueous capillary electrophoresis, *J. Pharm. Biomed. Anal.*, 13 (**1995**) 1473-1481.

12. Sahota R S and Khaledi M G, Nonaqueous capillary electrophoresis, *Anal. Chem.*, 66 (**1994**) 1141-1146.

13. Bjornsdottir I, Hansen S H, and Terabe S, Chiral separations in non-aqueous media by capillary electrophoresis using the ion-pair principle, *J. Chromatogr. A*, 745 (**1996**) 37-44.

14. Wang F and Khaledi M G, Chiral separations by non-aqueous capillary electrophoresis, *Anal. Chem.*, 68 (**1996**) 3460-3467.

15. Salimi-Moosavi N and Cassidy R M, Application of nonaqueous capillary electrophoresis to the separation of long-chain surfactants, *Anal. Chem.*, 68 (**1996**) 693-299.

16. Lu W, Poon G K, Carmichael P L, and Cole R B, Analysis of Tamoxifen and its metabolites by on-line capillary electrophoresis-electrospray ionization mass spectrometry employing non-aqueous media containing surfactants, *Anal. Chem.*, 68 (**1996**) 668-674.

17. Tjornelund J and Hansen S H, Determination of impurities in tetracycline hydrochloride by non-aqueous capillary electrophoresis, *J. Chromatogr. A*, 737 (**1996**) 291-300.

18. Hansen S H, Tjornelund J, and Bjornsdottir I, Selectivity enhancement in capillary electrophoresis using non-aqueous media, *Trends in Anal. Chem.*, 15 (**1996**) 175-179.

19. Valko I S, Siren H, and Riekkola M-J, Capillary electrophoresis in non-aqueous media: an overview, *LC GC Int.*, March (**1997**) 190-196.

20. Altria K D, Wallberg M, and Westerlund D, Determination of metal ions and small cations by capillary electrophoresis using indirect UV and non-aqueous solvents, *Chromatographia*, in press **1998**.

21. Johnson B D, Grinberg N, Bicker G, and Ellison D, The quantitation of a residual quaternary amine in bulk drug and amine in bulk drug and process streams using capillary electrophoresis, *J. Liq. Chromatogr. Rel. Techn.*, 20 (**1997**) 257-272.

14 The Use of Chemometrics and Experimental Designs in CE Method Development and Robustness Testing

14.1 Introduction

There are a great many experimental parameters that can be varied to influence the separation selectivity and performance achieved for a CE method. Conventionally when attempting to assess the impact of each of these parameters each factor may be varied sequentially. This "uni-variate" approach typically involves holding all parameters constant and varying one parameter at a time and measuring method responses such as resolution and analysis times. For example, a sequence of injections may be performed to assess the impact of pH, followed by a further injection series to assess impact of ionic strength and further sequences to assess the influence of factors such as buffer additives, organic solvent content, temperature, sample concentration etc. This step-by-step approach, although widely used, involves a large number of independent analyses, and could be replaced by statistically designed experimental protocols in which several factors are simultaneously varied. These multi-variate experimental design approaches have advantages in terms of reductions in the number of experiments, improved statistical interpretation possibilities, and reduced overall analysis time requirements.

A number of design types are available and have been used in CE applications (1–22). Table 14.1 shows that these designs have been used in both method development and method robustness studies. Chemometric method of data manipulation have also been used in identification of peaks in CE separations. General textbooks are available (23, 24) which cover chemometrics and experimental design procedures in greater depth. Experimental design approaches have been used in HPLC for method development (25–27) and robustness testing (28, 29).

Statistical evaluation of the data from suitably designed experimental sequences can often indicate whether parameters interact. An interaction between factors indicates that a simultaneous change in two factors together has a greater effect than the sum of each of the independent effects. For example, if the concentrations of both buffer and ion-pair reagent are simultaneously increased, they will both cause an increase in the current inside the capillary possibly changing selectivity or migration times. For example a 5°C rise in temperature may increase the current 5 µA and a voltage increase of 2 kV may also increase the current 5 µA. However as these factors interact a simultaneous increase of 5°C and 2 kV may result in a increase of 13 µA. These subtle interactions cannot be assessed in the uni-variate approaches as only one parameter is varied at a time. Therefore, the appropriate use of experimental designs can be extremely beneficial in evaluation of capillary electrophoresis methods.

Table 14.1 Applications of experimental designs and chemometrics in CE

Solute	Design	Application	Ref.
Acidic drugs	Central composite and fractional factorial	FSCE robustness study	1
Amlodipine	Central composite	Chiral method development and robustness	2
Amphetamine	Principle Component Analysis and Iterative Target Transform Factor Analysis	Peak identification in urine samples	3
Amphetamines	Central composite	Chiral method development	4
Antibacterial and related substances	Modified central composite	CEC method development	5
Basic drug and impurities	Central composite	FSCE robustness study	6
Basic drugs	Central composite and fractional factorial	FSCE robustness study	7
Clenbuterol	Plackett-Burman	Chiral method development	8
Dexfenfluramine	Plackett-Burman	Chiral method development	9
Drug residues	Central composite and fractional factorial	FSCE robustness study	10
Dyes	Multicomponent analysis (Kalman filtering)	Peak identification	11
Flavanoids	Overlapping resolution mapping	FSCE method development	12
Hapten-antibody complex	Various	FSCE method development	13
Fungal metabolites in penicillin commune	Fractional factorial	MECC method development	14
Mirtazapine, and five related substances	Various	FSCE method development	15
Potassium content in drug substance	Central composite	FSCE robustness study	16
Quinolone antibacterials	Overlapping resolution mapping	MECC method development	17
Ropivacaine	Full factorial	Robustness testing – chiral method	18
Sulphonamides	Overlapping resolution mapping	FSCE method development	19
Sulphonamides, dihydrofolate reductase inhibitors and beta-lactam antibiotics	Box-Behnen	Method development	20
Testerone esters	Plackett-Burman	MECC method development	21
Various	Uniform design	Method development	22

Various chemometric experimental designs have been employed (Table 14.1) for the optimisation of capillary electrophoresis (CE) methods. The designs employed include Central Composites, Fractional Factorials, Plackett-Burman, and Overlapping-Resolution Mapping. Optimisation studies have largely concentrated on the use of these designs on selection of the optimal electrolyte composition.

Similar designs have been utilised in the assessment of the robustness of CE methods. The robustness testing studies performed have involved the use of screening designs to identify the critical parameters affecting responses such as migration times and resolution. Further designs such as Central Composites can be then employed to set method limits following robustness studies.

The use of experimental designs and statistical data evaluation is particular attractive in CE as the autosamplers and equipment operation is entirely PC-controlled. Therefore vials containing a number of different electrolyte compositions can be placed on the autosampler and method development or robustness sequences can be programmed which vary parameters such as temperature injection time, voltage etc. for a range of different electrolyte compositions. This is also possible on HPLC but demands the use of sophisticated instruments whilst this can be performed on standard commercial CE equipment.

14.1.1 Full and Fractional Factorial designs

Factorial designs can be employed to screen the impact of all variables and potential interactions and, despite the potentially large number of experiments. To fully evaluate 5 factors (at 3 levels), 243 experiments would be required in a full factorial design. Therefore, Fractional Factorial designs and Plackett-Burman designs are employed to dramatically reduce the number of experiments whilst still maintaining the statistical ability to identify the influence of each parameter and to monitor possible identified interactions between factors.

14.1.2 Central Composite and Overlapping Resolution Mapping Designs

Having established the key factors affecting the performance of a method by using a screening design it may then be appropriate to optimise the method by obtaining response surfaces. These plots can provide a graphical representation of the data over the ranges studied and can be used to predict areas of optimal performance. Response surfaces can also show the impact of small changes away from a set value on the method performance which is valuable information relating to method robustness. Response surfaces can be obtained by employing designs such as Central Composites, Overlapping Resolution Mapping (ORM) and Box-Behnken.

14.1.3 Simplex optimisation

This is a popular approach in HPLC but has been of limited application in CE to-date. The only other report of the use of experimental design approaches for CE method development involves (30) the optimisation of an MECC method for the separation of 20 derivatised amino acids using a simplex scheme. In this approach the experimental results from each sequential injection are used to suggest the next combination of parameters. Subsequent experiments lead to the location of a localised optimal parameter combination. Percentage organic solvent, SDS concentration and pH were simultaneously investigated. An improved but incomplete separation of the 20 amino acids was obtained after only 10 analyses.

14.2 MECC Method Development

Electrolyte composition for the resolution of 7 corticosteroids was optimised (31) using a Central Composite design. The factors of pH, and concentrations of both SDS and borate were studied at 3 levels in an 19 experiment sequence. The mid-point conditions were randomly repeated throughout the sequence to assess variability. Desirability plots were constructed for all resolutions which indicated optimal conditions of pH 9.2 with 60 mM borate, and 10 mM SDS. These optimal conditions were experimentally confirmed resulting in clear baseline resolution of the components within 20 minutes.

A Plackett-Burman approach has been employed (21) in the optimisation of an micellar electrokinetic capillary chromatography (MECC) method for the separation of 4 testosterone esters. An initial set of operating conditions produced only a partial separation of a 4 component text mix with two peaks co-migrating. Five factors (pH, buffer concentration, acetonitrile content, SDS concentration and concentration of the co-micelle agent sodium heptyl sulphate (SHS) were investigated in an 8 injection sequence. Table 14.2 shows the design which includes 2 dummy factors. Statistical evaluation of the results indicated that resolution of the two peaks was decreased with % ACN and SHS concentration but improved by increasing pH and SDS concentration. An electrolyte containing no co-micelle agent, pH 9, 40 % acetonitrile, 50 mM SDS and 40 mM buffer concentration was shown to be close to the optimum and produced the required separation.

The effect of varying pH and SDS concentration in the ranges 6.5–8.5 and 10–50 mM respectively upon the separation of 14 quinolone antibacterials by MECC has been evaluated (12) using an ORM scheme. Seven experiments were performed covering concentration ranges for sodium heptanesulphonate, sodium cholate and acetonitrile. The overlapped resolution map obtained from analysis of the resolution data is given in Figure 14.1a. The most

Table 14.2 Plackett-Burman design for testosterone ester separation

buffer	dummy 1	pH	%SHS	%ACN	mM SDS	Dummy 2	mM buf.
1	+0	9	10	40	50	–0	20
2	–0	9	10	50	40	+0	20
3	–0	8	10	50	50	–0	40
4	+0	8	0	50	50	+0	20
5	–0	9	0	40	50	+0	40
6	+0	8	10	40	40	+0	40
7	+0	9	0	50	40	–0	40
8	–0	8	0	40	40	–0	20

Reproduced with permission from reference 21

suitable operating conditions predicted were confirmed (12) experimentally and the separation obtained is shown in Figure 14.1b.

Optimized separation conditions (14) for isolates of Penicillium commune were obtained by experimental designs. The effect of concentration of phosphate and borate, buffer pH, addition of sodium dodecyl sulphate (SDS), sodium deoxycholate (SCD), acetonitrile and methanol and voltage were examined using two-level fractional factorial design. Optimum separation conditions were obtained which gave good resolution of the components in the extract.

A Box-Behnken factorial design was used to optimise a 30 component MECC separation (20). The final optimised separation produced 25 peaks within a 25 minute analysis.

A statistical experiment-arrangement method called a uniform design has been employed (23) to systematically optimize MECC separations experiments. The uniform design utilises statistical regression analysis and can advantages over other optimization schemes such as overlapping resolution map. The design was used to successful predict optimal MECC separation conditions for seven barbiturates.

14.3 FSCE Method Development

The applicability of capillary electrophoresis (CE) for the determination (13) of association constants of an hapten-antibody complex with values as high as 10^7 mol l^{-1} was investigated. As a reference method the well known, enzyme-linked immunosorbent assay (ELISA) was selected. The study described the optimisation of the experimental conditions of the CE technique. The CE measurements were optimised according to an experimental design. The results of the CE and ELISA methods are compared giving consideration to the reproducibility (repeatability) of the two methods.

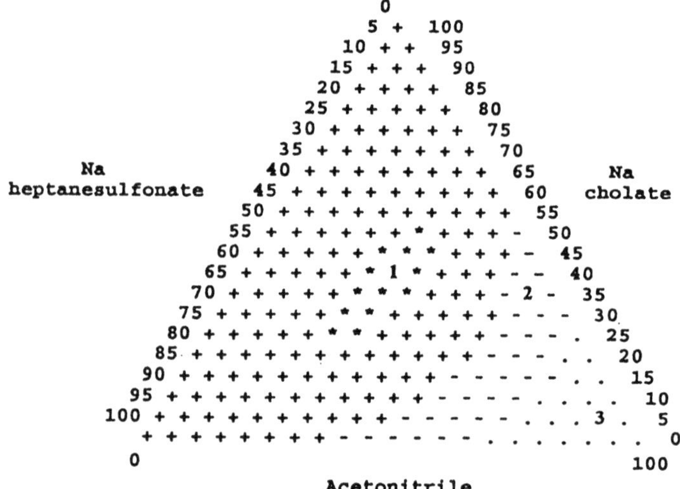

Figure 14.1a Overlapped resolution map for resolution of 14 quinolone antibacterials by MECC. Overlapped resolution map: Key: ● (Rs < 0.5), – (Rs 0.5–1.0), + (Rs 1.0–1.5), * (Rs greater than 1.5), # . Reproduced with permission from reference 12.

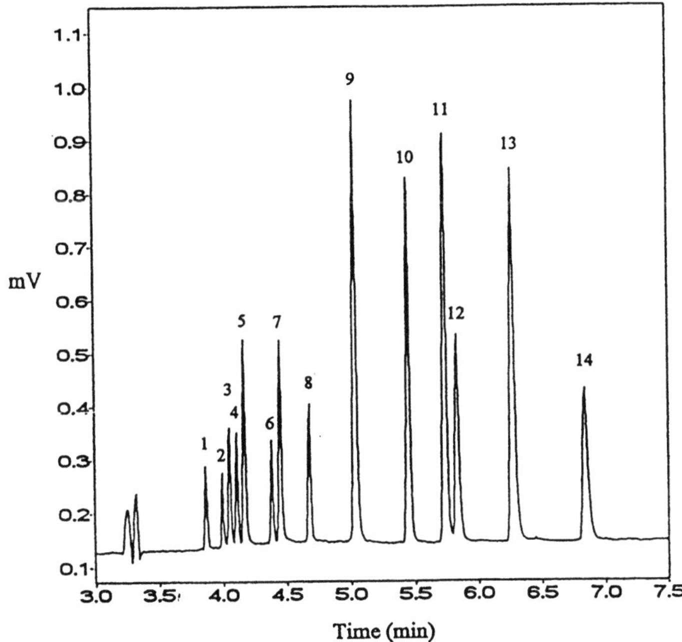

Figure 14.1b Optimised separation for resolution of 14 quinolone antibacterials by MECC. Optimised separation of 14 quinolone antibacterials by MECC. Separation conditions: 32 mM borate, 18 mM phosphate, 39 mM sodium cholate, 8 mM sodium heptansulphonate, 28 % v/v acetonitrile, 59 cm × 50 μm, 30 kV. Reproduced with permission from ref.12.

An overlapping resolution mapping scheme has been used (19) to optimise the separation of 8 sulphonamides. The buffer pH and β-cyclodextrin were varied to give an optimal separation shown in Figure 14.2.

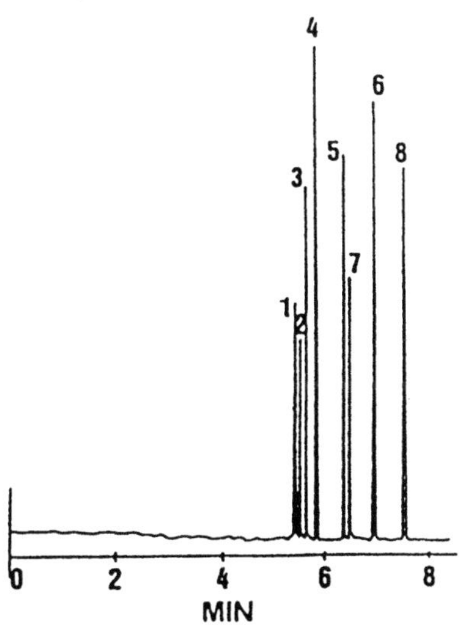

Figure 14.2 Separation of 8 sulphonamides using conditions optimised using an ORM scheme. Separation conditions: 50 mM phosphate-borate pH 6.4 with 2 mM β-cyclodextrin, 210 nm, 50 cm × 50 μm, 15 kV. Reproduced with permission from reference 19.

14.4 Chiral CE Method Development

A Plackett-Burman experimental design was used (9) to simultaneously optimise several operating conditions for a chiral CE method for the anorectic drug dexfenfluramine. The impact of concentration of the chiral selector dimethyl beta-cyclodextrin, concentration of methanol added to the buffer, pH of the background electrolytes, temperature of the capillary and applied voltage, was investigated on the resolution of the enantiomers, the analysis time and the peak symmetry. From these results, optimal values of the variables were selected for the development of a method capable of determining the levo-rotatory enantiomeric impurity of dexfenfluramine.

A Central Composite design has also been used (2) to optimise the chiral resolution of the drug amlodipine. The three factors studied were temperature, and concentrations of both electrolyte and hydroxypropyl-β-cyclodextrin. It was suggested that use of such designs in optimisation studies ensures method robustness at an early stage in the usage of a method.

A central composite design as been used (4) in the optimisation of the chiral separation of amphetamines using capillary electrophoresis. Five variables, i.e., buffer concentration, pH, chiral selector concentration, temperature, and applied voltage, were investigated. Enantiomeric resolutions as well as analysis time and generated power were established as responses for each experiment. A model of each response was obtained by multiple regression of a quadratic-degree mathematical expression. From the models the optimal separation conditions were obtained responses at threshold values. Results were compared (4) with a previous study in which a systematic investigation of the operating parameters was carried out. In order to visualize the robustness of the method, response surfaces were drawn for the significant variables. Figure 14.3 shows the response surface obtained for chiral resolution plotted against concentration of cyclodextrin concentration and temperature. This figure indicates that this resolution was highest at high cyclodextrin concentration and low temperature.

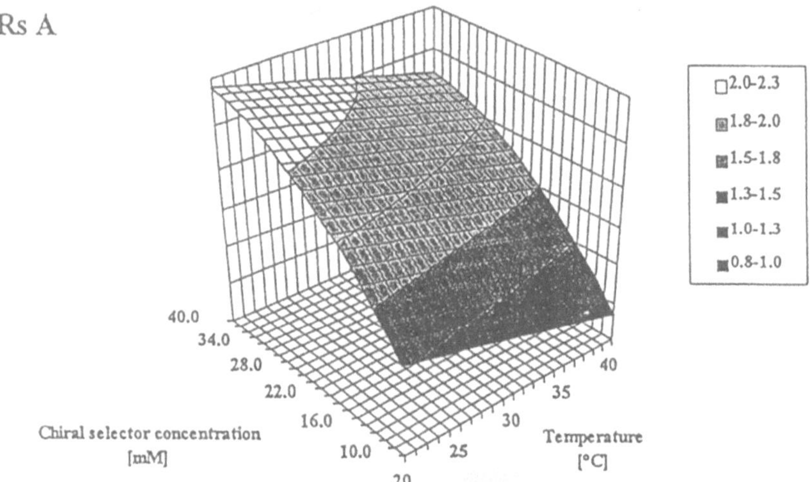

Figure 14.3 Response surface obtained for chiral resolution plotted against concentration of cyclodextrin and temperature. Reproduced with permission from reference 4.

A Plackett-Burman experimental design has also been employed in the optimisation of a chiral separation by CE (8). The influence of 5 parameters, each at 3 levels, upon the resolution of clenbuterol enantiomers was assessed in a 15 experiment study with duplicate injections of each experiment. The ranges investigated were pH (2.5, 4.0, 5.5), ionic strength (34, 67, 100 mM), CD concentration (6, 10, 16 mM), methanol content (0, 4, 8 %) and injection time (1, 5, 10 seconds). The data generated indicated that operation of pH 4, 100 mM electrolyte, 16 mM CD, 0 % methanol and a 1 sec injection time would produce the best resolution. This particular combination of parameters was not included in the initial design which highlights the power of this technique.

14.5 CEC Method Development

Capillary electrochromatography was employed (5) to separate the antibacterial 3-[4-(methylsulfinyl) phenyl]-5S-acetamidomethyl-2-oxazolidinone from its related S-oxidation products. The separation was optimized by a systematic search for the optimal separation conditions employing a modified central composite design. The variables examined in the optimization were in terms of applied potential, volume fraction of acetonitrile and buffer (Tris hydrochloride) concentration. A response surface mapping for selected variables at an applied potential of 25 kV was also performed. The optimised separation (Figure 14.4) gave resolution of the 4 components within 9 minutes.

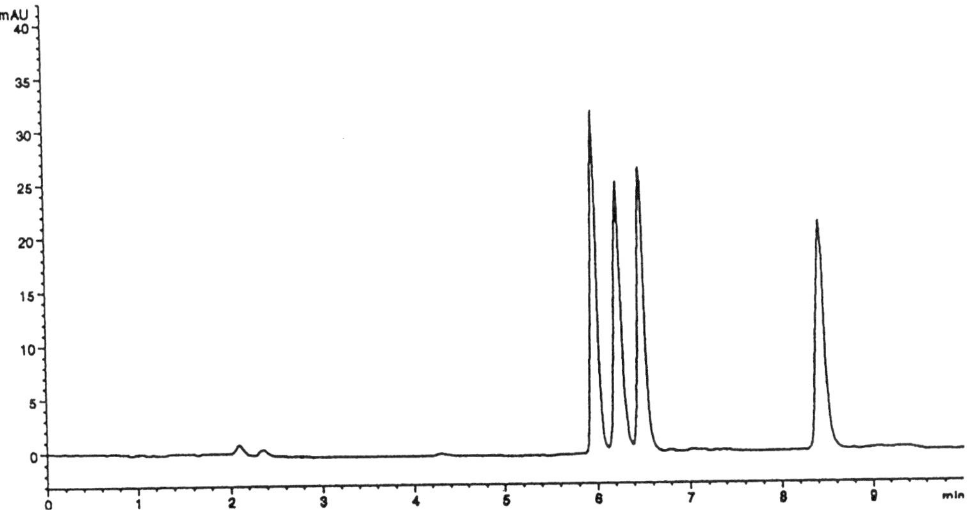

Figure 14.4 Optimised CEC separation of 4 related S-oxidation compounds. Operating conditions: Acetonitrile: 50 mM Tris HCl (60:40), 25 kV, 35 cm (25 cm packed) × 100 μm.

14.6 Robustness Testing

It is important that methods which are used routinely in fields such as pharmaceutical analysis are reliable as they will be used in many different laboratories by a variety of operators over a number of years. Method transfer and routine operation is improved for methods that have been developed and optimised with simultaneous regard for both performance and reliability. Robustness testing is an important aspect of method validation (Chapter 12) and can be conducted using experimental designs. If a method is robust then its performance will be unaffected by small deliberate change in the method parameters. Robustness testing involves making small deliberate changes to the operating conditions and measuring the effect upon a measured response such as resolution and analysis time. For example the buffer concentration may be set at 75 mM therefore in robustness testing studies the method will be run at 70 mM and 80 mM to check the performance. The number of parameters tested in robustness studies is high and include pH, electrolyte concentration, temperature, sample concentration, voltage etc. Use of a one-by-one approach to testing the parameters would result in a considerable workload. This can be considerably reduced using experimental designs. In addition use of experimental designs allows determination of any interaction effects.

Robustness studies indicate the factors having the most detrimental impact on analytical performance and allow tight limits to be set if necessary. Results from robustness testing can allow method limits to be appropriately selected for all operating parameters

The most appropriate scheme in the testing of robustness may be to use a fractional factorial design to identify the key method parameters and a further response surface generating design to enable method limits to be set together – with system suitability parameters.

This approach has been employed (6) in the evaluation of the robustness of a CE method used to determine levels of impurities in a basic drug. Initially, a Fractional Factorial design of 36 experiments was conducted to obtain the effect of 8 parameters upon resolution, % peak area and migration times. The parameters investigated are given in Table 14.3.

The method conditions are randomly repeated throughout the sequence to give an indication of the variability of the method. This variability is then included in the statistical evaluation of the data to indicate only those parameters that have a definite statistical effect on method performance.

Table 14.3 Fractional factorial design for related impurity method robustness study

Parameter	Method setting	Range investigated
Regeneration solution concentration	0.1 M NaOH	± 0.01M
Regeneration solution rinse time	1.0 minute	± 0.2
Electrolyte pH	2.1	± 0.2
Electrolyte concentration	50 mM	± 5
Electrolyte rinse time	2.0 minutes	± 0.2
Injection time	10 seconds	± 2
Sample concentration (mg/10ml)	10 mg	± 2
Applied voltage	10 kV	± 2

Reproduced with permission from ref. 6

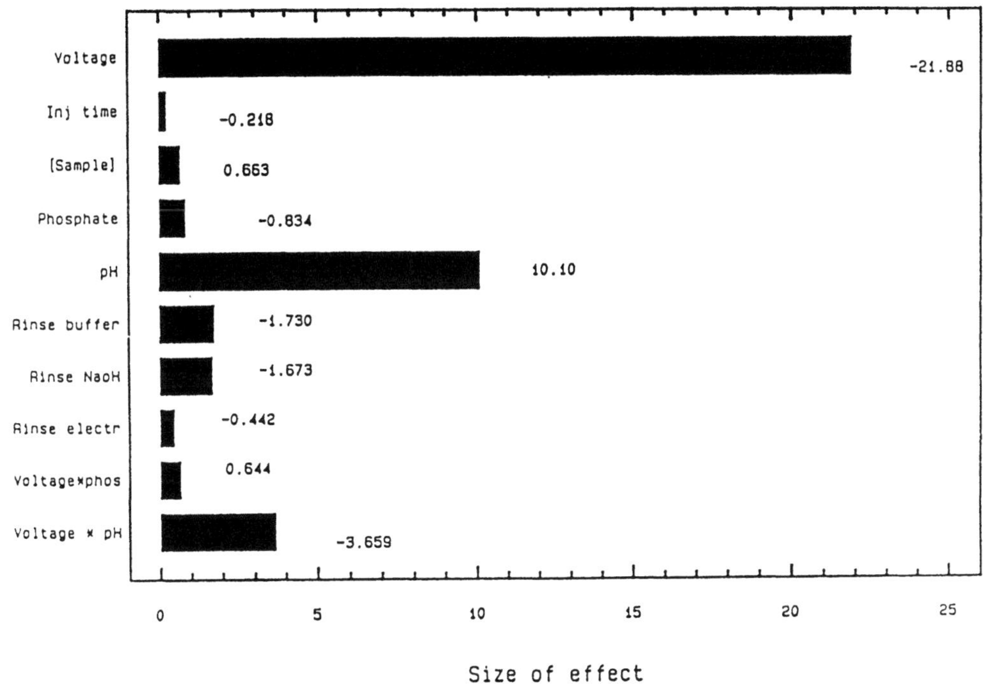

Figure 14.5 Pareto plot for migration time. Reproduced with permission from reference 6.

The effect of each parameter upon the various measured responses can be deduced by appropriate statistical examination of the data using a statistical evaluation computer software package. For example, Figure 14.5 shows a graphical representation (Pareto plot) of the 'size of effect' of each of the parameters investigated upon the migration time of the main peak. In this treatment a parameter is deemed to have a significant influence if the size of effect is greater than 2. Therefore, in Figure 14.5 voltage is shown to have a significantly negative effect upon migration time (i.e. an increase in voltage reduces migration time).

The overall analysis of the data set indicated that the only significant factors identified were sample concentration, injection time, voltage and pH. Increases in injection time or sample concentration had a detrimental effect on resolution of a closely migrating impurity from the main peak. Voltage and pH had more pronounced effects on several responses and were therefore assessed in a Central Composite design. A representation of the design is given in Figure 14.6. The voltage range employed was 8–12 kV whilst the effect of pH was assessed between 1.8–2.3 in a 16 analysis sequence. Analysis from these injections indicated operation at the method conditions of 10 kV and pH 2.1 were the most suitable.

An experimentally designed robustness evaluation has been performed (16) on a CE method employed for the determination of potassium drug counter-ion. The method employs an electrolyte comprised of formic acid and imidazole with indirect detection at 214 nm. Sodium is employed as an internal standard. A Fractional Factorial design was used to screen the effect of 7 parameters upon both resolution and migration times setting the minimum system

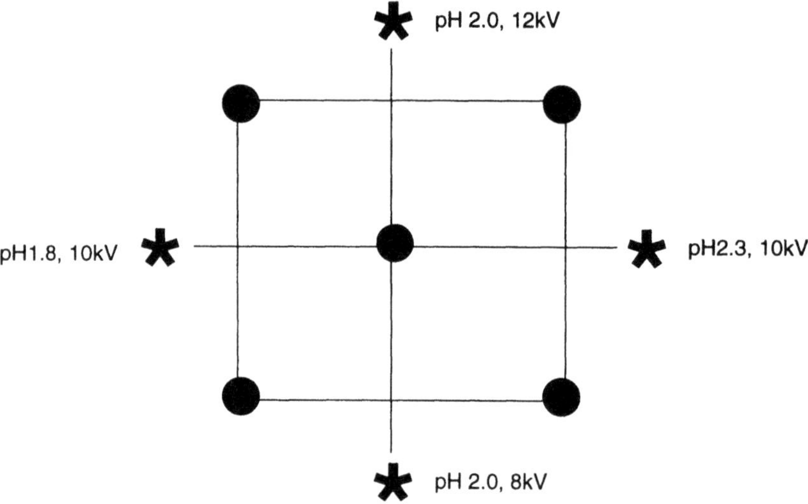

Figure 14.6   Schematic of the central composite design employed. Reproduced with permission from reference 6.

suitability criteria as being baseline resolution of potassium and sodium within 2 minutes. The parameters investigated were formic acid concentration, imidazole concentration, temperature, sodium concentration, potassium concentration, rinse time and applied voltage. The separation was shown to be robust within the ranges examined. The significant factors of voltage and formic acid were further assessed using a Central Composite design in the range 3–7 kV and 3.5–4.5 mM formic acid.

A capillary electrophoresis method for the determination of the enantiomeric purity of the local anaesthetic ropivacaine hydrochloride in injection solutions has (18) been validated. The method showed the required limit of quantitation of 0.1 % enantiomeric impurity. Good performances were shown for specificity, linearity, system repeatability, intermediate precision and accuracy. Robustness was tested via a full factorial design at two levels and the method proved to be robust. Figure 14.7 shows the response surface obtained for the chiral resolution of ropivacaine when varying temperature and pH during the robustness. The resolution is virtually unaffected by the change of pH or temperature which shows the method is highly robust in respect to these factors.

Similar approaches have been used in the evaluation of a CE method used to quantitatively determine a range of basic drugs (11). A Fractional Factorial and a Central Composite design were used to screen factors and to produce response surfaces as appropriate.

Table 14.4 provides an indication of the possible ranges that may be examined during robustness evaluation of CE methods. These ranges could be tightened if necessary, dependent upon the robustness of the method under evaluation.

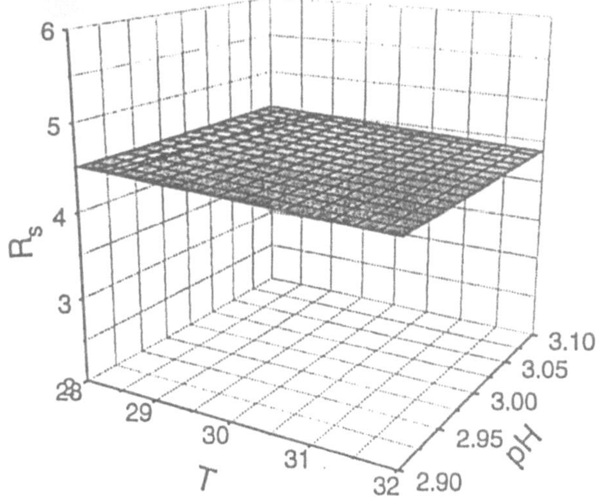

Figure 14.7 Response surface for chiral resolution of ropivacaine plotted against pH and temperature. Reproduced with permission from reference 18.

Table 14.4 Suggested ranges for robustness evaluation studies

Parameter	Approximate range
Buffer concentration	± 10 % of method value
Additive concentration(s)	± 10 % of method value
pH	± 0.2–0.5 from method value
Injection time	± 20 % of method value
Voltage	± 10–20 % of method value
Rinse time(s)	± 20 % of method value
Temperature	± 5°C from method value
Detection wavelength	± 5 nm from method value

14.7 Peak Identification

Drugs and their metabolites in biological specimens have been analyzed (3) by CE. The separations contained a large number of peaks which made peak identification difficult. This problem was addressed by use of a series of chemometric software tools. Principle Component Analysis and Iterative Target Transform Factor Analysis were used to inspect each electropherogram for spectral homogeneity of the peaks and to deconvolute comigrations. These algorithms are used to confirm the assay results and peak identifications. This approach was successfully demonstrated by the analysis of amphetamine and common interferences in human urine.

A chemometric method, Kalman filtering, has been applied (22) to deal with three-dimensional electropherograms, obtained by capillary electrophoresis connected to a charge-coupled device multiwavelength fluorescence detector. The capillary electrophoretic peaks of a mixture, which were not effectively separated, were resolved into electrophoretic peaks of individual components to obtain qualitative and quantitative information. The approach was applied to analyze a mixture of rhodamine fluorescent dyes with excellent results.

14.8 Selection of Experimental Design

Table 14.5 details the relative merits and application areas for the various previously discussed design types. It can be seen (32) that selection of the most appropriate design is largely dependent upon the requirements of the study. For example, if the purpose is simply to screen the effect of a large number of independent variables, then a Plackett-Burman design would be most appropriate giving a minimal number of experiments. However, if the intention is to fully characterise the effects of all parameters and all possible interactions, then a Factorial design would be appropriate with a considerably increased number of experiments. Therefore, often a compromise is made where Fractional Factorials are employed, with a reduction in the number of experiments, whilst only clearly identified interactions terms can be monitored.

Table 14.5 Summary of requirements and information obtained from various experimental designs

Design type	No. of expts.	Screening or optimisation	Response surface	Optimal no. of expts.	Interaction studies
Factorial	3 factors = 27 6 factors = 729	Both	Yes	less than 4	All
Fractional Factorial	3 factors = 9 6 factors = 27	Screening	No	less than 8	Selected
Central composite	3 factors = 20 6 factors = 53	Optimisation	Yes	less than 4	All
Plackett-Burman	8 factors = 9 12 factors = 13	Screening	No	less than 12	Selected
ORM	2 factors = 9	Optimisation	Yes	3	None
Simplex	3 factors = 10	Optimisation	No	3	None

References

1. Altria K D, Bryant S M, and Hadgett T, Validated capillary electrophoresis method for the assay of a range of acidic drugs and excipients, *J. Pharm. Biomed. Anal.*, 15 (**1997**) 1091-1101.

2. Small T S, Fell A F, Coleman M W, and Berridge J C, Central composite design for the rapid optimisation of ruggedness and chiral separation of amlodipine in capillary electrophoresis , *Chirality*, 7 (**1995**) 226-234.

3. Lilley K A and Wheat T E, Drug identification in biological matrices using capillary electrophoresis and chemometric software, *J. Chromatogr. B*, 683 (**1996**) 67-76.

4. Varesio E, Gauvrit J Y, Longeray R, Lantern P, and Veuthey J L, Central composite design in the chiral analysis of amphetamines by capillary electrophoresis, *Electrophoresis*, 18 (**1996**) 931-937.

5. Miyawa J H, Alasandro M S, and Riley C M, Application of a modified central composite design to optimize the capillary electrochromatographic separation of related s-oxidation compounds, *J. Chromatogr. A*, 769 (**1997**) 145-153.

6. Altria K D and Filbey S D, The application of experimental design to the robustness testing of a method for the determination of drug related impurities by capillary electrophoresis, *Chromatographia*, 39 (**1994**) 306-310.

7. Altria K D, Frake P, Gill I, Hadgett T, Kelly M A, and Rudd D R, Validated capillary electrophoresis method for the assay of a range of basic drugs and excipients, *J. Pharm. Biomed. Anal.*, 13 (**1995**) 951-957.

8. Rogan M M, Altria K D, and Goodall D M, Plackett-Burman experimental design in chiral capillary electrophoresis, *Chromatographia*, 38 (**1994**) 723-729.

9. Boonkerd S, Detaevernier M R, Heyden Y V, Vindevogel J, and Michotte Y, Determination of the enantiomeric purity of dexfenfluramine by capillary electrophoresis - use of a Plackett-Burman design for the optimization of the separation, *J. Chromatogr. A*, 736 (**1997**) 281-289.

10. Altria K D and Hadgett T, An evaluation of the use of capillary electrophoresis to monitor trace drug residue levels following the manufacture of pharmaceuticals, *Chromatographia*, 40 (**1995**) 23-27.

11. Xiong S X, Li J J, and Cheng J K, Multicomponent analysis of highly overlapped capillary electrophoretic peaks using multiwave-length charge-coupled devices detection, *J. Liquid Chromatogr. Rel. Techn.*, 19 (**1996**) 2129-2141.

12. Ng C L, Ong C P, Lee H K, and Li S F Y, Systematic optimisation of micellar electrokinetic chromatographic separation of flavanoids, *Chromatographia*, 34 (**1992**) 166-172.

13. Busch M H A, Boelens H F M, Kraak J C, Poppe H, Meekel A A P, and Resmini M, Critical evaluation of the applicability of capillary zone electrophoresis for the study of hapten-antibody complex formation, *J. Chromatogr. A*, 744 (**1996**) 195-203.

14. Nielsen M S, Nielsen P V, and Frisvad J C, Micellar electrokinetic capillary chromatography of fungal metabolites - resolution optimized by experimental design, *J. Chromatogr. A*, 721 (**1996**) 337-344.

15. Wynia G S, Windhorst G, Post P C, and Maris F A, Development and validation of a capillary electrophoresis method within a pharmaceutical quality control environment and comparison with high-performance liquid chromatography, *J. Chromatogr. A*, 773 (**1997**) 339-350.

16. Filbey S D and Altria K D, Robustness testing of a capillary electrophoresis method for the determination of potassium content in the potassium salt of an acidic drug, *J. Cap. Elect.*, 1 (**1994**) 190-195.

17. Sun S-W and Chen L-Y, Optimisation of capillary electrophoretic separation of quinolone antibacterials using the overlapping resolution mapping scheme, *J. Chromatogr. A*, 766 (**1997**) 215-224.

18. Sanger van de griend C E, Wahlstrom H, Groningsson K, and Widahlnasman M, A chiral capillary electrophoresis method for ropivacaine hydrochloride in pharmaceutical formulations – validation and comparison with chiral liquid chromatography, *J. Pharm. Biomed. Anal.*, 15 (**1997**) 1051-1061.

19. Ng C L, Lee H K, and Li SFY, Systematic optimisation of capillary electrophoresis of sulphonamides, *J. Chromatogr.*, 598 (**1993**) 133-138.

20. Hows M E P, Perrett D, and Kay J, Optimisation of a simultaneous separation of sulphonamides, dihydrofolate reductase inhibitors and beta-lactam antibiotics by capillary electrophoresis, *J. Chromatogr. A*, 768 (**1997**) 97-104.

21. Vindevogel J and Sandra P, Resolution optimisation in micellar electrokinetic chromatography: use of Plackett-Burman statistical design for the analysis of testerone esters, *Anal. Chem.*, 63 (**1991**) 1530-1536.

22. Guan F Y, Wu F, and Luo Y, A novel strategy for systematic optimization of micellar electrokinetic chromatography separations, *Anal. Chim. Acta*, 342 (**1997**) 133-144.

23. Massart D L, Nandeginste, Deming S N, Michotte Y, and Kaufman L, in *Chemometrics: a textbook*, published by Elservier Press, Amsterdam, **1988**.

24. Deming S N and Morgan S L, in *Experimental design: a chemometric approach*, published by Elsevier Press, Amsterdam, **1988**.

25. Vanbel P F, Gilliard J A, and Tilquin B, Chemometric optimisation in drug analysis by HPLC: a critical evaluation of the quality criteria used in the analysis of drug purity, *Chromatographia*, 36 (**1993**) 120-124.

26. Andersson A M, Karlsson A, Josefson M, and Gottfries J, Evaluation of mobile phase additives in LC-systems using chemometrics, *Chromatographia*, 38 (**1994**) 715-722.

27. Tucker R P, Fell A F, Berridge J C, and Coleman M W, Computer-aided models for optimisation of eluent parameters in chiral liquid chromatography, *Chirality*, 4 (**1992**) 316-322.

28. Mullholland M and Waterhouse J, Investigation of the limitations of saturated fractional factorial experimental designs, with confounding effects for an HPLC ruggedness test, *Chromatographia*, 25 (**1988**) 769-774.

29. Mullholland M, Ruggedness testing in analytical chemistry, *TRAC*, 7 (**1988**) 383-389.

30. Castagnola M, Rossetti D V, Cassiano L, Rabino R, Nocca G, and Giardina B, Optimisation of phenylhydantoinamino acid separation by micellar electrokinetic capillary chromatography, *J. Chromatogr.*, 638 (**1993**) 327-334.

31. Jumppanen J H, Wiedmer S K, Siren H, Riekkola M L, and Haario H, Optimized separation of seven corticosteroids by micellar electrokinetic chromatography, *Electrophoresis*, 15 (**1994**) 1267-1272.

32. Altria K D, Clark B J, Filbey S D, Kelly M A, and Rudd D R, Review of the application of experimental design approaches in capillary electrophoresis, *Electrophoresis*, 16 (**1995**) 2143-2148.

15 Forensic Applications of CE

15.1 Introduction

CE has developed a range of applications (1–23) in forensic studies (Table 15.1) and the subject has been recently reviewed (24, 25). The application areas mirror those of traditional pharmaceuticals and include identity confirmation, main component assay, related impurity determinations, clinical monitoring, chiral separations and inorganic impurity contents.

15.2 Identity Confirmation

General screening methods have been developed which allow simultaneous monitoring of a wide range of compounds of forensic interest. Often the identity of a component is confirmed by use of DAD UV detectors.

Figure 15.1 shows separation (14) of heroin and a wide range of heroin related species using a MECC electrolyte containing 50 mM cetyltrimethylammonium bromide and 10 % v/v acetonitrile with detection at 280 nm. The method had advantages over HPLC methods used in the author's laboratory (14) as CE allowed simultaneous results to be obtained for caffeine, narcotine, papaverine and acetylcodeine. Strychnine and morphine co-eluted on HPLC but were resolved by MECC.

Fourteen common basic drugs have been (15) screened using a 200 mM sodium phosphate run buffer. Linearity and reproducibility were shown for cocaine, heroin, methamphetamine, lysergic acid diethylamide (LSD), and phencyclidine (PCP). Known adulterants and impurities did not interfere with these drug compounds. Comparisons of CE quantitations with results from other laboratory techniques demonstrated (23) the reliable adaptation of CE to the forensic laboratory. Twenty five anabolic steroids have been (25) simultaneously resolved by an MECC method.

15.3 Assay

It is important to have quantitative methods in order to determine the strength of tablets and other seized materials. Heroin and derivatives were quantified using an MECC method. Table 15.2 shows a comparison of the data generated for 3 seizure samples by HPLC and MECC.

Table 15.1 Forensic applications of CE

Application	Electrolyte	Ref.	Comments
Amphetamine derivatives in "Ecstasy" tablets	pH 2	1	Identified by standard addition and UVDAD, good agreement with HPLC and GC
6 Amphetamines found in illicit drug seizures	pH 2	2	13 tablets profiled by CE and HPLC – no statistical difference between results
Amphetamine seizures	Various	3	Quantitative results generated
22 Basic drug stimulants in urine	Borate buffer	4	Injection precision less than 2%RSD
8 Cocaine related substances	CTAB, phosphate-borate pH 8.6 + 7.5% acetonitrile	5	Benzocaine, Procaine, Lignocaine, Cocaine, and Tetracaine separated
Dexamethasone and flumethasone in equine urine	MECC,SDS, MeOH	6	LOD of 1.1ng/ml, detection of dexamethasone 12 hours post-administration
18 Drugs of abuse	85 mM SDS 8.5 mM phosphate / 8.5 mM borate / 15 % acetonitrile pH 8.5	7	Good selectivity obtained
15 Diuretics in blood and urine	60 mM CAPS pH 10.6	8	SPE, relative migration times used for identification purposes
Drugs of abuse in urine	MECC, SDS	9	100 ppm detection at 195 nm
Drug overdoses – serum, urine	Phosphate pH 8, SDS MECC	10	Analysis of overdose samples with DAD
Ephedrine alkaloids in urine	pH 9.7 phosphate	11	2 µg/ml urine LOD, extensive validation data
Heroin and amphetamine seizures	Various	12	Impurity profiling and quantitative analysis
Heroin seizures	Various	13	Impurity profiling and quantitative analysis
Heroin seizures	Various	14	Impurity profiling and quantitative analysis
Illicit basic drugs including cocaine, heroin, lysergic acid diethylamide (LSD), and phencyclidine (PCP)	200 mM Sodium phosphate	15	14 drugs separated, cross-validation shown with other analytical techniques
Illicit drugs in hair	Borate pH 9.2	16	0.15 ng/mg LOD, 90 %+ recoveries
LSD (and nor-LSD, iso-LSD and iso-nor-LSD) in blood	Citrate-methanol buffer, pH 4.0	17	Laser-induced fluorescence detection and electrokinetic injection allowed an LOD of 0.1–0.2 ng LSD per ml blood, methylergometrine as IS
Metal ions and anions in illicit heroin	Various indirect UV detection CE methods	18	Range of anions and cations determined in heroin exhibits at sub 1 % levels.

Table 15.1 continued

Application	Electrolyte	Ref.	Comments
Methadone and metabolites in urine	MECC, SDS	19	20 ppm LOD, 6 minute assay, good correlation of sample test results with immunoassay and CG-MS data
Methadone and metabolites in urine, serum and urine	Low pH with DIMEB	20	Chiral separation of methadone and metabolites was achieved using a cyclodextrin additive
Opium alkaloids	Low pH MECC	21	Zwitterionic surfactant used for selectivity
Opiates in urine including pholcodine, 6-monoacetyl-morphine, morphine, hero-in, codeine and dihydro-codeine	100 mM phosphate at pH 6	22	Detection limits in the region of 10 ng per ml with electrokinetic injection. Data in agreement with HPLC results.
Range of drugs in forensic samples	CE and MECC conditions	23	Quantitative data compared to other techniques

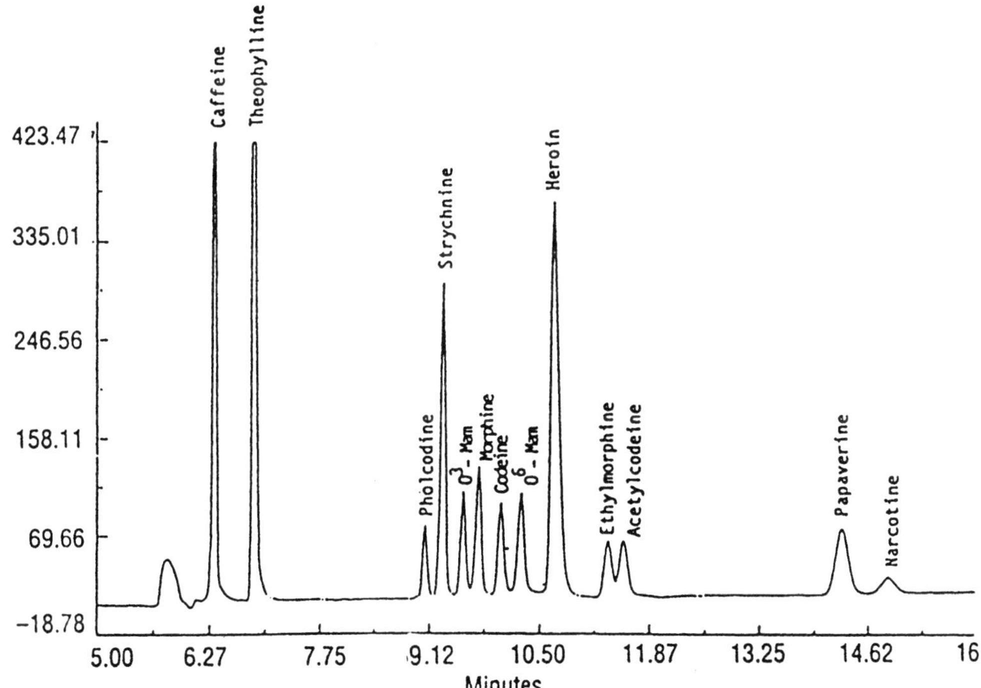

Figure 15.1 Separation of heroin and a wide range of heroin related species. Separation conditions : 10 mM KH_2PO_4, 10 mM borate, 50 mM cetyltrimethylammonium bromide and 10 % v/v acetonitrile, detection at 280 nm, 72 cm × 75 μm capillary, 15 kV. Reproduced with permission from reference 14.

Table 15.2 Comparison of the data generated for 3 seizure samples by HPLC and MECC.

Sample	MAM (%)		DAM (%)		AC(%)	
	MECC	HPLC	MECC	HPLC	MECC	HPLC
1–68	1.0	0.8	90.0	88.0	4.0	4.1
AFP/1	0.5	0.7	28.9	28.3	2.0	2.1
AFP/2	0.5	0.7	29.3	30.0	2.0	2.1
AFP/3	0.5	0.7	31.2	30.0	2.0	2.1

MAM = monoacetylmorphine, DAM = diacetylmorphine [heroin], AC = acetylcodeine
Reproduced with permission from reference 14

Six commonly determined amphetamines were quantified (2) by CE with good linearity, precision, accuracy and limits of detection. The CE method was applied to the analysis of 13 seized tablets. The results of this testing were statistically similar to those obtained by a GC and HPLC procedure

CE with diode array detection, has been used (26) in the quantitative analysis of ephedrine and 9 amphetamines. The method offered advantages over the conventional technique of GC in terms of elimination of the need for sample derivatisation. Detection limits at 214 nm were 13 and 68 μg/ml for the individual compounds which is similar to that obtained by GC.

Amphetamine and amphetamine derivatives have been quantified (1) in "Ecstasy" tablets. The tablets were dissolved in 0.01 N hydrochloric acid and diluted with an aqueous solution of the internal standard phenylephrine. Separation was achieved with a pH 2.2 phosphate buffer. Identification verification was also achieved by comparing UV DAD with spectra obtained from standards. Good agreement was achieved between CE and HPLC for the amphetamine or amphetamine derivatives content in 56 "Ecstasy" tablets and powder.

15.4 Clinical Applications

Forensic analysis of biofluids is conducted to confirm consumption of narcotics by determining the levels of narcotics or their metabolites in the body fluids of suspects. Alternatively biofluid analysis from athlete's may be conducted to monitor levels of narcotics or banned substances such as stimulants or steroids. This activity is also of concern in areas such as horse racing where the levels of stimulants are routinely monitored. There a number of advantages of using CE for bioassays which are covered in greater details in Chapter 10. These advantages include the ability to directly inject the sample into the capillary with no sample pretreatment.

Drug abuse can be confirmed (16) by the analysis of hair samples. Drugs are deposited into the hair as the unmetabolised species and the timescale and level of drug ingestion can be calculated based on average growth rate of hair. Hair analysis is therefore a useful technique as drugs are normally rapidly metabolised and would disappear from blood and urine within a few days. Hair samples were (16) extracted with diethyl ether and 0.01 M HCl

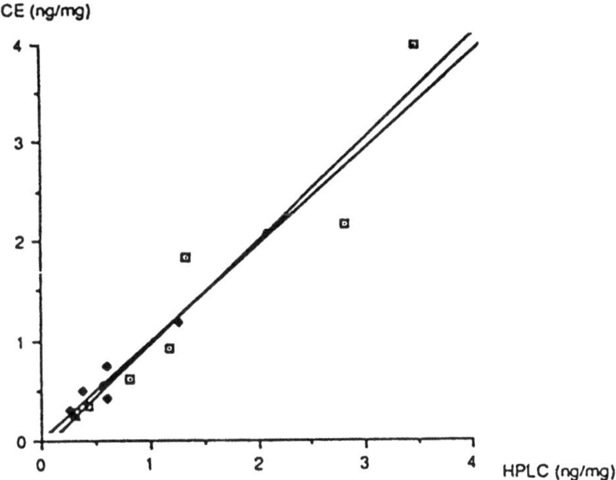

Figure 15.2 Correlation of CE and HPLC results obtained for the determination of cocaine levels in 8 samples. Filled-in diamonds denote cocaine results and squares indicate morphine results. Reproduced with permission from reference 16.

and reconstituted with buffer. Detection limits of 600 ng/ml, equivalent to 0.15 ng/mg of hair were obtained for cocaine. Tetracaine was used as an internal standard to give acceptable injection precision. Recovery data for cocaine spiked into hair was in the region of 85–98 %. Figure 15.2 shows the good correlation of CE and HPLC results obtained for the determination of cocaine levels in 8 samples.

Diuretics such as amiloride, triamterene, bendroflumethiazide and bumetanide have been included (26) in the list of banned substances in since 1988. Levels of diuretics in urine have been determined (4) by CE with fluorescence detection and a pH 8 buffer. Direct injection of the urine was possible (4) as no endogenous urine components interfered with the assay. Limits of detection range between 0.5 fmol for triamterene and 21.6 fmol for bumetanide within an 8 minute analysis

The use of a sensitive HeCd laser-induced fluorescence (LIF) detector enabled D-lysergic acid diethylamide (LSD) to be determined (17) in blood samples at the 0.4–0.5 ng/ml level. Sample preparation involved precipitation of proteins, solvent extraction, evaporation of the solvent, and reconstitution of the residue with methanol. The method used a citrate-methanol buffer, pH 4.0 electrolyte and gave a 8 minute analysis time with methylergometrine used as an internal standard. Results obtained by the CE method were in-line (17) with those generated using a radioimmunoassay.

100 mM disodium hydrogenphosphate at pH 6 has been used (22) to resolve a range of opiates including pholcodine, 6-monoacetylmorphine, morphine, heroin, codeine and dihydrocodeine. The method was then applied to determine opiate content in urine with a 10 ng/ml limit of detection. Values of urine levels of dihydrocodeine and pholcodine were found to be comparable to those obtained by HPLC. Figure 15.3 shows analysis of urine spiked with the six opiates (the inserted figure shows the blank urine containing an endogenous peak [E] and the internal standard [IS], levallorphan). Limits of detection of 4–8 ng/ml were obtained with 200 nm detection and DAD was used for identity confirmation.

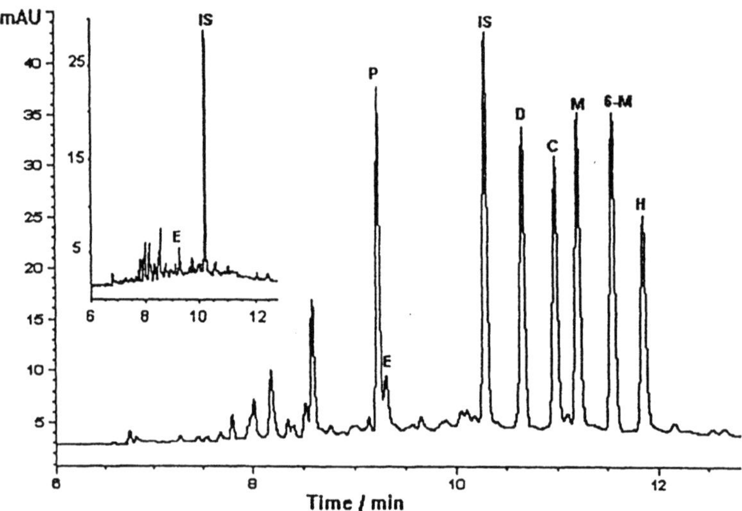

Figure 15.3 Analysis of urine spiked with the six opiates (the inserted figure shows the blank urine containing an endogenous peak [E] and the internal standard [IS], levallorphan). P = pholcodine, 6-M = 6-monoacetylmorphine, H = heroin, C = codeine, M = morphine, D = dihydrocodeine. Separation conditions: 100 mM Na_2HPO_4 pH 6, 200 nm detection, 20 kV, 65 cm × 50 μm

15.5 Purity Determination

Purity determinations are especially important in forensic science as the profile of impurities in the drug can be indicative (13, 29) of the source of the material and its possible origins. Invariably narcotics are not sold as pure substances but are diluted (cut) with other cheaper materials such as aspirin and quinine. The ability to determine the presence and levels of these adulterants can provide useful information in forensic investigations.

A low pH phosphate buffer has been used (15) to analyse the purity of cocaine and heroin samples. Figure 15.4 shows application of the method to samples of seized cocaine and heroin. The analysis of these samples confirmed their adulteration with caffeine, quinine, paracetamol and aspirin. Good correlation was obtained between CE and GC for the analysis of 21 illicit drug samples. The analysis of LSD blotted onto paper patches was easier by CE than HPLC as CE produced considerable lower number of interfering peaks.

A range of charged and neutral opiates and adulterants were separated (28) using a pH 4 50 mM 6-aminocaproic acid containing 50 mM MAPS (a zwitterionic surfactant 3-N, N-dimethylmyristylammoniopropanesulfonate), 5 mM 1-heptanesulfonic acid and 10 % acetonitrile. The method was successfully applied to a number of heroin seizure samples.

MECC was used (7) to separate 18 common drugs of abuse. A 25 cm × 50 μm capillary with 20 kV and 85 mM SDS 8.5 mM phosphate / 8.5 mM borate / 15 % acetonitrile pH 8.5 with detection at 210 nm. CE gave twice as many impurity peaks as HPLC.

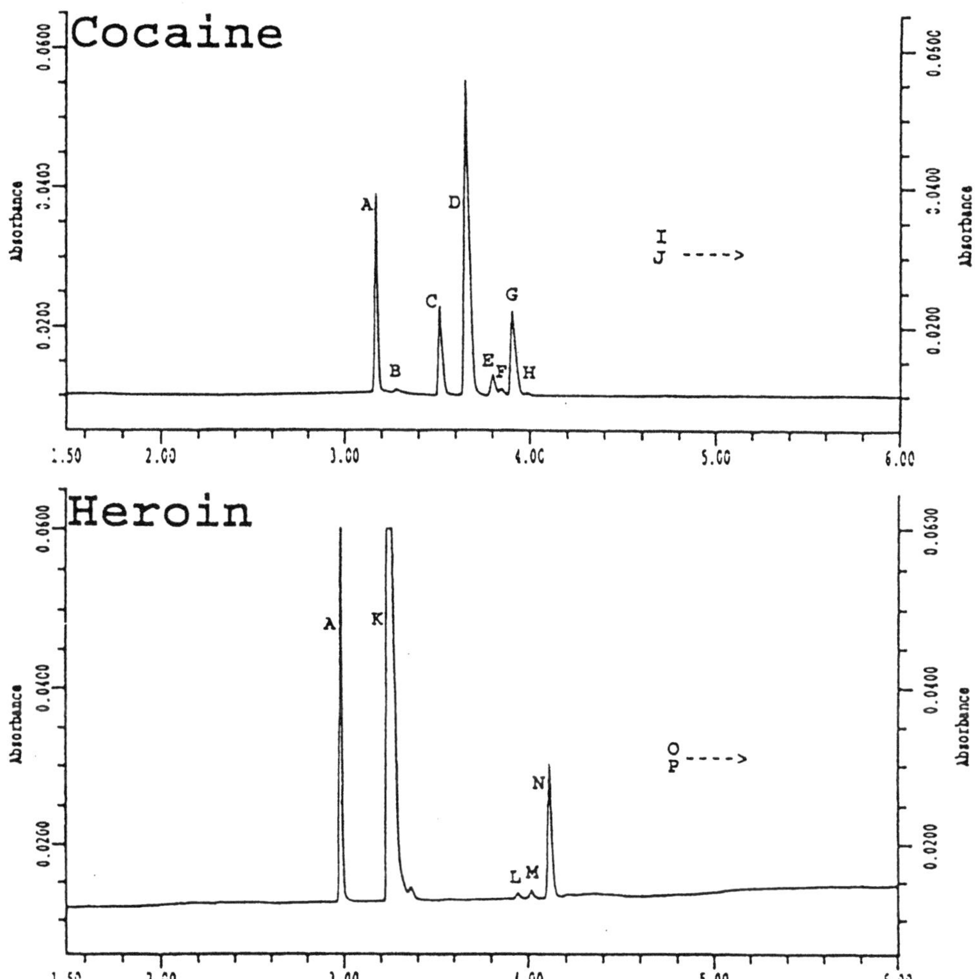

Figure 15.4 Impurity profiling of samples of seized cocaine and heroin. A = naphazoline, B = ephedrine, C = procaine, D = cocaine, E = lidocaine, F = ciscinnamoylcocaine, G = tetracaine, H = trans-cinnamoylcocaine, I = benzocaine, J = caffeine, K = quinine, L = acetylcodeine, M = acetylmorphine, N = heroin, O = acetaminophen, P = aspirin. Separation conditions : 200 mM phosphate pH 4.5, 37 cm × 50 µm, 20 kV, detection at 230 nm.

An extensive study on the use of CE for impurity profile has been reported (13). Figure 15.5 shows the separation of a heroin exhibit using a MECC buffer containing both cyclodextrin and acetonitrile with detection at 210 nm. DAD was used to confirm peak identities in samples. The use of laser-induced fluorescence (LIF) detection improved sensitivity considerably. For example the LOD for acetylthebaol using LIF was 2 ng/ml which was 500 times better than UV detection.

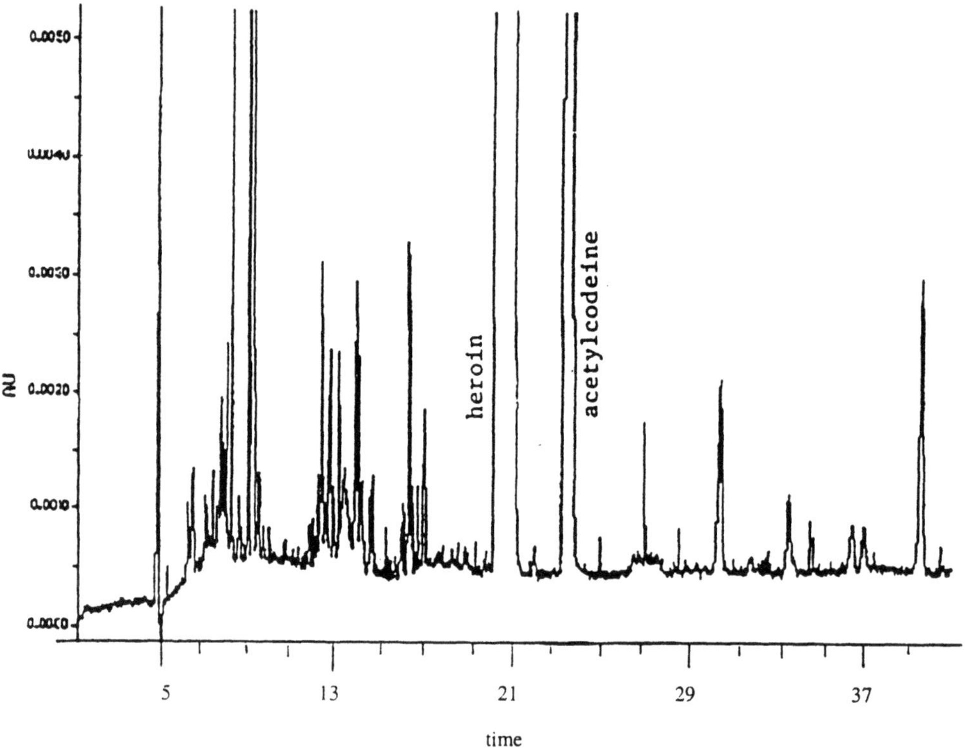

Figure 15.5 Separation of a heroin exhibit using a MECC buffer. Operating conditions: 57 cm × 50 μm, 9 mM phosphate and 9 mM borate, 45 mM SDS, 6.9 mM SBE (sulphobutyl-beta cyclodextrin) and 10 % v/v acetonitrile, detection at 210 nm, 21 kV. Reproduced with permission from reference 13.

15.6 Inorganic Ions

The levels of small cations and anions in illicit drugs is importance for intelligence purposes. CE has been used (18) to determine levels of inorganic cations and both inorganic and organic anions in illicit heroin using CE with indirect UV detection. The methods used were similar to those used in stoichiometric determinations (Chapter 5). Lithium and nitrate were used as internal standards in the cation and anion methods respectively. Detection limits for the cations were in the order of ng/ml whilst this was high ng/ml for the anions. Linearity data covering ranges of 0.5–60 mg/l gave correlation coefficient's of 0.997–0.999 for both anions and cations. Figure 15.6 shows separation of anions in a south-west Asian heroin hydrochloride exhibit which contains tartrate, citrate, chloride and acetate. The same sample was shown to contain sodium, calcium and magnesium at the 0.42, 0.11 and 0.02 % w/w level respectively.

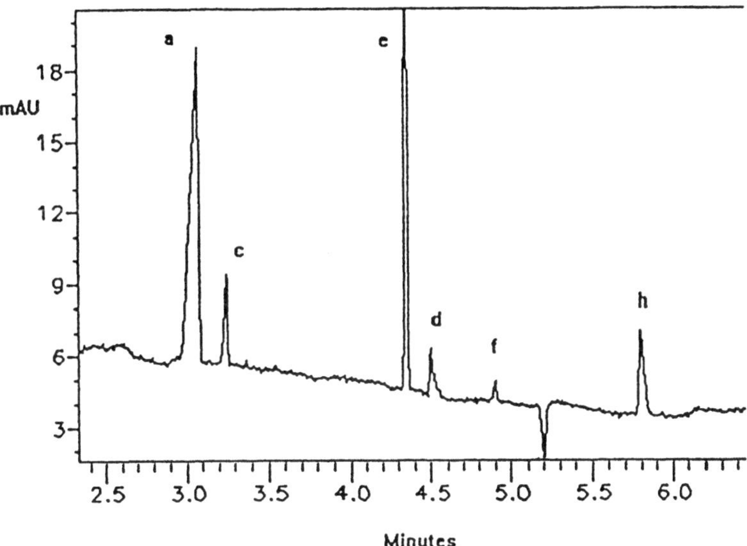

Figure 15.6 Separation of anions in a heroin exhibit by CE with indirect UV detection. a = chloride, c = nitrate (internal standard), d = citrate, f = phosphate, h = acetate. Separation conditions: 2.25 mM pyromellitic acid, 6.5 mM NaOH, 0.75 mM hexamethonium hydroxide and 1.6 mM treithanolamine pH 7.7, 72 cm × 50 μm, −30 kV, indirect UV detection at 230 nm

15.7 Non-Aqueous CE

Many narcotics have limited water solubility and therefore the use of non-aqueous separation conditions offers useful separation possibilities. Chapter 13 covers the use of non-aqueous electrolyte systems in CE.

The separation selectivity of 6 opiates was investigated (30) in a range of organic solvents including acetonitrile, methanol, formamide and n-methylformamide. Optimal conditions were found to be acetonitrile:methanol 50:50 containing 250 mM acetic acid and 25 mM ammonium acetate. Opium tincture was analysed using these conditions. Sample preparation involved a simple 100 fold dilution and direct injection onto the capillary - no interferences from the tincture were obtained.

Aqueous and non-aqueous electrophoresis media, respectively, have been compared (31) for the quantitative determination of morphine in pharmaceutical preparations. In the aqueous system the separation from other opium alkaloids was achieved using 2,6-di-O-methyl-beta-cyclodextrin as an additive to the electrophoresis buffer. In the non-aqueous system no other additives than the electrolytes were necessary in order to achieve separation of the opium alkaloids. The two methods have been partially validated and compared with a currently used high-performance liquid chromatography method. From the overall point of view the validations show that the three methods are equivalent in performance and that they are appropriate for the purposes they are intended for.

15.8 Chiral

The ability to chirally resolve enantiomeric resolve compounds by CE is one of the most popular applications of CE (Chapter 4). Many narcotics and banned drugs are chiral and CE has been used in their analysis.

Methadone and its main metabolite 2-ethylidene-1,5-dimethyl-3,3-diphenylpyrrolidine (EDDP) have been chirally resolved (20) using a low pH buffer containing heptakis-(2,6-di-O-methyl)-beta-cyclodextrin (DIMEB) as the chiral selector. The enantioselective metabolism of methadone was followed by analysis of the urine and serum samples. Figure 15.7 shows analysis of a urine sample following administration of a dose of methadone. Serum and urine samples were extracted at pH 9–10 with n-hexane. A limit of detection of 2ng/ml in serum and 10ng/ml in urine for each enantiomer. Table 15.3 shows recovery data for R-methadone enantiomer spiked into urine and serum. Similar results were obtained in spiking experiments for S-methadone and both R- and S-EDDP.

Table 15.3 Recovery data for R-methadone enantiomer spiked into urine and serum.

Amount added ng/ml		Amount determined by CE (ng/ml)	
Serum	Urine	Serum	Urine
5	50	6	58
50	500	60	510
250	2500	239	2389

Reproduced with permission from reference 20.

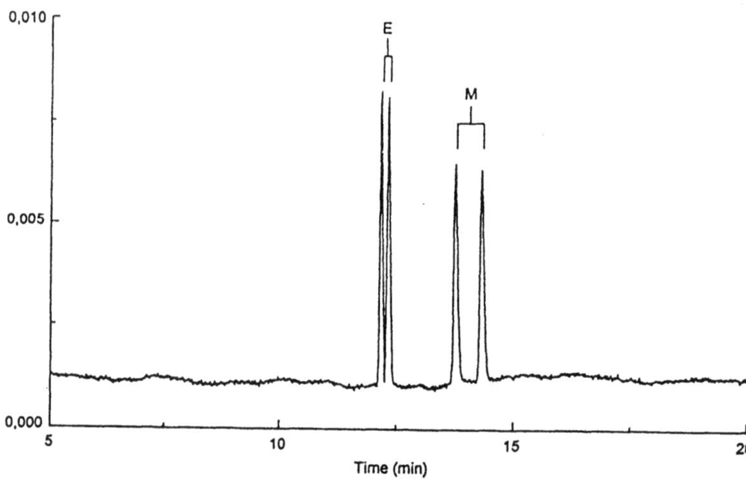

Figure 15.7 Analysis of a urine sample following administration of a dose of methadone. M = methadone, E = EDDP. Separation conditions: 100 mM phosphate pH 2.2 with 2 mM DIMEB and 10% methanol, 47 cm × 50 μm, detection at 200 nm. Reproduced with permission from reference 20.

References

1. Frost M, Kohler H, and Blaschke G, Analysis of ecstasy by capillary electrophoresis, *Int. J. of legal medicine*, 109 (**1996**) 53-57.

2. Sadeghipour V, Varesio E, Giroud C, Rivier L, and Veuthey J L, Analysis of amphetamines by capillary electrophoresis and liquid chromatography – application to drug seizures and cross-validation, *Forensic Sci. Int.*, 86 (**1997**) 1-13.

3. Trenerry V C, Robertson J, and Wells R J, Analysis of illicit amphetamine seizures by capillary electrophoresis, *J. Chromatogr. A*, 708 (**1995**) 169-176.

4. Gonzalez E and Laserna J J, Capillary zone electrophoresis for the rapid screening of banned drugs in sport, *Electrophoresis*, 15 (**1994**) 240-243.

5. Trenerry V C, Robertson J, and Wells R J, The determination of cocaine and related substances by micellar electrokinetic capillary chromatography, *Electrophoresis*, 15 (**1994**) 103-108.

6. Gu X, Meleka-Boules M, and Chen C-L, Micellar electrokinetic capillary chromatography with immunoaffinity chromatography for identification and determination of dexamethasone and flumethasone in equine urine, *J. Cap. Elec.*, 3 (**1996**) 43-49.

7. Weinberger R and Lurie I S, Micellar electrokinetic capillary chromatography of illicit drug substances, *Anal. Chem.*, 63 (**1991**) 823-827.

8. Jumppanen J, Siren H, and Riekkola M L, Screening for diuretics in urine and blood serum by capillary zone electrophoresis, *J. Chromatogr.*, 652 (**1993**) 441-450.

9. Wernly P and Thormann W, Drug of abuse confirmation in human urine using stepwise solid-phase extraction and micellar electrokinetic capillary chromatography, *Anal.Chem.*, 64 (**1992**) 2155-2159.

10. Caslavska J, Lienhard S, and Thormann W, Comparative use of 3 electrokinetic capillary methods for the determination of drugs in body fluids – Prospects for rapid determination of intoxications, *J. Chromatogr.*, 638 (**1993**) 335-342.

11. Chicharro M, Zapardiel A, Bermejo E, Perez J A, and Hernandez L, Direct determination of ephedrine and norephedrine in human urine by capillary zone electrophoresis, *J.Liq. Chromatogr.*, 18 (**1995**) 1363-1381.

12. Krogh M, Brekke S, Tonnesen F, and Rasmussen K E, Analysis of drug seizures of heroin and amphetamine by capillary electrophoresis, *J. Chromatogr. A*, 674 (**1994**) 235-240.

13. Lurie I S, Chan K C, Spratley T K, Casale J F, and Issaq H J, Separation and detection of acidic and neutral impurities in illicit heroin via capillary electrophoresis, *J. Chromatogr. B*, 669 (**1995**) 3-13.

14. Trenerry V C, Wells R J, and Robertson J, The analysis of illicit heroin seizures by capillary zone electrophoresis, *J. Chromatogr. Sci.*, 32 (**1994**) 1-6.

15. Walker J A, Marche H L, Newby N, and Bechtold EJ, A free zone capillary electrophoresis method for the quantitation of common illicit drug samples, *J. Forensic Sci.*, 41 (**1996**) 824-829.

16. Tagliaro F, Poiesi C, Aiello R, Dorizzi R, Ghielmi S, and Marigo M, Capillary electrophoresis for the investigation of illicit drugs in hair – Determination of cocaine and morphine, *J. Chromatogr.*, 638 (**1993**) 303-309.

17. Frost M, Kohler H, and Blaschke G, Determination of LSD in blood by capillary electrophoresis with laser-induced fluorescence detection, *J. Chromatogr. A*, 693 (**1997**) 313-319.

18. Lurie I, The analysis of cations and anions in illicit heroin using capillary electrophoresis with indirect UV detection, *J. Cap. Elec.*, 5 (**1996**) 237-242.

19. Molteni S, Caslavska J, Allemann D ,and Thormann W, Determination of methadone and its primary metabolite in human urine by capillary electrophoretic techniques, *J. Chromatogr. B*, 658 (**1994**) 355-367.

20. Frost M, Kohler H, and Blaschke G, Enantioselective determination of methadone and its main metabolite 2-ethylidene-1,5-dimethyl-3,3-diphenylpyrrolidine (EDDP) in serum, urine and hair by capillary electrophoresis, *Electrophoresis*, 18 (**1997**) 1026-1034.

21. Bjornsdottir I and Hansen S H, Determination of opium alkaloids in opium by capillary electrophoresis, *J. Pharm. Biomed. Anal.*, 13 (**1995**) 687-693.

22. Taylor RB, Low AS, and Reid RG, Determination of opiates in urine by capillary electrophoresis, *J. Chromatogr. B*, 75 (**1996**) 213-223.

23. Tagliaro F, Smith F P, Turrina S, Equisetto V, and Marigo M, Complementary use of capillary zone electrophoresis and micellar electrokinetic capillary chromatography for mutual confirmation of results in forensic drug analysis, *J. Chromatogr. A*, 735 (**1996**) 227-235.

24. Tagliaro F, Turrina S, and Smith F P, Capillary electrophoresis – principles and applications in illicit drug analysis, *Forensic Sci. Int.*, 77 (**1996**) 211-229.

25. Lurie I S, Application of capillary electrophoresis to the analysis of seized drugs *Int. Lab.*, 26 (**1996**) 21-28.

26. Esseiva P, Lock E, Goueniat O, and Cole MD, Identification and quantification of amphetamine and analogues by capillary zone electrophoresis, *Science and Justice*, 37 (**1997**) 113-119.

27. Ventura R and Segura J, Detection of diuretic agents in doping control, *J. Chromatogr. B*, 687 (**1997**) 127-144.

28. Naess O and Rasmussen KE, Micellar electrokinetic chromatography of charged and neutral drugs in acidic running buffers containing a zwitterionic surfactant, sulfonic acids or sodium dodecyl sulphate – separation of heroin, basic by-products and adulterants, *J. Chromatogr. A*, 760 (**1997**) 245-251.

29. Flurer C L and Wolnik K A, Chemical profiling of pharmaceuticals by capillary electrophoresis in the determination of drug origin, *J. Chromatogr. A*, 674 (**1994**) 153-163.

30. Bjornsdottir I and Hansen S H, Determination of opium alkaloids in crude opium using non-aqueous capillary electrophoresis, *J. Pharm. Biomed. Anal.*, 13, (**1995**) 1473-1481.

31. Bjornsdottir I and Hansen S H, Comparison of aqueous and non-aqueous capillary electrophoresis for quantitative determination of morphine in pharmaceuticals, journal of *J. Pharm. Biomed. Anal.*, 15 (**1997**) 1083-1089.

16 Determination of Radioactive Compounds by CE

16.1 Introduction

The use of radiolabelled drugs is common practise in areas of pharmaceutical analysis such as drug metabolism, quality control for radiopharmaceuticals and method validation. Typically analysis of radiolabelled drugs is conducted by HPLC using radioactivity detectors. These detectors often contain scintillation materials that emit a photon when impacted by an appropriate radiowave. The photons are then detected by photomultiplier tubes which generate the detector response signal. Radioactivity detectors are more sensitive than UV absorbance detectors as the natural background radioactivity signal is low. This low background signal enables the signal to be highly amplified without generation of high levels of noise. The UV signal cannot be so highly amplified due to stray light effects which would cause excessive noise levels. The background radiation signal can be further reduced by shielding of the detector with lead or other appropriate material. The use of radioactivity detectors can be beneficial when the sample contains many other UV active components which would interfere with the detection of the peak using UV detection. The radioactivity detector will only record peaks that are radioactive so the separation profile obtained is much cleaner. This selective detection ability is especially useful in bioassays where there are often many non-radioactive endogenous species present in the sample solution.

Radioactive reagents can be used to produce a radioactive final drug substance. Generally a final product is prepared which emits beta radiation. The adsorption, metabolism, distribution and excretion of the drug in the patient can be assessed by a radioactivity based analysis of biofluids such as saliva, tears, sweat, urine, blood and plasma. The sensitivity of the radioactive detector is of particular importance when low doses of the drug are employed in these studies.

Radioactive metal ions (such as $_{43}Tc^{99m}$) can be complexed with organic compounds to prepare radiopharmaceuticals which are widely used in hospitals as tracer compounds. The organic compounds used are generally polycarboxylic acids such as 1,2-diaminopropane-N,N,N',N'-tetraacetic acid (DPTA) and (+/-)-trans-1,2-diaminocyclohexane-N,N,N',N'-tetraacetic acid (DCTA) (1). Ethylene dicysteine is used (2) a tracer for renal functions following chelation with $Tc^{99m}O_4^-$. Figure 16.1 shows the radiolabelling scheme, $SnCl_2$ is added in the reaction mixture.

The radiopharmaceutical is administered to the patient and is then selectively concentrated in particular areas of the body such as the liver, heart, bone etc. For example (3) Tc^{99m}-MDP is a bone-seeking agent. Generally the radiopharmaceutical emits gamma rays which

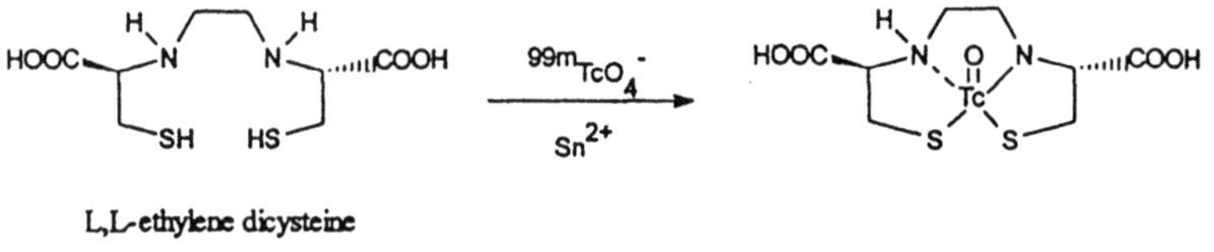

L,L-ethylene dicysteine

Figure 16.1 Radioactive labelling of L,L-ethylene dicysteine with radioactive pertechnetate. Reproduced with permission from reference 2.

can penetrate through the body. The body area of interest can then be visualised using a gamma ray camera (1) in order to diagnose disease or to monitor the transport function of specific organs. Quality control methods are needed to support this manufacture. Non-radioactivity monitoring methods are used to determine the amount and purity of the chelating material. Radioactive detection methods are needed to determine the concentration of the radioactive metal ion in solution. Radioactive methods are also necessary to determine the purity and concentration of the radiolabelled complex – this is important as the complexation may not be entirely selective or complete. Assay of the radiopharmaceutical solution content is important as this information is used by the clinician to the measure the radiation dose administered to the patient.

Radiolabelled drugs are also employed in recovery experiments in method validation studies. Typically this aspect is addressed in HPLC method validation studies. In HPLC there is the possibility that a fraction of the drug (and/or impurities) injected onto the column can become irreversibly adsorbed onto the stationary phase resulting in loss of sample. This possibility can be assessed by analysis of radiolabelled drug substance. A volume of radiolabelled drug solution can be injected onto the HPLC column and the analysis performed. If the volume injected and solution concentration are accurately known the expected radiation intensity of the eluent collected from the column can be calculated. This calculated amount can be compared to the eluent collected from the system. For additional verification the column can be unpacked and the radioactivity signal determined for the stationary phase material. Table 16.1 gives a listing of the radioactivity related applications of CE

16.2 Radioactivity Detectors

Radioactivity detectors have been developed for use in CE (4–10) but due to the limited use of radioactivity detection methods there has been limited commercial interest in developing these detectors by the major CE instrument manufacturers. The majority of attention has been paid to detector capable of detecting γ and high-energy β emissions. This is largely due to the difficulty of low energy β- emission to penetrate through the capillary walls. However a capillary electrophoresis (CE) postcolumn radionuclide detector for low-energy beta emitters has been developed based on a phosphor-imaging detector (10). The limit of detection (LOD) for S-35-labelled analytes was 0.13 amol (8.7 pM or 0.007 Bq), while the LOD for

Table 16.1 Radioactivity related applications of CE

Analyte	Ref.	Comment
Complexes of radioactive metal ions with polycarboxylic acids such as EDTA and DCTA	4	Monitoring of radioactive waste
Diatrizoic acid and its impurities	5	Determination in radiopaque Injection solutions
DMSA-Tc complex	6	Analysis of input pertechnetate and radiopharmaceutical with γ detector
Ethylene dicysteine	2	Determination of the radiopharmaceutical purity and assay
P^{32} labelled TTP, ATP and CTP	7	Applicable to high energy β-emitters and γ-emitters
Radiolabelled heavy metals	8	proton-induced X-ray emission
Range of radiolabelled metal ions	9	Fission products analysed and compared to results from CE analysis with indirect UV detection.

ATP = adenosine 5'[αP^{32}]triphosphate, CTP = cytidine 5'[αP^{32}]triphosphate, DCTA = (+/-)-trans-1,2-diaminocyclohexane-N,N,N',N'-tetraacetic acid, DMSA = dimercaptosuccinic acid, TTP = thymidine 5'[αP^{32}]triphosphate

P-32-labelled analytes was 4.9 zmol (0.33 pM or 0.002 Bq). A linear range from 1.5 amol to 1.5 fmol was demonstrated for S-35-labelled compounds. Use of another CE detector for low-energy beta emitters has been reported (11).

Figure 16.2 shows a schematic of a self constructed on-line β and γ radioactivity detector. The detector consisted (9) of conical plastic scintillating material with the capillary passing through the centre to provide a 4 pi detection geometry. The sintillator material was dye-doped polyvinyltoluene. This geometric arrangement provides the optimal (80 % efficient) collection of signal resulting in improved sensitivity. The wide end of the cone is optically coupled to a photomultiplier tube. Radioactivity detection of Eu-152, and Cs-137 was possible at the nanocurie level for 80-100 nL injections.

A semiconductor radioisotope detector was used (7) to analyse a range of labelled compounds. Figure 16.3 shows separation of P-32-labelled ATP and related compounds.

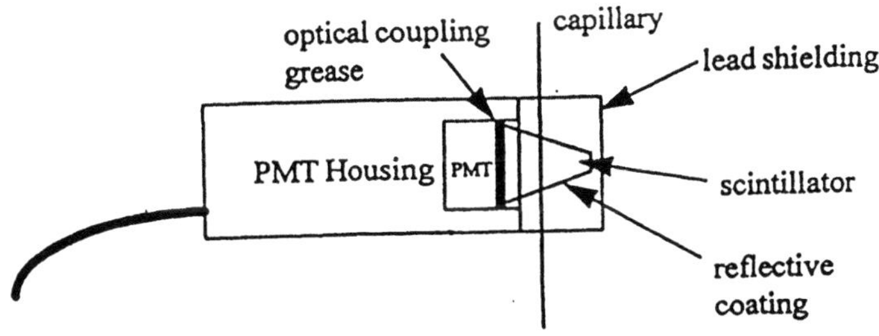

Figure 16.2   Schematic of an on-line β and γ radioactivity detector. Reproduced with permission from reference 9.

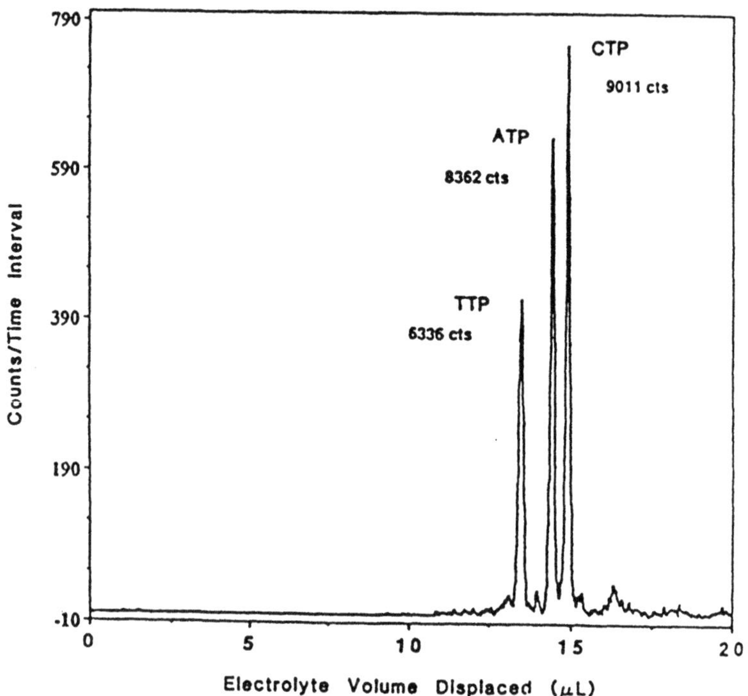

Figure 16.3 Separation of P^{32} labelled ATP and related compounds with radioactivity detection. Separation conditions : plastic scintillator detector, 100 cm × 100 μm, 20 kV, 200 mM borate. TTP = thymidine 5'[αP^{32}]triphosphate, ATP = adenosine 5'[αP^{32}]triphosphate, CTP = cytidine 5'[αP^{32}]triphosphate. Reproduced with permission from reference 7.

CE with laser induced resonance energy transfer detection has been used (4) to analyse complexes of terbium (Tb) with polycarboxylic acids such as ethylenediamine tetraacetic acid (EDTA), 1,2-diaminopropane-N,N,N',N'-tetraacetic acid (DPTA) and (+/-)-trans-1,2-di-aminocyclohexane-N,N,N',N'-tetraacetic acid (DCTA). Detection limits in the 1×10^{-7} M range were achieved. The method was applied to monitoring these compounds in complex radioactive waste.

16.3 Radiopharmaceutical Purity and Assay Determinations

Separation and determination of diatrizoic acid (DTZA) and its four mono- and diiodo degradation products (2-iodo,4-iodo, 2,4-diiodo, and 2,6-diiodo-3,5-diacetamidobenzoic acid) in radiopaque solution for injection (RSI) has (5) been achieved by CE. An RSD of 1.7 % was obtained for method was peak area precision at high sample concentration. Detector linearity for DTZA concentration over the range of 5–60 mg/mL gave a correlation coefficient of 0.997. Samples of RSI were shown to contain levels of each of the mono and diiodo impurities.

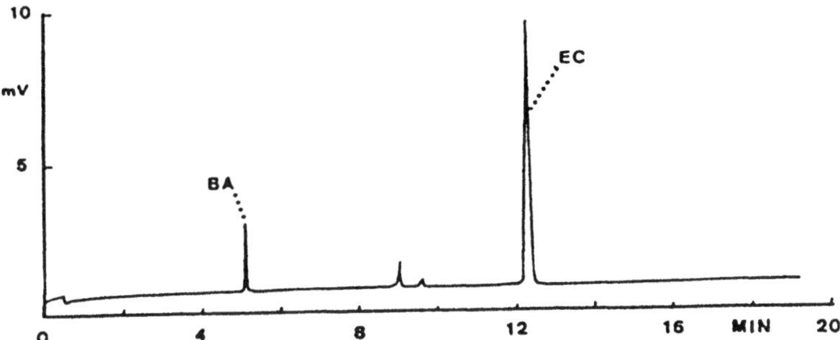

Figure 16.4   Separation of ^{99m}Tc-ethylene dicysteine Injection solution following storage for 2 hours. Separation conditions : 70 cm × 75 µm, detection at 235 nm, 20 kV, 10 mM borate pH 10. Reproduced with permission from reference 2.

A pH 10 borate buffer has been used (2) to determine the purity of ethylene dicysteine batches with UV detection at 235 nm. Selectivity of the method was demonstrated for all the potential degradants such as thiaproline and cysteine derivatives. Figure 16.4 shows separation of a ^{99m}Tc-ethylene dicysteine Injection solution following storage for 2 hours. The complex is unstable in the high pH phosphate buffer used to reconstitute the dried powered complex material. Figure 16.4 shows the occurrence of 2 impurities (between 9 and 10 minutes) which were not visible in the freshly prepared solution. Benzyl alcohol (BA) was used to indicate the EOF rate.

A gamma-ray detector has been used (6) to determine sodium pertechnetate $NaTc^{99m}O_4$ The detector was shown to have a linear response from the limit of detection 10 to 500 $Bqcm^{-3}$ corresponding to 5.1×10^{-17} to 2.55×10^{-15} g cm^{-3} $_{43}Tc^{99m}$. The method was used to analyse Tc^{99m}-dimercaptosuccinic acid complex which is a radiopharmaceutical used to image tumours and kidneys. The ability to quantify uncomplexed pertechnetate would allow the method to monitor the degree of complexation.

References

1. Mohammad AA, Laser induced resonance energy transfer – a novel approach towards achieving high sensitivity in capillary electrophoresis part B – applications for environmental testing, *J. Chromatogr. A*, 767 (**1997**) 217-221.

2. Van Schedael A, Verbeke K, Van Nerom C, Hoogmartens J, and Verbruggen A, Capillary electrophoretic analysis of ethylene dicysteine a precursor of the radiopharmaceutical ^{99m}Tc ethylene dicysteine, *J. Chromatogr. B*, 697 (**1997**) 251-254.

3. Denotaristefani F, Pani R, Scopinaro F, Barone L M, Blazek K, Devincentis G, Malatesta T, Maly P, Pellegrini R, Pergola A, Soluri A, and Vittori F, First results from a YAP-CE gamma camera for small animal studies, *IEEE Transactions on nuclear science*, 43 (**1996**) 3264-3271.

4. Bissell M G, Okorodudu A O, Petersen J R, and Mohammad A A, Laser induced resonance energy transfer – a novel approach towards achieving high sensitivity in capillary electrophoresis part B – applications for environmental testing, *J. Chromatogr. A*, 767 (**1997**) 217-221.

5. Farag S A and Wells C E, Capillary electrophoresis determination of diatrizoic acid and its impurities in diatrizoate radiopaque solutions, *Mikrochimica Acta*, 126 (**1997**) 141-145.

6. Altria K D, Simpson C F, Bharij A K, and Theobald A E, A gamma-ray detector for Capillary Zone Electrophoresis and its use in the analysis of some radiopharmaceuticals, *Electrophoresis*, 11 (**1990**) 732-734.

7. Pentoney S L, Zare R N, and Quint J F, Semiconductor radioisotope detector for capillary electrophoresis, *J. Chromatogr.*, 480 (**1989**) 259-271.

8. Vogt C, Vogt J, and Wittrisch H, Element-sensitive X-ray detection for capillary electrophoresis, *J. Chromatogr. A*, 727 (**1996**) 301-310.

9. Kunder G L, Andrews J E, Grant P M, Andresen B D, and Russo R E, Analysis of fission products using capillary electrophoresis with on-line radioactivity detection, *Anal. Chem.*, 69 (**1997**) 2988-2993.

10. Tracht S E, Cruz L, Stobbawiley C M, and Sweedler J V, Detection of radionuclides in capillary electrophoresis using a phosphor-imaging detector, *Anal. Chem.*, 68 (**1996**) 3922-3927.

11. Tracht S, Toma V, and Sweedler J V, Postcolumn radionuclide detection of low-energy beta emitters in capillary electrophoresis, *Anal. Chem.*, 66 (**1984**) 2382-2389.

17 Miscellaneous Pharmaceutical Analysis Related Areas of CE

17. 1 Drug Diet Determinations

In toxicology trials drug may be administered to the animals by incorporation into their feed. The analysis of drug in feedstuff can be complicated as the matrix is complex and a large number of interfering peaks can be obtained. Use of CE can be of benefit in reducing the number of interferences and CE has been successfully applied (1–3) in this area. The ability to inject sample solutions with complicated matrices can lead to reductions in sample preparation requirements – this is highlighted in clinical applications of CE (Chapter 10). Sensitivity requirements are often not stringent in drug diet analysis and CE is therefore consider to be useful in this area.

The chiral purity and content of various ephedrine compounds in nutritional supplements have been determined (1) by FSCE. Low pH buffers containing cyclodextrins gave highly repeatable separations. The methods were also applied to the chiral analysis of a number of amphetamines. The extraction procedure was relatively simple and comprised of addition of water adjusted to pH 2 followed by 30 minutes sonication. Sample solutions were filtered and directly injected onto the CE system. Figure 17.1 shows separation of a 1:100 acidic extract of a nutritional supplement which shows the presence of norephedrine (Nor), ephedrine (E), methylpseudoephedrine (MeE) and pseudoephedrine (Ψ–E).

Dansylated amino acids have been determined (2) in feedstuffs by MECC methods using SDS or sodium cholate as the micellar agents.

A combination of ITP and CE enabled (3) detection limits of 10^{-8} M for halofuginone with UV detection at 254nm. Three hundred micron capillaries were used for separation and 25 µl of sample solution was used for analysis. Precision data of 1 % RSD was reported with detector linearity of 0.9999

17.2 Regulatory Aspects

The attitude of regulatory authorities towards a new technology is of paramount concern to workers in the field of pharmaceutical analysis. Reluctance to accept a new technique can limit its general development. It is generally agreed that CE has been well received by regulatory bodies and this has served to stimulate growth and interest in CE within the pharmaceutical industry.

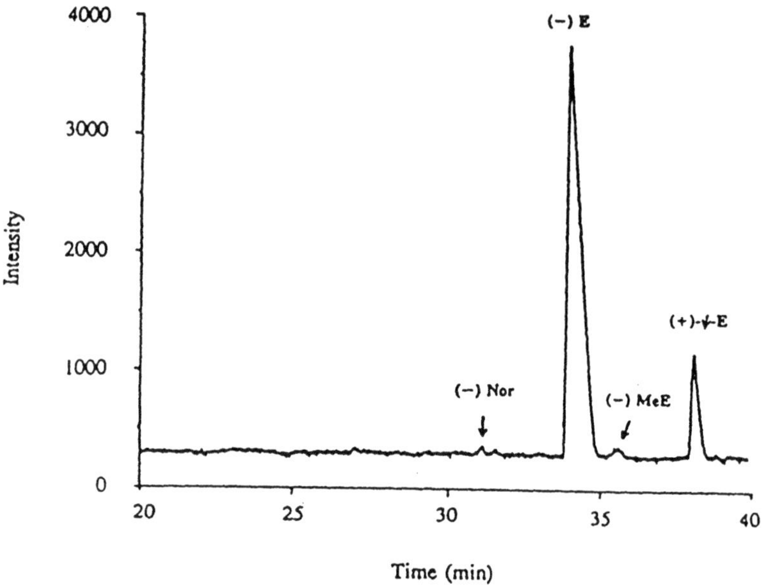

Figure 17.1 Separation of a 1:100 acidic extract of a nutritional supplement. Separation conditions: 90 cm × 50 µm, 70 mM hydroxypropyl-beta-cyclodextrin, 30 mM tetramethylammonium chloride, 10 mM SDS pH 2, 28 kV. Reproduced with permission from reference 1.

The acceptance of CE methods by regulatory authorities was highlighted (4) in a confidential survey of a number of major UK and US pharmaceutical companies. Table 17.1 shows the data obtained relating to regulatory submissions. All CE methods submitted had been accepted (4) without technical query. All companies surveyed indicated that they had no reluctance to submit appropriate CE data in submissions. The survey was conduced in mid 1994 and the number of submitted methods has certainly increased. For example, a stability indicating MECC method for the analysis of BMS-188484 has been successfully included in a regulatory submission from Bristol-Myers-Squibb. (5).

Table 17.1 Regulatory acceptance of CE methods

Number of companies who have submitted regulatory documents containing CE data

	Included	Intend	Would	Would not
UK	1	4	11	0
US	3	2	5	0

Acceptability of CE data in Regulatory Submissions

	Accepted	Awaiting	No data	Queried
UK	1	0	14	0
US	3	1	4	0

Reproduced with permission from reference 4.

A number of companies at a recent CE conference (6) indicated that they have included CE in New Drug Applications (NDA's).

Pharmacopoeia have also recognised the advancing application of CE within pharmaceutical companies. The savings of solvent purchase and disposal have been highlighted (7) particularly from an environmental concern viewpoint. A draft USP general chapter on CE has been published (8) in anticipation of future monographs containing CE analytical methods. The first USP monograph containing CE is due in 1998 (6) for the drug ethambutol.

A number of researchers within regulatory authorities are actively investigating applications of CE. These include studies by FDA researchers into batch profiling of pharmaceuticals from different suppliers (9), impurity content and assay of cephalosporins (10), enantiometric determination of ephedrine compounds in nutritional supplements (1) and impurity contents of submitted samples (11).

17.3 Biopharmaceuticals

The majority of pharmaceuticals are small synthetic organic molecules and the previous chapters cover application of CE to these types. However there is a increasing tendency in many pharmaceutical companies to develop biopharmaceuticals such as proteins, peptides DNA and oligonucleotides. The analysis of many of these compound types is often accomplished by using isoelectric focusing and capillary gel electrophoresis. For completeness some examples of CE in the analysis of biopharmaceuticals are included. For more complete details of this type of analysis readers are directed to other reference sources (12, 13).

17.3.1 Proteins

A CE method has been developed (14) for the quality control of insulin formulations. An extensive method development exercise resulted in the use of a 50mM sodium acetate buffer containing 85 mM of zwitterionic buffer 2-(N-cyclohexylamino) ethanesulphonic acid (CHES) and 10 % acetonitrile (pH 7.75). Figure 17.2 shows separation of the deamidation products from insulin in a formulated product by both CE and ion-exchange chromatography (IEC). Results for the deamidation products were generated using CE and IEC and good agreement was obtained (Table 17.2). The CE method also showed good recovery, repeatability and linearity results.

Peptic digests are also used (15) to monitor quality control of biopharmaceuticals and CE has been used in this application. Impurities were detected at 0.01 % of the active compound using (15) laser-induced fluorescence detection.

Recombinant human erythropoietin (rhEPO) has been analysed by CE (16) in preparations formulated with large amounts of a protein excipient (human serum albumin, HSA). Addition of 1 mM nickel chloride to the electrophoretic buffer was required to separate rhEPO from HSA. Validation of the method include linearity range of 0.03–1.92 mg/ml, and LOD and LOQ values of 0.01 and 0.03 mg/ml, respectively. The method was successfully

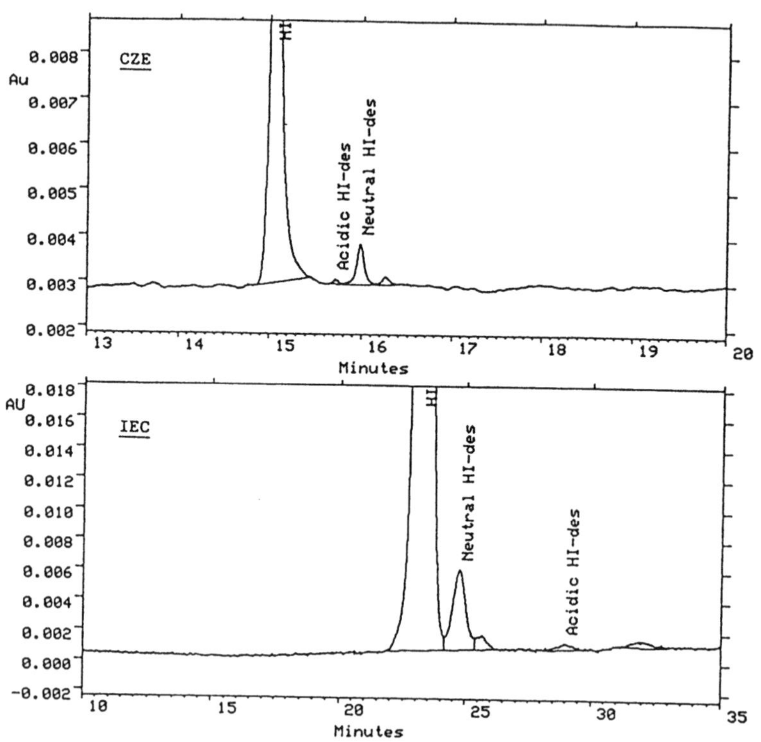

Figure 17.2   Separation of formulated insulin sample by both CE and IEC. Reproduced with permission from reference 14.

applied to a range of products. The method developed was applied to HSA formulations containing rhEPO. Electrospray ionisation mass spectrometry was used to determine the heterogeneous nature of HSA.

A low pH phosphate buffer containing 0.25% hydroxypropylmethylcellulose (HPMC) was used (17) to analyse acidic fibroblast growth factor (aFGF). The addition of HPMC reduced peak tailing effects but excessive HPMC led to a reduction in separation performance. The method was applied to pharmaceutical purity analysis (Figure 17.3). An alternative approach to the reduction of protein-capillary wall interactions is the addition of diamines such as 1,4-diaminobutane (18) to the electrolyte.

Table 17.2   Comparison of insulin purity results by CE and IEC.

Technique	Sample	% area/area results		
		HI	AD-I	ND-I
CE	1	98.41	0.24	1.35
	2	95.53	0.29	3.69
IEC	1	97.90	0.25	1.82
	2	95.36	0.30	3.65

Reproduced with permission from reference 14.

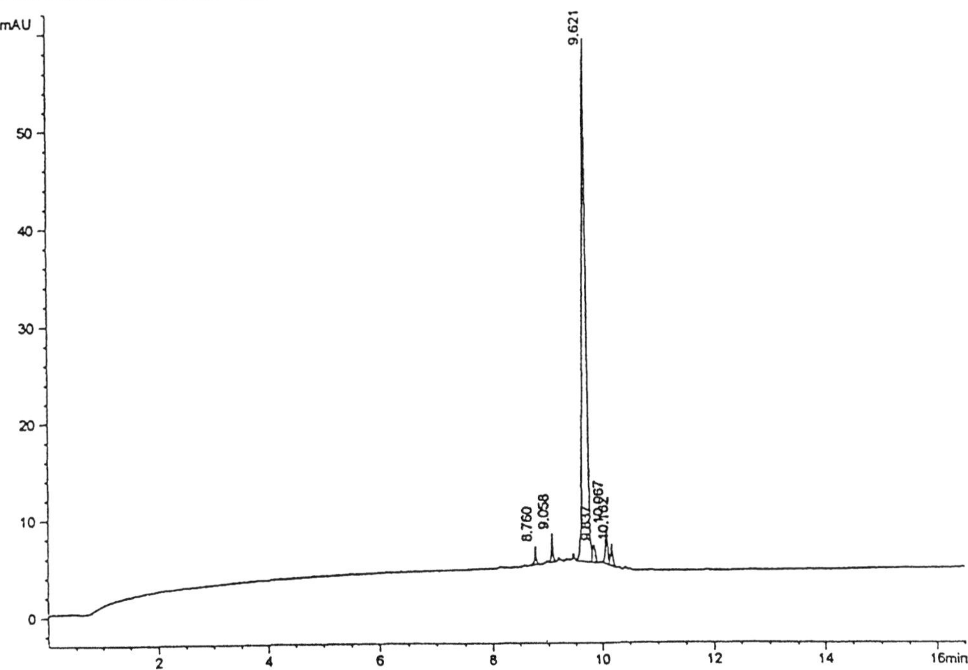

Figure 17.3   Separation of acidic fibroblast growth factor by CE for purity control. Operating conditions: 50 mM phosphate pH 2.5 with 0.25 % HPMC, 15 kV, 1.0 mg/ml, 200 nm, 40 cm × 75 µm.

A variety of CE approaches have been used (19) to analyse the glycoprotein, recombinant tissue plasminogen activator (rtPA). A free Solution CE method using pH 5 buffers and coated capillaries was used to separate rtPA and related impurities. Capillary isoelectric focusing (CIEF) was used to provide an alternative means of protein profiling. Proteins strongly interact with SDS and it was possible (19) to obtain unique separations with gels containing SDS (SDS-CGE). Workers from the same group have also (20) used similar methods to replace and compliment slab gel techniques with capillary electrophoresis in a quality control environment. In particular for the analysis of recombinant humanized monoclonal antibody HER2 (rhuMAbHER2). The CIEF method gave good injection precision and linearity data with a 2 ppm detection limit for the protein. A capillary isoelectric focusing (cIEF) methods has also been validated (21) to monitor charge heterogeneity in a recombinant glycoprotein. Recovery data was >100 % and an acceptable sensitivity of 25 ng of protein was obtained. Linearity for detector response with protein content was shown over the range 50–1000 µg/mL.

17.3.2 Peptides

A combination of CE and HPLC have been used (22) to monitor the stability of a basic hexapeptide (antagonist G). MS detection was used to provide structural information on the degradants. A 50 mM phosphate pH 5.5 has been used (23) to separate the anticoagulant peptide MDL28050 from its deletion by-products.

Substance P (SO) and 8 its metabolites have been separated (24) by capillary electrophoresis at neutral pH. Phytic acid was employed as a run buffer additive to eliminate the interaction of SP and its cationic N-terminus metabolites with ionized silanol groups. Additional selectivity was obtained by the addition of sulfobutyl ether beta-cyclodextrin.

Chiral separation of peptides is of importance and CE has been used for this purpose employing various additives such as vancomycin (25) and cyclodextrins (26).

17.3.3 Oligonucleotides

Chemically modified phosphorothioate oligodeoxynucleotides (ODNs) have become critical tools for research in the fields of gene expression and capillary gel electrophoresis (CGE) has been applied to monitor the content and purity of these compounds.

A CGE method was used (27) for the quantitative assay of a phosphorothioate oligonucleotide in pharmaceutical formulations. Migration time repeatability of less than 0.1 % RSD was reported. An internal standard was used to give injection precision values of around 2 % RSD using electrokinetic injections. The method was applied to formulated product and gave results in line with label claim (1.008 mg/ml for a 1 mg/ml product and 0.331 mg/ml for a 0.33 mg/ml product). Similar CGE approaches has been used by other workers (28, 29) to assess the content and stability of antisense oligonucleotides.

20-mer ODNs have been determined (30) in plasma and urine by both CGE and HPLC. CE gave increased basepair resolution compared to HPLC in these quantitative studies.

17.4 Combinatorial Libraries

The use of combinatorial libraries to synthesise assortments of potential drug candidates is an increasing development area in many pharmaceutical companies. Often robotic systems are used (31) to mix reagents at different ratios to produce mixtures of new lead compounds. The solutions generated often contain 10–1000 new compounds which are then screened for biological activity. Combinations of HPLC and CE has been used (31) in the analysis of these complex samples.

It is possible (32, 33) to use CE separations to identify likely drug candidates from within these mixtures. Figure 17.4 shows a schematic of the process that occurs within the capillary during separation. For example the mixture of compounds is injected at the end of the capillary further from the detector and the pH is selected such that these compounds are uncharged and therefore not resolved. A negatively charged compound such as enzyme or other target compound is injected at the detector end of the capillary. If a compound from within the mixture has an interaction with the enzyme then it will be retained (ie peaks 1–4 in Figure 17.4). The peak with the most interaction (binding) will be detected last. The identity of the peaks can be obtained by utilising MS detection. The compounds having an interaction can then be prepared in a pure form ready for further biological activity testing.

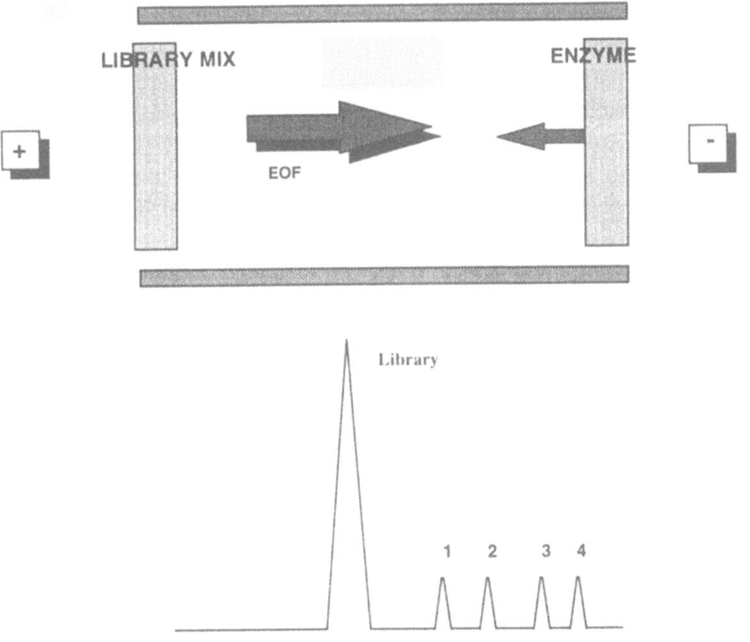

Figure 17.4   Use of CE in compound selection from combinatorial libraries

17.5 Physicochemical Property Determinations Using CE

17.5.1 Binding constants

The extent of binding/interaction of a compound with a host molecule is characterised by its binding constant. For example in Figure 17.4 peak 4 would have a stronger binding constant than peaks 1–3. CE can be used to obtain binding constants (34, 35) based on migration time data. Broadly speaking the migration time of the compound is determined in the electrolyte containing no binding agent. The migration time is then determined in the electrolyte containing the binding agent. A change in migration time can be related to the binding constant associated between the test solute and the binding agent. This form of analysis has been termed (34) affinity capillary electrophoresis (ACE) and has been used (34) to measure binding constants for a range of compound types include solute-antibody, solute-oligosaccharides and solute-protein. Binding constants have also been determined by ACE for beta-cyclodextrin inclusion complex constants for 3,4-dihydro-2-h-1-benzopyran enantiomers (35), peptides with vancomycin (36). The use of ACE to measure binding constants has recently been reviewed (37, 38).

17.5.2 Dissociation constants

The dissociation constant of a compound can also be determined (39, 40) from migration time data. The solute is analysed using electrolytes covering a range of pH values. The mobility of the solute can be calculated directly from the migration time and EOF times. A plot of the mobility versus pH is constructed (Figure 17.5) and the pH at a point corresponding to 50 % of the mobility values equates to the pKa value. For example methylbenzylamine was calculated to have a dissociation constant of pH 6.76 in Figure 17.5. The pKa results from CE have been found to correlate well with literature values. the advantages of using CE for pKa measurements include automated operation and low sample amount requirements. It is also possible (41) to measure dissociation constants of in water insoluble or sparingly soluble compounds using CE by the addition of solvents such as methanol into the electrolyte.

17.5.3 Partition coefficients

The hydrophobicity of a drug is an important property as it governs the transport properties of the drug within the body. The hydrophobicity of a compound can be expressed as its partition coefficient, P, between water and an immiscible, non-polar solvent. Octanol is a typical solvent used log P_{ow} values are determined by a "shake-flask technique". Data from MECC migration times has been correlated (42-45) with log P_{ow} values as a highly hydrophobic compound will be strongly retained by the micelles used in MECC. For example (45) when the log P_{ow} values for 100 compounds were statistically compared to MECC data a correlation coefficient of 0.98 was obtained.

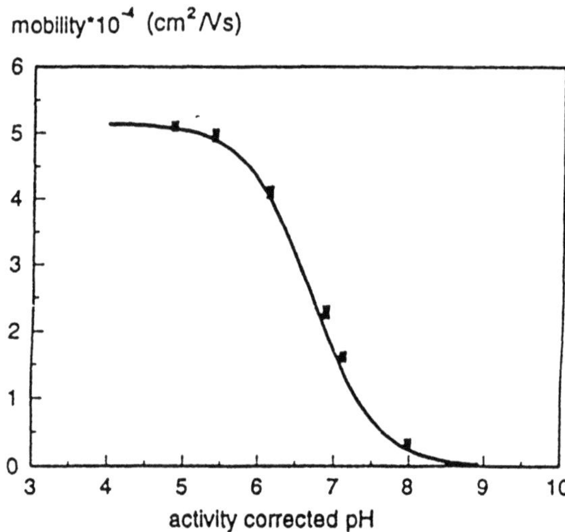

Figure 17.5 Mobility plot versus pH for methylbenzylamine. Reproduced with permission from reference 39.

17.5.4 Isoelectric point determinations

The isoelectric point (pI) of a zwitterionic compound is a pH value where the number of negatively charged groups exactly equals the number of positively charged groups. This property of zwitterionic compounds is exploited in isoelectric focusing (see Chapter 1 for further details). The pI of a compound can be calculated (46, 47) from CE migration time (mobility) data obtained covering a range of electrolyte pH values.

A plot can be constructed similar to Figure 17.5. At a pH above the pI value the mobility will be negative whilst the mobility will be positive at a pH value below the pH. The intercept value of the plot where the mobility is zero is the pI value for the compound.

References

1. Flurer C L, Lin L A, Satzger R D, and Wolnik K A, Determination of ephedrine compounds in nutritional supplements by cyclodextrin-modified capillary electrophoresis, *J. Chromatogr. B*, 669 (**1995**) 133-139.

2. Michaelsen S, Moller P, and Sorensen H, Analysis of dansyl amino acids in feedstuffs and skin by micellar electrokinetic capillary chromatography, *J. Chromatogr. A*, 680 (**1994**) 299-310.

3. Krivankova L, Foret F, and Bocek P, Determination of halofuginone in feedstuffs by the combination of capillary isotachophoresis and capillary zone electrophoresis in a column-switching system, *J. Chromatogr.*, 545 (**1991**) 307-313.

4. Altria K D and Kersey M, Capillary Electrophoresis and Pharmaceutical Analysis: a survey of the industrial application and their status in the United Kingdom and United States, *LC-GC Int.*, 8, April. (**1995**) 201-208.

5. Bretnall A E, Hodgkinson M M, and Clarke G S, Micellar electrokinetic chromatography stability indicating assay and content uniformity determination for a cholesterol-lowering drug product, *J. Pharm. Biomed. Anal.*, 15 (**1997**) 1071-1075.

6. Weinberger R, The seventh annual Frederick conference on capillary electrophoresis, *Amer. Lab.*, Jan. (**1997**) 38-38.

7. Cohen EM and Bell RG, *Pharm. Form.*, 20 (**1994**) 7870-7870.

8. *Pharm.Forum*, Jan-Feb (**1996**) 1727-1735.

9. Flurer C L and Wolnik K A, Quantitation of gentamicin sulfate in injectable solutions by capillary electrophoresis, *J. Chromatogr.*, 663 (**1993**) 259-263.

10. Sciacchitano C J, Mopper B, and Specchio J J, Identification and separation of five cephalosporins by micellar electrokinetic capillary chromatography, *J. Chromatogr.*, 657 (**1994**) 395-399.

11. Flurer C L and Wolnik K A, Chemical profiling of pharmaceuticals by capillary electrophoresis in the determination of drug origin, *J. Chromatogr.*, 674 (**1994**) 153-163.

12. Separation of proteins and peptides by capillary electrophoresis, *Beckman Educational Primers on Capillary Electrophoresis Volume V*, Beckman Part Number 727484.

13. Heller C (ed.), Analysis of nucleic acids by capillary electrophoresis, Vieweg Publishing, Wiesbaden, **1997**, ISBN 3-528-06871-X.

14. Mandrup G, Rugged method for the determination of deamidation products in insulin solutions by free zone capillary electrophoresis using an untreated fused-silica capillary, *J. Chromatogr.*, 604 (**1992**) 267-281.

15. Lee T T, Lillard S J, and Yeung E S, Screening and characterization of biopharmaceuticals by high-performance capillary electrophoresis with laser-induced native fluorescence detection, *Electrophoresis*, 14 (**1993**) 429-438.

16. Bietlot H P and Girard M, Analysis of recombinant human erythropoietin in drug formulations by high-performance capillary electrophoresis, *J. Chromatogr. A*, 759 (**1997**) 177-184.

17. Roddy T P, Molnar T E, Mckean R E, and Foley J P, Method of analysis of recombinant acidic fibroblast growth factor by capillary electrophoresis, *J. Chromatogr. B*, 695 (**1997**) 49-58.

18. Girard M, Bietlot H P, and Cyr T D, Characterisation of human serum albumin heterogeneity by capillary zone electrophoresis and electrospray ionization mass spectrometry, *J. Chromatogr. A*, 772 (**1997**) 235-242.

19. Thorne J M, Goetzinger W K, Chen A B, Moorhouse K G, and Karger B L, Examination of capillary zone electrophoresis, capillary isoelectric focusing and sodium dodecyl sulfate capillary electrophoresis for the analysis of recombinant tissue plasminogen activator, *J. Chromatogr. A*, 744 (**1996**) 155-165.

20. Hunt G, Moorhouse K G, and Chen A B, Capillary isoelectric focusing and sodium dodecyl sulfate capillary gel electrophoresis of recombinant humanized monoclonal antibody HER2, *J. Chromatogr. A*, 744 (**1996**) 295-301.

21. Moorhouse K G, Rickel C A, and Chen A B, Electrophoretic separation of recombinant tissue-type plasminogen activator glycoforms – validation issues for capillary isoelectric focusing methods, *Electrophoresis*, 17 (**1996**) 423-430.

22. Reubsaet J L E, Beijnen J H, Bult A, Teeuwsen J, Koster E H M, Waterval J C M, and Underberg W J M, Reversed-phase high-performance liquid chromatography and capillary electrophoresis in the stability study of the neuropeptide growth factor antagonist [arg(6), d-trp(7,9), me-phe(8)]-substance P{6-11}: a comparative study, *Anal. Biochem.*, 220 (**1994**) 98-102.

23. Chen T M, George R C, and Payne M H, Separation of the anticoagulant peptide MDL28050 from its deletion by-products by capillary zone electrophoresis, *J. High Res. Chromatogr.*, 13 (**1990**) 782-784.

24. Kstel K L, Freed A L, and Lunte S M, Complete capillary electrophoretic separation of substance P and its metabolites at neutral pH using ionic run buffer additives, *J. Chromatogr. A*, 744 (**1996**) 241-248.

25. Wan H and Blomberg L G, Enantiomeric separation by capillary electrophoresis of di- and tri-peptides derivatized with 9-fluorenylmethyl chloroformate using vancomycin as chiral selector, *J. Micro. Sepn.*, 8 (**1996**) 339-344.

26. Skanchy D J, Wilson R, Poh T, Xie G H, Demarest C W, and Stobaugh J F, Resolution of acylated dipeptide stereoisomers by capillary electrophoresis using sulfobutylether derivatized beta-cyclodextrin, *Electrophoresis*, 18 (**1997**) 985-995.

27. Srivatsa G S, Batt M, Schuette J, Carlson R H, Fitchett J, Lee C, and Cole D L, Quantitative capillary gel electrophoresis assay of phosphorothioate oligonucleotides in pharmaceutical formulations, *J. Chromatogr. A*, 680 (**1994**) 469-477.

28. Bruin G J M, Bornsen K O, Husken D, Gassmann E, Widmer H M, and Paulus A, Stability measurements of antisense oligonucleotides by capillary gel electrophoresis, *J. Chromatogr. A*, 709 (**1995**) 181-195.

29. Vilenchik M, Belenky A, and Cohen A S, Monitoring and analysis of antisense DNA by high-performance capillary gel electrophoresis, *J. Chromatogr A*, 663 (**1994**) 105-113.

30. Chen S H, Qian M X, Bnnan J M, and Gallo J M, Determination of antisense phosphorothioate oligonucleotides and catabolites in biological fluids and tissue extracts using anion-exchange high-performance liquid chromatography and capillary gel electrophoresis, *J. Chromatogr. B*, 692 (**1997**) 43-51.

31. Boutin J A, Hennig P, Lambert P H, Bertin S, Petit L, Mahieu J P, Serkiz B, Volland J P, and Fauchere J L, Combinatorial peptide libraries: Robotic synthesis and analysis by nuclear magnetic resonance, mass spectrometry, tandem mass spectrometry, and high-performance capillary electrophoresis techniques, *Anal. Biochem.*, 234 (**1996**) 126-141.

32. Chu Y H, Kirby D P, and Karger B L, Free solution identification of candidate peptides from combinatorial libraries by affinity capillary electrophoresis/mass spectrometry, *J. Am. Chem. Soc.*, 117 (**1995**) 5419-5420.

33. Chu Y-H, Dunayevskiy, Kirkby D P, Vouros P, and Karger B L, Affinity capillary electrophoresis-mass spectrometry for screening combinatorial libraries, *J.Am.Chem.Soc.*, 118 (**1996**) 7827-7835.

34. Takeo K, Advances in affinity electrophoresis, *J. Chromatogr. A*, 698 (**1995**) 89-105.

35. Baumy P, Morin P, Dreux M, Viaud M C, Boye S, and Guillaumet G, Determination of beta-cyclodextrin inclusion complex constants for 3,4-dihydro-2-h-1-benzopyran enantiomers by capillary electrophoresis, *J. Chromatogr. A*, 707 (**1995**) 311-326.

36. Chu Y H and Whitesides G M, Affinity capillary electrophoresis can simultaneously measure binding constants of multiple peptides to vancomycin, *J. Org. Chem.*, 57 (**1992**) 3524-3525.

37. Winzor D J, Measurement of binding constants by capillary electrophoresis, *J. Chromatogr. A*, 696 (**1995**) 160-163.

38. Oravcova J, Bohs B, and Lindner W, Drug-protein binding studies – new trends in analytical and experimental methodology, *J. Chromatogr. B*, 677 (**1996**) 1-28.

39. Gluck S J and Cleveland J A, Capillary zone electrophoresis for the determination of dissociation constants, *J. Chromatogr. A* 680 (**1994**) 43-48.

40. Ishihama Y, Oda Y, and Asakawa N, Microscale determination of dissociation constants of multivalent pharmaceuticals by capillary electrophoresis, *J. Pharm. Sci.* 83 (**1994**) 1500-1507.

41. Bellini S, Uhrova M, and Deyl Z, Determination of the thermodynamic equilibrium constants of water-insoluble (sparingly soluble) compounds by capillary electrophoresis, *J. Chromatogr. A*, 772 (**1997**) 91-101.

42. Herbert B J and Dorsey J G, n-octanol water partition coefficient estimation by micellar electrokinetic capillary chromatography, *Anal. Chem.*, 67 (**1995**) 744-749.

43. Ishihama Y, Oda Y, Uchikawa K, and Asakawa N, Correlation of octanol-water partition coefficients with capacity factors measured by micellar electrokinetic chromatography, *Chem. Pharm. Bull.*, 42 (**1994**) 1525-1527.

44. Smith J T and Vinjamoori D V, Rapid determination of logarithmic partition coefficients between n-octanol and water using micellar electrokinetic capillary chromatography, *J. Chromatogr. B*, 669 (**1995**) 59-66.

45. Garcia M A, Diez-Masa J C, and Marina M L, Correlation of the logarithm of capacity factors for aromatic compounds in micellar electrokinetic chromatography and their octanol-water coefficients, *J. Chromatogr. A*, 742 (**1996**) 251-256.

46. Kleparnik K, Slais K, and Bocek P, Determination of the isoelectric points of low and high molecular mass ampholytes by capillary electrophoresis, *Electrophoresis*, 14 (**1993**) 475-479.

47. Yao Y J, Khoo K S, Chung M C M, and Li S F Y, Determination of isoelectric points of acidic and basic proteins by capillary electrophoresis, *J. Chromatogr. A*, 680 (**1994**) 431-435.

Subject Index

Capillary Electrophoresis
Methods and Potentials

by Heinz Engelhardt, Wolfgang Beck,
and Thomas Schmitt

1995. x, 215 pp. Hardcover DM 79,50
ISBN 3-528-06668-7

Capillary electrophoresis combines the analytical separation technique of classical electrophoresis with the instrumental potential of modern chromatographic detection and automation. It eminently supplements chromatography for the separation of polar and water-soluble substances. The extent of its applications is extremely broad, spanning the range of separation from small cations to the highest molecular weight ionic biopolymers. The breadth of its application potential has now made capillary electrophoresis the fastest growing area of instrumental analysis. This book provides a practical introduction to capillary electrophoresis separation techniques. Particular value is placed on the developing and optimizing a separation.

Stand 1.4.98
Änderungen vorbehalten.
Erhältlich im Buchhandel
oder beim Verlag.

Abraham-Lincoln-Str. 46, Postfach 1547, 65005 Wiesbaden
Fax: (06 11) 78 78-4 00, http://www.vieweg.de

Analysis of Nucleic Acids by Capillary Electrophoresis

by Christoph Heller (Ed.)
1997. x, 313 pp. (Chromatographia CE-Series;
ed. by Altria, K. D.) Vol. 1. Hardcover DM 168,00
ISBN 3-528-06871-X

Part 1: Basic Concepts: Separation Matrix - Electrophoresis theories - Microscopic studies / *Part 2: Factors affecting the separation:* Electric field and polymer concentration - Sample matrix and injection - Agglomeration / Part 3: *Development in instrumentation:* Pulsed field CE - New type of separation matrix - Blotting - DNA separation / *Part 4: Applications:* Separation of restriction fragments - Analysis of oligonucleotides - DNA sequencing - Antisense DNA - Mutational analysis

This book on capillary electrophoresis is unique in its focus on the separation of nucleic acids. The importance of electrophoretic separation in every molecular biology laboratory justifies this specialization, which is also reflected in the development of instrumentation. This book is aimed to help to implement this rather new and promising technology in the biological laboratories and to help to overcome typical problems that can occur when starting with a new technique. It should also trigger further development in the field. It covers the theoretical background as well as practical examples of usual applications.
The authors are all experts in their field, having many years of experience with capillary electrophoresis and their application to nucleic acids.

Stand 1.4.98
Änderungen vorbehalten.
Erhältlich im Buchhandel
oder beim Verlag.

Abraham-Lincoln-Str. 46, Postfach 1547, 65005 Wiesbaden
Fax: (06 11) 78 78-4 00, http://www.vieweg.de

If you have any concerns about our products,
you can contact us on
ProductSafety@springernature.com

In case Publisher is established outside the EU,
the EU authorized representative is:
Springer Nature Customer Service Center GmbH
Europaplatz 3, 69115 Heidelberg, Germany

Printed by Libri Plureos GmbH
in Hamburg, Germany